# Sustainable Materials and Technology

**Series Editors**

**Sustainable Materials and Technology (SMT)** book series publishes research monographs (both edited and authored volumes) showcasing the latest developments in the field and comprehensively covering topics such as:

- Recycling of waste into useful material and their energy applications
- Catalytic action of Nano oxides for efficient carbon reforming process
- Sustainable technologies for plastic transformation
- Bifunctional nanoparticles for sustainable water splitting applications
- Sustainable dying and printing
- New materials from waste
- Sustainable Manure Management and Technology: Potentials, Uses and limitations
- Sustainable Mechanical Engineering Approach
- Sustainable biochemistry for the improvement of health
- Sustainable development of Mechanical recycling of automotive components
- Sustainable -waste recycling and conversion in useful materials for different applications
- Sustainable development of inexpensive Nano-photo catalysts
- Sustainable development of recycling of discarded lithium ion batteries
- Modern sustainable cement and concrete
- Sustainable adsorbent for hazardous removal
- Sustainable superior electromagnetic shielding materials
- Excellent sustainable nanostructured materials for energy storage device
- Sustainable development of heavy metal detoxification from water
- Carbon dioxide utilization for sustainable energy
- Sustainable development in green syntheses of materials
- Environment friendly and sustainable cloth for garments application
- Sustainable design and application of eco-materials
- Nanoparticles for sustainable environment applications
- Sustainable remediation of industrial contaminated water towards potential industrial applications
- Biomaterials for sustainable bioremediations

Siti Noorbaini Sarmin · Mohammad Jawaid ·
Rob Elias
Editors

# Wood Waste Management and Products

Springer

*Editors*
Siti Noorbaini Sarmin
Department of Wood Industry, Faculty of Applied Sciences
Universiti Teknologi MARA
Pahang, Malaysia

Mohammad Jawaid
Laboratory of Biocomposite Technology
Universiti Putra Malaysia
Serdang, Malaysia

Rob Elias
University of Wales Bangor
Bangor, UK

ISSN 2731-0426 ISSN 2731-0434 (electronic)
Sustainable Materials and Technology
ISBN 978-981-99-1904-8 ISBN 978-981-99-1905-5 (eBook)
https://doi.org/10.1007/978-981-99-1905-5



This Springer imprint is published by the registered company Springer Nature Singapore Pte Ltd.
The registered company address is: 152 Beach Road, #21-01/04 Gateway East, Singapore 189721, Singapore

*To all readers,*
*Find out how to make small changes that are environmentally friendly and have a long-term impact*

# Preface

Wood is our preferred material for environmentally friendly product construction and manufacturing. Wood has excellent properties for reuse, which can be realized and implemented through cascading utilization, introducing intermittent product lives.

Wood waste has the potential to be used to manufacture a wide range of products, including engineered wood products, energy generation (heat and electricity), mulching, and animal bedding. These low-cost, underutilized feedstock has the potential to boost the added value of wood waste.

The massive amount of wood waste generated from forest biomass and the furniture industry raised a number of environmental concerns. From an ecological standpoint, the sustainability of manufacturing processes is a critical global concern, particularly in the wood-based industry, where wood waste is a major issue. This is a sector that manufactures products from renewable forest resources. As a result, it is well-positioned to produce items that must contribute to environmental sustainability.

The biomass material used in the forestry sector is derived from natural and plantation forests, as well as the wood-based industry. They build up during the production of logs, primary wood processing, and wood processes. Sustainable wood and paper products may be preferable to other materials because they are made from renewable resource trees that are grown with sunlight, soil nutrients, and water. Second, they store carbon: most trees extract carbon dioxide from the atmosphere and replace it with oxygen during photosynthesis, lowering greenhouse gas emissions. Carbon emitted during harvesting is offset by carbon taken up during regeneration and regrowth in sustainably managed forests, making them carbon neutral. Third, they have the ability to store carbon for decades, if not centuries. Carbon can be stored in solid wood and paper-based materials for decades, if not centuries. Finally, because they are recyclable, they can be repurposed into new products, extending their useful life, and contributing to the pool of wood fiber resources.

The purpose of this book is to examine the potential utilization and conversion of wood-related waste materials into value-added products.

Pahang, Malaysia — Siti Noorbaini Sarmin
Serdang, Malaysia — Mohammad Jawaid
Bangor, UK — Rob Elias

# Contents

# About the Editors

**Siti Noorbaini Sarmin** is Senior Lecturer at Malaysia's Universiti Teknologi MARA Pahang. She was awarded a postdoctoral scholarship from the Malaysian Ministry of Higher Education from 2021 to 2022 at INTROP, Universiti Putra Malaysia. She received her B.Tech. in Bioresources, Paper, and Coating from Universiti Sains Malaysia in 2007, her M.Sc. from the same university in 2009, and her Doctor of Natural Sciences from the University of Hamburg in Germany in 2017. Natural fiber composites, conventional wood composites, timber trade management, and wood sciences are among her research interests. She has produced or co-authored over 50 publications in international journals, book chapters, and conference proceedings/seminars to date. She currently holds many national research grants for related studies.

**Dr. Mohammad Jawaid** is currently working as Senior Fellow (Professor) at Biocomposite Technology Laboratory, Institute of Tropical Forestry and Forest Products (INTROP), Universiti Putra Malaysia (UPM), Serdang, Selangor, Malaysia, and also has been Visiting Professor at the Department of Chemical Engineering, College of Engineering, King Saud University, Riyadh, Saudi Arabia, since June 2013. He has more than 20 years of experience in teaching, research, and industries. His area of research interests includes hybrid composites, lignocellulosic-reinforced/filled polymer composites, advanced materials: graphene/nanoclay/fire retardant, modification and treatment of lignocellulosic fibers and solid wood, biopolymers and biopolymers for packaging applications, nano-composites and nanocellulose fibers, and polymer blends. So far, he has published 40 books, 65 book chapters, more than 350 peer-reviewed international journal papers, and several published review papers under top 25 hot articles in science direct during 2013–2019. He also obtained 2 patents and 6 copyrights. H-index and citation in Scopus are 52 and 12521, and in Google Scholar, H-index and citation are 61 and 17402. He is founding Series Editor of Composite Science and Technology Book Series from Springer Nature, also Series Editor of Springer Proceedings in Materials, Springer Nature, and also International Advisory Board Member of Springer Series on Polymer and Composite Materials. He worked as Guest Editor of special issues of SN-applied Science, Frontiers in

Sustainable Food Systems, Current Organic Synthesis, and Current Analytical Chemistry, International Journal of Polymer Science, and IOP Conference Proceeding. He is also Editorial Board Member of Journal of Polymers and The Environment, Journal of Plastics Technology, Applied Science and Engineering Progress Journal, Journal of Asian Science, Technology and Innovation, and the Recent Innovations in Chemical Engineering. Besides that, he is also Reviewer of several high-impact international peer-reviewed journals of Elsevier, Springer, Wiley, Saga, ACS, RSC, Frontiers, etc. Presently, he is supervising 12 Ph.D. students (5 Ph.D. as Chairman and 7 Ph.D. as Member) and 6 master's students (1 master as Chairman and 5 master as Member) in the fields of hybrid composites, green composites, nano-composites, natural fiber-reinforced composites, nanocellulose, etc. Twenty-six Ph.D. and 13 master's students graduated under his supervision in 2014–2020. He has several research grants at university, national, and international levels on polymer composites of around 3 million Malaysian ringgits (USD 700,000). He also delivered plenary and invited talks in international conferences related to composites in India, Turkey, Malaysia, Thailand, the UK, France, Saudi Arabia, Egypt, and China. Besides that, he is also Member of technical committees of several national and international conferences on composites and material science. Recently, Dr. Mohammad Jawaid received Excellent Academic Award in the Category of International Grant-Universiti Putra Malaysia-2018 and also Excellent Academic Staff Award in Industry High-Impact Network (ICAN 2019) Award. Besides that, he received Gold Medal-Community and Industry Network (JINM Showcase) at Universiti Putra Malaysia. He also received Publons Peer Review Awards, 2017 and 2018 (Materials Science), and Certified Sentinel of Science Award Recipient, 2016 (Materials Science) and 2019 (Materials Science and Crossfield). He is also Winner of Newton-Ungku Omar Coordination Fund: UK-Malaysia Research and Innovation Bridges Competition 2015. Recently, he got Fellow and Chartered Scientist from the Institute of Materials, Minerals and Mining (IOM), UK. He is also Life Member of Asian Polymer Association and Malaysian Society for Engineering and Technology. He has professional membership of American Chemical Society (ACS) and Society for Polymers Engineers (SPE), USA.

**Dr. Rob Elias** is Director of the BioComposites Centre at Bangor University with a staff of 25 scientists. The center was established in 1989 to work with companies to develop new technologies based on sustainable materials such as wood. The center has state-of-the-art facilities including pilot-scale equipment that enables companies to demonstrate their ideas by developing prototype materials. Rob has a major interest in the development of bio-derived materials that reduce global warming potential. He has an industrial and academic background in natural fiber production. His expertise includes wood-based panel production, biomass extraction/chemical composition, and product development. His current research interests include biorefining, the production of bioplastic products, extraction of value-added molecules from plant materials, utilization of wastes, and agricultural co-products for construction applications.

# Challenges and Opportunities in Wood Waste Utilization

**Nurul Huda Abu Bakar and Nurjannah Salim**

**Abstract** Wood wastes have the potential to be utilized in the manufacturing of a wide range of products, such as engineered wood products, energy generation, and additive manufacturing. These low-cost biomasses that are only partially exploited have the potential to increase the value that can be added to waste wood products. The objective of this chapter is to address the challenges encountered as well as the opportunities presented by the utilization of wood waste. With the aid of this knowledge, the right approach can be identified for the development of wood waste in the future, which will result in the most long-term benefits for both the environment and the economy. The lack of adaption of more sophisticated technology and the absence of organizations concerned with the potential advantages of making use of such wastes is the source of the problem with wood waste. From this review, it is indicated that wood waste has the potential to be used as a source for the manufacture of a variety of materials; therefore, in order to make the most of the value of wood waste resources, the government should implement efficient guidelines for wood waste management.

## 1 Overview of the Wood Industry and Its Residue

Back closer to the middle of the 1800s, humans mainly utilized wood for cooking, heating, and lighting. It is a crucial fuel, notably for cooking and heating in several underdeveloped nations, and the main energy source in many nations. The materials that come from logging, the manufacture of wood, and materials that are reused or recycled all fall under the category of wood waste. Wood residues have many advantages, such as fuel in manufacturing facilities where the waste is burnt or used as raw material sources for bioenergy processes. For the pulp and paper industries, wood waste might also be converted into pellets or smaller wooden objects. According to recent studies, wood and wood waste, such as bark, sawdust, wood chips, wood

N. H. Abu Bakar · N. Salim (✉)
Faculty of Industrial Sciences and Technology, Universiti Malaysia Pahang, Lebuhraya Tun Razak, Gambang, 26300 Kuantan, Pahang, Malaysia
e-mail: njannah@ump.edu.my

 
S. N. Sarmin et al. (eds.), *Wood Waste Management and Products*, Sustainable Materials and Technology, https://doi.org/10.1007/978-981-99-1905-5_1

scrap, and paper mill byproducts, accounted for around 2.1% of the country's yearly energy consumption [1].

The use of wood, however, can also be a source of some difficulties. For instance, there are several steps in the production of timber goods, from the removal of logs to the creation of the finished goods, all of which have the potential to contaminate the land, the air, and the water. About 50% of wood is transformed into useful items, and the remaining 50% is thrown [2]. Bark, slabs, sawdust, chips, coarse residues, planer shavings, peeler log cores, and end trimmings are a few examples of wood waste produced by significant industrial processes. As a result, efficient use of wood waste significantly reduces environmental repercussions without endangering global forests [2].

Since the evolution of wood technology, it is crucial to understand the properties of wood, its processing method, and the potential utilization of wood waste. One can overcome the wood industry challenges by comprehending the potential applications and finding new opportunities to expand the applications. Therefore, this overview of current challenges and applications of wood residues seems to be one of the most important developments and a novel topic in the vast field of wood technologies. The discussion is also delivered with proper cohesiveness between sub-chapters for better insight.

## 2 Classification of Wood Waste

Wood waste is a byproduct of wood-related activities, such as from the forest or agriculture sector and mill operations, as well as non-commercial wood resources. Wood waste classification can be clarified as follows [3, 4]:

(1) Material left in the forest in the form of undesired, underutilized species, tops, high stumps, cull logs or bolts, undersized yet damaged trees, shattered material, etc.
(2) The byproducts of mechanically processing wood, including sawdust, shavings, bark, slabs, edgings, trims, faulty pieces, and veneer log cores.
(3) The waste wood products from several industries, including the railroad construction business, individual houses, and the wood packaging, demolition, and building industries.
(4) The extractives, lignin sugars, hemicelluloses, and cellulose that are not recovered in pulp production; the chemicals and gases that are not recovered in the production of charcoal; and the bark that is taken from wood prior to pulping.
(5) The enormous waste is caused by the destruction of wood by fire, insects, and decay in the forest, during storage and use, and by mechanical wear and breakage during service or storage.

Applications for both energy- and non-energy-related uses can be made from this waste wood. The creation of composite boards, surface products, composting, and cement boards are examples of non-energy uses for wood waste, as are the combustion, cogeneration, pellet, and briquette processes.

## 3 Opportunities for Utilization of Wood Waste

### *3.1 Biomass for Energy*

It is common to practice using logging-related wood debris as a source of biomass for the creation of energy [5]. The stems, branches, and other leftover parts of trees and other plants were utilized to create the woody biomass. Typically, these remnants of forestry operations are burned or left on the ground, which pollutes the environment. The above-ground dry weight of wood in the stem and limbs of live trees with a minimum diameter of 2.5 cm (1 in) was therefore suggested as a method for obtaining woody biomass [6]. Dead plants, foliage, seedlings, and other forest-growing vegetation are removed before harvesting the woody biomass. For instance, most developing countries, such as Nigeria and Indonesia, are actively developing their forest and collecting wood waste to be used as an affordable fuel supply to prevent the higher cost of fossil fuels [7]. Aside from collecting the wood waste from the forest, they also utilize sawdust from sawmills as an investment asset to generate energy. In this context, sawdust has been fabricated into the form of compacted briquettes or pellets by the densifying process [8]. Niño et al. demonstrated that the mechanical properties of the briquettes are highly dependent on the type of wood waste used in the processes, thereby affecting the quality of the energy produced [9]. In addition, the timber industry is also reported to produce sawdust briquettes to generate energy such as heat and electricity. The sawdust briquette from the timber industry successfully supplied large mass quantities, thereby minimizing waste disposal problems.

Other commercial producers of wood products and paper in the United States used the waste from lumber mills and paper mills to generate steam and energy. It is observed that using wood waste has lowered the manufacturing cost instead of purchasing electricity to operate the facilities. In 2020, it was reported that about 2.2 million households in the U.S. utilized wood waste as the primary energy source for space heating fuel [10]. Recent studies demonstrate that wood waste from the bark can also be used as an effective energy generation by gasification. When the synthetic gas from the gasification of wood waste is injected into an internal combustion engine, heat and electricity are produced. In comparison to raw sawdust and wood pellets, it was claimed that bark waste might improve the calorific value in the form of briquettes [11].

Crushed wood waste can be burned in boilers to provide thermal energy in the form of steam or hot water at the interval. In this regard, a device transports the crushed wood waste into a combustion chamber and burns it in a fuel bunker, supplying fuel continuously and regularly [12]. The boilers are typically finished with a solid fuel boiler, ash removal equipment, and chimneys fitted with safety systems and combustion gas purification systems. Another study reported that wood waste in pellet form is much more suitable for combustion using boilers compared to gasification and is increasingly used in many countries [13]. The generated heat energy is sufficient for maintaining industrial enterprises, social objects, and household consumers.

In accordance with the European Commission's waste management guidelines policy, one example of energy recovery is the combustion of wood waste. Since most wood products use resins or adhesives, the wood waste may occasionally contain glues [14]. The combustion emissions will contain one of the following compounds: $CO_2$, $H_2O$, $SO_2$, $N_2/NO_2$, and $O_2$ after the full oxidation of all fuel components made up of carbon, hydrogen, sulfur, nitrogen, and oxygen. Contrarily, insufficient air-fuel mixing in the combustion chamber will result in incomplete combustion, which will result in the production of carbon monoxide (CO), polycyclic aromatic hydrocarbons (PAHs), polychlorinated dibenzo-p-dioxins/dibenzofurans (PCDD/F), and polychlorinated biphenyls (PCB) [15]. According to a review by Cesprini et al., several adhesive resins with high nitrogen content may have a negative impact on $NO_x$ production during combustion [16]. The production of $NO_x$ has been significantly impacted by the use of timber biomass and adhesives as solid biofuels. Particularly, the quantities of oxides in the exhaust emissions are increased by the nitrogen-based resins. Since it is a challenge to reduce greenhouse gases emission during waste combustion of wood waste high in carbon, oxygen, and hydrogen, other alternatives to replace traditional adhesives have been developed. In this regard, wood glues were derived from bio-based molecules such as proteins, polysaccharides, lignin, and tannin to ensure reusable wood waste energy. The bio-based adhesives were also reported could overcome the emission of pollutants during combustion.

On the other hand, pollutants mean that certain wood waste has limited potential as reusable biomass. Waste wood burning in combined heat and power (CHP) facilities is becoming a routine practice in many European nations due to the low pre-treatment costs [17]. For instance, the European Union (EU28) produced about 48.46 Mt of wood wastes (including lumber residues), of which 48% were burned with energy recovery, 49% were recycled, 2% were burned, and 1% were dumped [18]. However, it was found that contaminants in the form of deposits and alkali chlorides occurred during the combustion of waste wood. These contaminants raise the possibility of boiler corrosion, slagging, and fouling in biomass CHP plants [19]. Consequently, a small number of studies discovered that using additions such as coal fly ash and halloysite lessens the environmental effects of creating heat and electricity from wood wastes. The additives replace other environmentally damaging energy-producing methods [20]. According to estimates, the use of additives in the combustion of scrap wood might bring potential greenhouse gas emissions down to 2.03 Gt $CO_2$ eq/yr.

In addition to using wood waste to generate energy, the pyrolysis process may also turn lignocellulosic materials derived from wood into biofuel [21]. Harvested crops and forest trash are gathered, and energy is recovered through thermal breakdown, which results in the breakdown of polymeric chain components without oxygen and the production of char, gaseous chemicals, and oil [22]. A recent study showed that pyrolysis has become a successful method for converting biomass materials derived from wood into liquid fuels. Compared to normal diesel oil, pyrolysis oil often has a higher energy density and is easier to transport.

As was mentioned, it appears that the non-fossil energy market for wood waste is beginning to take off. In conclusion, there are two ways that wood waste can be transformed into biomass: directly through combustion and indirectly through gasification and pyrolysis. These procedures for recovering the materials demonstrated that they may be used as fuels or to provide electrical and thermal energy. Additionally, it is discovered that recycling the urban wood waste produced by trimming bushes and trees in park areas and city streets offers significant potential for energy production [11]. Other waste materials selection and preparation factors, such as particle size and moisture content, would also determine whether the pellets and briquettes produced will provide adequate fuel. Overall, proper wood waste recovery management is critical in raising sustainability requirements.

## *3.2 Secondary Products for the Construction Industry*

Woody biomass from construction and demolition (C&D) is the second-largest component that produces wood waste. In 2018, it was estimated that the C&D component contributes 20–30% of waste; from that, approximately 10 percent of the amount was deposited into landfills. The Construction Materials Recycling Association (CMRA) also reported that 29 million tons of waste were produced and available for recycling after recovery or combustion. It shows that the wood waste from C&D experiences rapid expansion of recycling activities. However, recycling alone might not be enough to maximize wood waste utilization from C&D activities as it might produce a lesser quality product. Therefore, recovery by upcycling into secondary products seems more trustworthy. It allows the substitution of virgin resources as well as minimizes the environmental burdens in the scope of industrial ecology.

The 2030 Agenda for Sustainable Development Goals, which has 169 linked targets, was introduced by the United Nations in 2015. This agenda's primary goal is to offer a solution to the global $CO_2$ emissions produced by the construction industry as well as other detrimental environmental effects. The need for a better solution that involves building construction using wood waste is essential because construction activities are unavoidable due to the growing human population.

In this context, many researchers have dedicated the use of wood waste to the construction industry by producing particleboard [23, 24]. The high-quality particleboard was obtained from the wood waste formwork and further transformed into cement-bonded particleboard using magnesium oxide cement (MOC). The product was also reported to use a green cementitious binder which exhibits practical and eco-friendly management options for the construction industry [25]. The use of green cement and recycled wood waste has been proven to be superior to that of their conventional and/or virgin counterparts in terms of the life cycle environmental benefits as tested by Life cycle assessment (LCA; ISO 14040-14,044) [26, 27]. The cement-bonded particleboard is well known for its lightweight interior panels, noise barriers, partition walls, and ceilings [28]. It is worth mentioning that the addition of the MOC and wood waste successfully transformed the conventional ordinary Portland cement (OPC) and virgin wood into the production of novel cement-bonded particleboard. On the other hand, rapid-shaping cement-bonded particleboards can be obtained by adding magnesia-phosphate cement (MPC) with timber formwork made from Masson pine [29]. The work presented that MPC particleboards are more advantageous than OPC-based particleboards due to the fast-setting property and short compression time (5 min) [25].

Meanwhile, various types of post-industrial wood waste, including mixed hardwood/softwood powder and bark, can be liquefied and applied as bio-adhesives for particleboards [30]. It was reported that the particleboards produced from recycled wood bonded with urea-formaldehyde resin-modified liquefied wood (LW) approximately 20% qualified the requirements of European standards concerning particleboards. The application of LW in particleboard production significantly increases the use of recycled wood as raw material and expands the exploitation of lignocellulosic biomass [31] reported that a similar liquefied process was utilized on waste pine wood to produce polyurethane foam. The liquefied pine-based polyol was applied to synthesize melamine phosphate modified wood type polyurethane for flame retardant properties. The properties are essential to decreasing fire-related casualties and property loss in construction industries.

Previous studies found that most particleboards were produced by binding the waste wood particles with a thermosetting binder such as formaldehyde, phenol formaldehyde, phenol-resorcinol formaldehyde, and melamine–formaldehyde [31–33]. However, formaldehyde has adverse effects on human health, thereby different solutions to replace it are widely researched, such as employing expanded polystyrene (EPS) waste as a binder [34]. Masri et al. [35] reported that particleboards from date palm and EPS wastes exhibit reliable fiber-matrix interface adhesion with excellent bending strength and stress values of 0.78 GPa and 2.84 MPa [35]. It was also found that particleboards from sawdust and low-density EPS foam waste could form water-resistant particleboards [36]. It is due to the high mobility of EPS resin on the wood surface that forms hydrophobic films within the cells. In addition, formaldehyde-free particleboards were investigated using high-density wood–polyethylene composites from rubberwood flour and sludge [37]. Song et al. [38] also synthesized formaldehyde-free and water-resistant boards with high physicomechanical plywood using polypropylene film as a binder [38]. The results showed that the polypropylene film bonded plywood more than the urea formaldehyde bonded plywood.

Aside from particleboard, wood waste can be mixed into concrete or mortar for building materials. For instance, wood waste aggregation in gypsum or ash wood waste is introduced in mortars or concrete mixture [39–41]. Ramos et al. [42] first reported that wood waste ash was successfully utilized as a pozzolanic partial substitution material for cement, with minimal strength loss. The durability of the product is significant and thus contributes to sustainable construction. It is also found that using the optimal percentage of biochar in the concrete mixture would positively affect the concrete and subsequently aid in the development of building materials for a sustainable economy [43]. Wood waste powders and fibers are also recommended as raw materials for manufacturing cement mortars [44]. They support delivering a more environmentally friendly alternative to the current waste management choices for recycling industrial trash and conserving natural resources.

It's interesting to note that wood waste is also discussed in the formulation of low-strength concrete materials [45]. Research has already been done to create a novel material called "Wood-crete," which uses concrete products and wood waste as fillers for wall panels and hollow blocks as a thermal insulators [46]. With a density of 473 kg/m$^3$, the "Wood-crate" forms thin blocks that act as insulating materials. Its thermal conductivity ranges from 0.046 to 0.069 W/m K, and its tolerable compressive strength is 0.08. Overall, using various wood wastes in cementitious composite successfully improves the thermal insulation properties of the building construction element.

## *3.3 Additive Manufacturing (AM)*

The possibility of using wood waste from primary and secondary wood processing in additive manufacturing technology is higher. Using additive manufacturing (AM), which has cheap manufacturing costs and little waste, it is possible to create objects with complicated shapes more automatically [47]. Wood waste is thought to be one of the most effective natural resources for lowering the use of petroleum resources [48]. Wood waste is included in AM because it has better characteristics and a smaller environmental impact. The capacity to be upcycled or downcycled at their end of life is improved when using wood waste with biodegradable or recycled polymers.

The potential of wood waste as additive material in various 3D-printing technologies is reviewed by Krapež et al. [47]. It has been found that a variety of performance levels in the compounds can be achieved by combining wood waste particles with polymers in the proper proportions in terms of size, distribution, and content. In this sense, wood fibers have a low aspect ratio (length/width), whereas wood flour contains random shapes and dimensions that are roughly the same. Therefore, when the polymer matrix is reinforced with distinct wood types, different mechanical strengths would be attained. It is anticipated that the total strength of wood-plastic composites made from wood fiber will be higher than that of wood flour. Wood fiber is preferable because 3D printing comprises methods for binding the material and deposition of the material [49]. Sawdust was mentioned as another inexpensive filler and reinforcement for bending and tensile strength for AM use in addition to wood fiber [50].

Fused Deposition Modelling (FDM), which has been extensively studied in relation to 3D printing, is now the production process utilized for desktop 3D printers. To create wood-polymer filaments, the wood particles are frequently included with FDM materials. The foundation of long-used wood-plastic composites (WPC) serves as the basis for the technology of wood-plastic filaments. WPC is well known for being used as an alternative to real wood for door and window profiles, packaging, automotive dashboard components, and other applications. As its creation is based on melting thermoplastics mixed with wood particles and extruding via a matrix or in molds, WPC exhibits favorable hybrid qualities between wood and plastic. Given that the pre-preparation of WPC entails grinding the wood particles to a specified size before mixing them with polymers and additives, it is important to note that a variety of wood waste is viable for this use.

Therefore, wood industry sawdust appears to be a valuable waste source for producing high-quality 3D polymer materials, as Narlıoğlu et al. [51] reported. Waste pine sawdust was employed in the study as reinforcement and was extruded with polylactic acid (PLA) polymer. Other than that [52], reported on the use of wood flour, lignin, and cellulose nanofibers as functional additives and reinforcements in thermoplastic and thermoset matrices used in 3D printing. Waste pine sawdust was employed in the study as reinforcement and was extruded with polylactic acid (PLA) polymer [52].

Furniture trash may be recycled and used as a feedstock for wood-plastic filaments, according to a study by Pringle et al. [53]. In order to make wood mixable with PLA, it has been found that solid slabs and sawdust from medium-density fiberboard (MDF), low-density fiberboard (LDF), and melamine wood waste can be milled and ground into powder. The author claim that the mixture with 30% wood waste is the most usable but did not take environmental implications into account.

Wood waste could also be used as panel plywood or veneers by laminating object manufacturing (LOM) technology [54]. LOM is considered an AM technology where complex geometries and higher strength of wood are achieved by cutting and laminating veneers. The process is more advantageous if compared with subtractive manufacturing techniques. Furthermore, lesser wood defects can be achieved by manipulating the orientation of the layers, thereby producing higher-strength products.

Overall, employing wood waste in AM technology, particularly 3D printing, has a wide range of applications and motivations. Products manufactured from renewable, biodegradable, and non-petroleum-based materials, such as wood waste, are becoming more and more in demand from both consumers and businesses.

## 4 Challenges in Wood Waste Utilization

There are still some challenges to overcome despite the opportunities presented by wood waste. It frequently has to do with wood waste that has been chemically preserved, such as waste wood that has been preserved with older preservatives or formaldehyde adhesives. Because recycling this wood debris has a negative impact on the environment, it could have major health consequences. In addition, when nonbiodegradable binders are applied, the secondary product of wood waste with binder could not be completely biodegradable [55].

A variety of environmental effects could be led by wood products at different stages of the production process, from harvesting to disposal. Environmental effects are mostly caused by the energy required to make wood products and the greenhouse gas emissions (GHG) that occur throughout the manufacturing process from raw materials to finished goods. Despite the fact that forests and wood function as carbon sinks by absorbing carbon dioxide from the atmosphere through carbon sequestration, the production of wood products emits carbon [56]. However, up to 17% of GHG emissions into the atmosphere are attributed to the forestry sector as a whole and tree removal due to deforestation [57]. Other types of environmental effects associated with wood products include the transportation of timber commodities, the use of chemicals, and wood waste [58].

In the process of making wood, a variety of chemicals are used, mainly for coating the finished product, applying adhesive, and preservation treatment. Although these substances have increased the longevity of wood products, they can also have a negative influence on the environment because of the components they contain. Another significant environmental concern, for instance, is the disposal of wood from demolished construction sites that still employ high quantities of preservatives. There are laws against the use of dangerous chemicals in several countries.

Although adhesives, which are materials made of both natural and manmade substances, are necessary for gluing wood components into wood products, they may have negative environmental impacts [59]. Phenol-formaldehyde (PF) and urea-formaldehyde (UF) are the two adhesives most frequently employed in outdoor settings because of their excellent weather- and water-resistance properties [60]. Even cured adhesives that are thought to be safe and innocuous might release chemicals that are bad for the environment and people [59].

Wood coatings shield the wood from harmful environmental elements such as moisture radiation, mechanical, chemical, and biological deterioration. However, they have the ability to produce volatile organic compounds and contain organic solvents or liquid (VOC). From the viewpoint of air pollution, human health, and safety, VOCs, such as those containing chlorofluorocarbon, are seen as a major environmental issue [61].

The disposal of wood products has several negative environmental repercussions, particularly in urban settings. Commercial and industrial garbage, construction and demolition work, pallets and packaging, and utilities are the main sources of urban wood waste [62]. When products are thrown away rather than reused, recycled, or repaired, they contribute to toxic waste that can leach from landfills, they take up a lot of space in landfill sites, and they necessitate the construction of new waste disposal facilities. This results in outside pollution and GHG emissions from transport from the source to the landfill site.

Similar to this, burning used materials results in the release of gases, pollution, and smoke into the environment. For example, solid contamination causes disposal issues by decreasing burning efficiency and producing waste, whereas too much chlorine during burning also reduces burning efficiency and can lead to the production of dioxins [63]. It appears that additional research is required to study the engineered wood products made by using various types and sizes of wood waste in order to gain a better understanding of the performance, given the identification of potential sources of environmental impact and the difficulties in utilizing wood waste. The utilization of thermal energy produced by sawmill byproducts, ecologically friendly chemical preservatives, adhesives, coatings, and integrated industrial sites are just a few strategies that can be used to address the related problems.

## 5 Future Outlook

According to a prediction made in 2010, Malaysia's forest area will produce less wood as a result of the shrinking size of forested regions and the designation of the majority of the land for development. Based on these assumptions, the Ministry of Plantation Industries and Commodities constructed a number of forest plantations to address deforestation issues. Several other governments forbid the export of lumber in order to protect their remaining forests. Additionally, the availability of raw materials in the wood sector may be negatively impacted by rising costs for raw materials as a result of greater competition for them on the global market. The local industry should adjust its practices in this regard to boost domestic wood production and decrease imports [64].

The sustainability of raw materials has become a concern in the wood industry. Numerous agricultural wastes, such as oil palm trunk (OPT), an oil palm frond (OPF), empty fruit bunch (EFB), palm fruit bunch (PPF), pineapple leaves (PALF), kenaf, bamboo, and jute, remain limited unutilized. Some of these products, such as OPT, have shown potential as alternative materials for the wood sector despite being considered raw material wastes. Due to the lengthy maturation period of forest trees, forest plantations produce few resources slowly. Alternative sources for the wood-based industry may include waste from plantations and other commodities. As a result, the government is launching a variety of initiatives to find alternate resources for the wood processing industry [64, 65].

The Malaysian government has launched many programs to look for alternate materials derived from other sources because it is aware of the limitations of the timber industry. One example of a substitute raw material that the wood processing sector might use is the employment of various chemicals that can produce agricultural waste. Southeast Asia's biocomposite market is still growing, and the shift to this substitute material is expected to limit the availability of skilled workers and the local labor force. The government developed numerous wood training institutions in Malaysia and abroad to support the biocomposite sector. Resource conservation and environmental protection have been hotly contested issues among the global community for more than a decade. Global demand for "green" or environmentally friendly items has increased, particularly in poorer nations. These changes put pressure on Asian wood producers and their goods [64, 66].

The issue with the wood market is that it could limit the availability of natural resources while also increasing the cost of those resources. To maintain ongoing growth, stability, and competitiveness, the government should promote the use of substitute wood products. Therefore, scientists are researching biocomposite materials that could replace the wood products now on the market. Biocomposite materials, which provide wood product manufacturers an alternative by combining biomaterials or natural fibers, have been well embraced on the international market. Additionally, biocomposite materials need to be readily available globally and able to meet market demand. Manufacturers are becoming more confident in biocomposite-based materials, which are innovative and could be useful for furniture and other uses. In addition, the most crucial considerations for manufacturers when looking at biocomposite as an alternative material for the manufacturing industry are the availability of natural fiber-like resources from forests and the use of agricultural wastes in biocomposite materials [34, 64].

On the market for agro-based biocomposites, one might anticipate fierce competition from overseas companies [64]. Future challenges could affect the market for composite-based goods, including strict requirements, purchasing rules and policies, environmental issues, and law enforcement. Southeast Asia has a solid history in the global wood market, but its market has grown competitive. Each country must provide a range of creative answers to these problems in order to stay competitive in the business.

## References


1. Monthly Energy Review (2022). https://www.eia.gov/totalenergy/data/monthly/pdf/mer.pdf
2. Pandey S (2022) Wood waste utilization and associated product development from under-utilized low-quality wood and its prospects in Nepal. SN Appl Sci 4(6):168
3. Besserer A et al (2021) Cascading recycling of wood waste: a review. Polymers 13(11):1752
4. Kong H (2000) Current status of biomass utilization in Malaysia. Forest Res Inst Malays 1:15
5. Fernandes U, Costa M (2010) Potential of biomass residues for energy production and utilization in a region of Portugal. Biomass Bioenergy 34(5):661–666
6. Morhart C et al (2016) Above-ground woody biomass allocation and within tree carbon and nutrient distribution of wild cherry (Prunus avium L.)—A case study. Forest Ecosyst 3(1):1–15
7. Shafie SM, Othman Z, Hami N (2017) Potential utilisation of wood residue in Kedah: a preliminary study. J Technol Oper Manag 60–69
8. de Jong J et al (2017) Realizing the energy potential of forest biomass in Sweden-how much is environmentally sustainable? For Ecol Manag 383:3–16
9. Niño A, Arzola N, Araque O (2020) Experimental study on the mechanical properties of biomass briquettes from a mixture of rice husk and pine sawdust. Energies 13(5):1060
10. Residential Energy Consumption Survey (2020) Housing Characteristics. https://www.eia.gov/consumption/residential/data/2020/hc/pdf/HC%201.1.pdf
11. Khudyakova G, Danilova D, Khasanov R (2017) The use of urban wood waste as an energy resource. In: IOP Conference series: earth and environmental science. IOP Publishing
12. Juszczak GR, Miller M (2016) Detour behavior of mice trained with transparent, semitransparent and opaque barriers. PloS One 11(9):e0162018
13. Gehrig M et al (2016) Influence of firebed temperature on inorganic particle emissions in a residential wood pellet boiler. Atmos Environ 136:61–67
14. Zhao L-F et al (2011) State of research and trends in development of wood adhesives. For Stud China 13(4):321–326
15. Wei S et al (2012) Tolerant mechanisms of Rorippa globosa (Turcz.) Thell. hyperaccumulating Cd explored from root morphology. Bioresour Technol 118:455–459

16. Cesprini E et al (2020) Energy recovery of glued wood waste—A review. Fuel 262:116520
17. Vis M, Mantau U, Allen B (2016) CASCADES: study on the optimised cascading use of wood. Publications Office
18. Eurostat (2019). Treatment of waste by waste category, hazardousness and waste management operations. https://ec.europa.eu/eurostat/web/waste/data/database
19. Åmand L-E et al (2006) Deposits on heat transfer tubes during co-combustion of biofuels and sewage sludge. Fuel 85(10):1313–1322
20. Corona B et al (2020) Consequential life cycle assessment of energy generation from waste wood and forest residues: the effect of resource-efficient additives. J Clean Prod 259:120948
21. Rao YK et al (2022) Investigation on forestry wood wastes: pyrolysis and thermal characteristics of Ficus religiosa for energy recovery system. Adv Mater Sci Eng 2022
22. Sheth PN, Babu B (2008) Differential evolution approach for obtaining kinetic parameters in nonisothermal pyrolysis of biomass. Mater Manuf Process 24(1):47–52
23. Pędzik M, Janiszewska D, Rogoziński T (2021) Alternative lignocellulosic raw materials in particleboard production: a review. Ind Crops Prod 174:114162
24. Hossain MU et al (2018) Environmental and technical feasibility study of upcycling wood waste into cement-bonded particleboard. Constr Build Mater 173:474–480
25. Wang L et al (2016) Value-added recycling of construction waste wood into noise and thermal insulating cement-bonded particleboards. Constr Build Mater 125:316–325
26. Ruan S, Unluer C (2016) Comparative life cycle assessment of reactive MgO and Portland cement production. J Clean Prod 137:258–273
27. Dylewski R, Adamczyk J (2014) Life cycle assessment (LCA) of building thermal insulation materials. Eco-efficient construction and building materials. Elsevier, pp 267–286
28. Karade S (2010) Cement-bonded composites from lignocellulosic wastes. Constr Build Mater 24(8):1323–1330
29. Wang L et al (2018) Upcycling wood waste into fibre-reinforced magnesium phosphate cement particleboards. Constr Build Mater 159:54–63
30. Janiszewska D, Frąckowiak I, Mytko K (2016) Exploitation of liquefied wood waste for binding recycled wood particleboards. Holzforschung 70(12):1135–1138
31. Yue D et al (2017) Liquefaction of waste pine wood and its application in the synthesis of a flame retardant polyurethane foam. RSC Adv 7(48):30334–30344
32. Mamza PA et al (2014) Comparative study of phenol formaldehyde and urea formaldehyde particleboards from wood waste for sustainable environment. Int J Sci Technol Res 3(9):53–61
33. Tay CC, Hamdan S, Osman MSB (2016) Properties of sago particleboards resinated with UF and PF resin. Adv Mater Sci Eng 2016
34. Keskisaari A, Kärki T (2017) Raw material potential of recyclable materials for fiber composites: a review study. J Mater Cycles Waste Manag 19(3):1136–1143
35. Masri T et al (2018) Characterization of new composite material based on date palm leaflets and expanded polystyrene wastes. Constr Build Mater 164:410–418
36. Akinyemi BA et al (2019) Durability and strength properties of particle boards from polystyrene—wood wastes. J Mater Cycles Waste Manag 21(6):1541–1549
37. Homkhiew C et al (2018) Potential utilization of rubberwood flour and sludge waste from natural rubber manufacturing process as reinforcement in plastic composites. J Mater Cycles Waste Manag 20(3):1792–1803
38. Song W et al (2017) Utilization of polypropylene film as an adhesive to prepare formaldehyde-free, weather-resistant plywood-like composites: process optimization, performance evaluation, and interface modification. BioResources 12(1):228–254
39. del Río-Merino M et al (2022) A review of the research about gypsum mortars with waste aggregates. J Build Eng 45:103338
40. Tamanna K et al (2020) Utilization of wood waste ash in construction technology: a review. Constr Build Mater 237:117654
41. Ayobami AB (2021) Performance of wood bottom ash in cement-based applications and comparison with other selected ashes: overview. Resour Conserv Recycl 166:105351

42. Ramos T, Matos AM, Sousa-Coutinho J (2013) Mortar with wood waste ash: mechanical strength carbonation resistance and ASR expansion. Constr Build Mater 49:343–351
43. Sirico A et al (2021) Biochar from wood waste as additive for structural concrete. Constr Build Mater 303:124500
44. Ince C, Tayançlı S, Derogar S (2021) Recycling waste wood in cement mortars towards the regeneration of sustainable environment. Constr Build Mater 299:123891
45. Batool F et al (2021) Effectiveness of wood waste sawdust to produce medium-to low-strength concrete materials. J Build Eng 44:103237
46. Aigbomian EP, Fan M (2013) Development of Wood-Crete building materials from sawdust and waste paper. Constr Build Mater 40:361–366
47. Krapež Tomec D, Kariž M (2022) Use of wood in additive manufacturing: review and future prospects. Polymers 2022, 14, 1174. 2022, s Note: MDPI stays neutral with regard to jurisdictional claims in published
48. Mazzanti V, Malagutti L, Mollica F (2019) FDM 3D printing of polymers containing natural fillers: a review of their mechanical properties. Polymers 11(7):1094
49. Gao W et al (2015) The status, challenges, and future of additive manufacturing in engineering. Comput Aided Des 69:65–89
50. Ayrilmis N, Kariž M, Kitek Kuzman M (2019) Effect of wood flour content on surface properties of 3D printed materials produced from wood flour/PLA filament. Int J Polym Anal Characteriz 24(7):659–666
51. Narlıoğlu N, Salan T, Alma MH (2021) Properties of 3D-printed wood sawdust-reinforced PLA composites. BioResources 16(3)
52. Obielodan J et al (2019) Characterization of PLA/lignin biocomposites for 3D printing. In: 2019 International solid freeform fabrication symposium. University of Texas at Austin
53. Pringle AM, Rudnicki M, Pearce JM (2018) Wood furniture waste—Based recycled 3-D printing filament. For Prod J 68(1):86–95
54. Tao Y, Yin Q, Li P (2020) An additive manufacturing method using large-scale wood inspired by laminated object manufacturing and plywood technology. Polymers 13(1):144
55. Adhikari S, Ozarska B (2018) Minimizing environmental impacts of timber products through the production process "From Sawmill to Final Products." Environ Syst Res 7(1):6
56. Le Quéré C et al (2009) Trends in the sources and sinks of carbon dioxide. Nat Geosci 2(12):831–836
57. Baccini A et al (2012) Estimated carbon dioxide emissions from tropical deforestation improved by carbon-density maps. Nat Clim Chang 2(3):182–185
58. Wootton JT (2012) Effects of timber harvest on river food webs: physical, chemical and biological responses
59. Yang M, Rosentrater K (2015) Environmental effects and economic analysis of adhesives: a review of life cycle assessment (LCA) and techno-economic analysis (TEA). In: 2015. ASABE annual international meeting. https://core.ac.uk/download/pdf/38931621.pdf
60. Zhang W et al (2013) Lignocellulosic ethanol residue-based lignin–phenol–formaldehyde resin adhesive. Int J Adhes Adhes 40:11–18
61. de Meijer M (2001) Review on the durability of exterior wood coatings with reduced VOC-content. Prog Org Coat 43(4):217–225
62. Taylor J, Warnken M (2008) Wood recovery and recycling: a source book for Australia
63. Taylor J et al (2005) Recycling and end-of-life disposal of timber products
64. Suhaily S et al (2012) A review of oil palm biocomposites for furniture design and applications: potential and challenges. BioResources 7(3):4400–4423
65. Salim N, Abdullah Y, Hashim R (2021) Potential of Seaweed (Kappaphycus alvarezii) as a Particleboard. In: Materials science forum. Trans Tech Publ
66. Keya KN et al (2019) Natural fiber reinforced polymer composites: history, types, advantages and applications. Mater Eng Res 1(2):69–85

# Life Cycle Assessment of Wood Waste

**Siti Noorbaini Sarmin, Mohammad Jawaid, and Rob Elias**

**Abstract** Wood waste can be decreased without significantly impacting the world's forests by enhancing the effectiveness of primary wood consumption and utilising raw wood supplies produced by sustainable forest administration. Wood waste can be utilised to create a wide range of goods, including engineered wood products, energy production (heat and electricity), mulch, and animal bedding. These low-cost, underutilised feedstocks have the potential to boost the added value of wood waste. Life cycle assessment (LCA) is a method for examining the environmental impact of materials, products, and services, and it is intended to aid in the development of sustainable decisions. This chapter discusses important difficulties concerning the life cycle assessment (LCA) of wood waste products. We looked at the process by which LCA evaluates the whole environmental effects of wood output, whether as input or output, over the course of a product's life, from raw material to end-of-life disposal or rebirth as a new product.

**Keywords** Life cycle assessment · Wood waste · Environmental sustainability

## 1 Introduction

Timber products are regarded to be natural resources that are renewable and sustainable. Similar to other products, timber products can have a variety of negative environmental effects at different points throughout the supply chain, from harvesting to disposal [1]. The consumption of energy necessary to manufacture timber goods, as

S. N. Sarmin (✉)
Department of Wood Technology, Faculty of Applied Science, University Teknologi MARA, 26400 Jengka, Pahang, Malaysia
e-mail: baini@uitm.edu.my

M. Jawaid
Department of Bio-Composite Laboratory, Institute of Tropical Forestry and Forest Products, University Putra Malaysia, 43400 UPM Serdang, Selangor, Malaysia

R. Elias
The Biocomposites Centre, Bangor University, Bangor LL57 2UW, UK



S. N. Sarmin et al. (eds.), *Wood Waste Management and Products*, Sustainable Materials and Technology, https://doi.org/10.1007/978-981-99-1905-5_2

well as the creation of greenhouse gases (GHG) throughout the production process after raw materials to final products, are significant resources of environmental effects [2, 3]. The importance of satisfying this demand while minimising environmental repercussions is being recognised more and more as environmental awareness and demand for forest products expand.

Most wood product manufacturing processes produce wastes such as sander dust, bark, chips, sawdust, hog fuel, trimmings, and shavings. Depending on the manufacturing stage, these can be either green (usually around 50% moisture content, wet foundation) or dry [3]. Higher value residues are sold as pulp chips, bark for landscaping, sawdust for animal bedding, or in an assortment of types to wood composite manufacturers for products like particleboard and fiberboard [4]. In the manufacture of wood, the wood residue is frequently used as a source of steam or direct heat for equipment like dryers [5]. One way for comprehending the complex interaction of components involved in wood waste utilisation, recycling and reused is life cycle assessment. LCA gives info regarding the current environmental influences of wood waste management, as well as the upstream environmental loads of wood waste materials and products.

The life cycle of wood begins with raw material extraction and ends with end-of-life dumping, and it is a linear method that generates a significant quantity of waste (Fig. 1) [6]. As a result, it is critical to adopt the circular economy throughout the entire life cycle of wood in order to achieve sustainable development. From log extraction to finished product, the production process for timber products contains multiple stages that might pollute the surrounding environment through land, air, and water contamination [2]. With the use of the LCA technique, it is possible to quantify the state of the environment now and identify areas that could be improved to lessen potential environmental effects in the future [6]. LCA assessments often distinguish the most significant drivers of environmental consequences, allowing for more focused efforts to reduce those influences. Energy and greenhouse gas balances have been shown to be critically dependent on how wood-based products are managed towards the end of their useful lives. Retrieval of post-use material for use as bioenergy is advantageous, whereas landfill disposal often has more negative consequences.

## 2 Recycled Wood Waste

Unwanted wood products from a variety of sources, including private homes, railroad construction, and the destruction and packaging of wood products, are known as "wood wastes" [7]. There are two sorts of wood trash: industrial garbage produced within the industry and final rubbish generated after the items have been consumed. This waste can be used as a secondary raw material source for providing energy as well as the development of innovative possibility goods such as chemicals, ethanol, and other lignocellulosic materials (Fig. 2). To meet the growing demand for wood

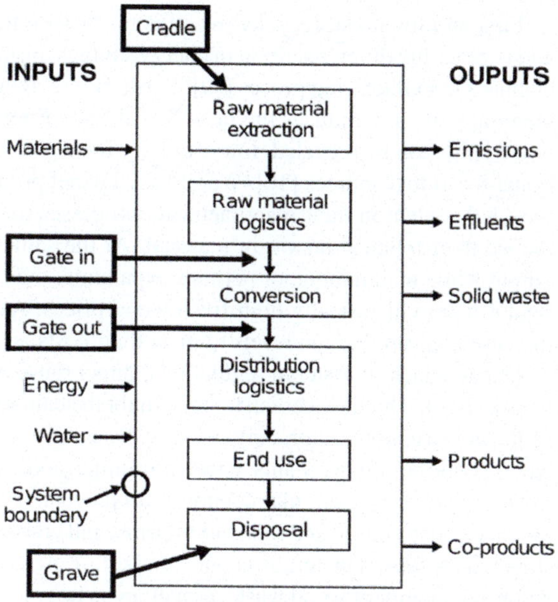


**Fig. 1** Life cycle of trees from the regeneration to the disposal of wood

without further destroying the world's forests, the timber industry may be able to reduce its environmental impact with the help of the prevention of wood waste [8].

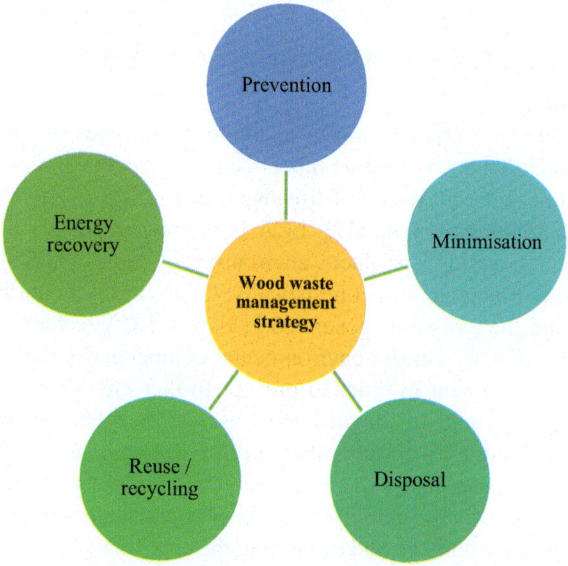


**Fig. 2** Wood waste management strategy

First and foremost, recycled wood is up to 40% less expensive than fresh wood and is generally dryer, resulting in energy savings during the drying stage of the particleboard manufacturing process [9]. Furthermore, recycling scrap wood extends the material's life and reduces the need for additional lumbering. While higher quality wood chips can be recycled, lower quality waste wood can be used to replace traditional fossil fuel sources [10]. Wood-based panel producers are working to enhance recycled content in their manufacturing processes to assure economic viability and reduce their reliance on virgin material. At the same time, they want to maximise output while maintaining, or perhaps expanding, their margins, which have become tighter due to historically high fresh wood prices. As a result, they are required to increase their production levels to meet their costs and function financially.

Due to a lack of material regulations, direct reuse of wood waste may not always be permitted, which makes customers doubt the caliber of repurposed materials [11]. In the construction sector, certain wood-based products such as formwork, pallets, wood-frame buildings, beams, window framing, and doors are recycled. Furthermore, in order to execute a circular economy, all players, including contractors, engineers, architects, demolition and renovation firms, and consumers, must collaborate to set a standard for reused materials and the market. A greater understanding of the composition and quality of wood waste, as well as an upgrade in present sorting procedures, are required. Trash could be effectively categorised based on chemical composition using automated sorting methods based on spectrometric detectors, particularly those that use medium and near-infrared light [3, 8]. Optimising recycling pathways may be aided by a stage of purification for the wood fibers or particles.

## 3 Life Cycle Assessments Add Value

The most efficient way to examine the full environmental impact of all energy and material flows, whether input or output, over the course of a product's life cycle, from raw materials to end-of-life disposal or rebirth as a new product, is through life cycle assessment (LCA) [12]. LCA is the internationally recognised standard approach for assessing a product's environmental impact (Fig. 3). LCA calculates a product's environmental effect holistically, taking into account both the resources consumed and the emissions generated [13]. The LCA technique not only makes it possible to quantify current environmental competencies, but also to find opportunities for improvement in order to prevent further environmental consequences. LCA findings normally distinguish the most significant drivers to environmental degradation, allowing for a concentrated effort to decrease such impacts.

LCA evaluates the environmental consequences of a product system or activity holistically by quantifying the energy and materials used, the wastes discharged into the environment, and estimating the environmental impacts of the energy, materials, and wastes. The aim and scope phase, the life cycle inventory phase, the life cycle impact assessment phase, and the interpretation phase are the four stages of LCA [9, 14]. The second step of LCA is the Life Cycle Inventory (LCI), which includes

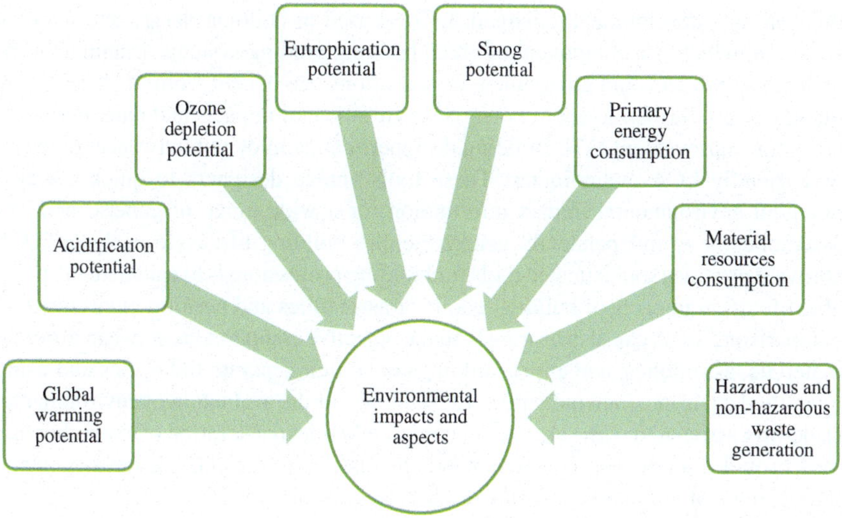


**Fig. 3** LCA on potential environmental impacts and product

activities related to the search, collection, and interpretation of data necessary for measuring environmental impacts. This study makes use of gate-to-gate LCI data. Product input included raw materials, assembly, material packaging, and electricity. The process's output comprises both product and non-product output (NPO). NPO are sawdust, wood chips, and carbon emissions. Energy consumption is measured by logging all machines employed in manufacturing by-product inside the boundary of the production system under consideration. The power on each machine (kw) is then identified and multiplied by the product's time (hour) by the machine to obtain the machine's energy consumption to create the product in kwh [3].

Types and levels of environmental impact are calculated in Life Cycle Impact Assessment (LCIA) utilising the software Simapro Ecoinvent database v 7.1 and 2.0, as well as the impact assessment technique Eco costs 2012 v 3.03 [5, 8]. LCIA calculation in the Eco costs technique consists of many stages, including characterisation, normalisation, and a single score based on the resulting environmental impacts. The results of this indicator category are generated by assigning the Life Cycle Inventory data for emissions to soil, water, and air of the substances used to group representing environmental issues, the so-called impact category. The final stage of the LCA process is life cycle interpretation. Life cycle interpretation is the systematic identification, measurement, examination, and evaluation of information from LCI and LCIA data [15].

Several European countries, including Germany, Zurich, and Brussels, have made LCA necessary before issuing building permission [16]. Building products were initially analysed in order to construct and populate entire building tools. As the green building movement progressed over the last few decades, developing wood building products, as well as other forest-based products that incorporated LCA in their product development, emerged. LCA is used in building design and material selection, which is a component of green building grading systems. Manufacturers, architects, builders, and government organisations can benefit from LCA by giving quantitative information about potential environmental impacts and data to suggest areas for improvement [17]. Building designers can employ publicly accessible and user-friendly LCA technologies. These tools enable designers to quickly access potential environmental impact information for a wide range of generic building assemblies or to independently generate entire building life cycle analyses. LCA software provides sophisticated tools for building professionals to calculate the probable life cycle impacts of building goods or assemblies and perform environmental comparisons. LCA can also be used to make objective comparisons between different materials, assemblies, and entire buildings over their separate life cycles and using quantified environmental indicators. LCA allows for the evaluation of environmental trade-offs associated with selecting one material or design solution over another and thus provides an appropriate framework for assessing the relative environmental implications of alternative building design scenarios [8, 15, 16].

The Life Cycle Sustainability Analysis (LCSA) framework serves as a foundation for future LCA. It broadens the scope of current LCA from focusing solely on environmental consequences to include all three elements of sustainability (people, planet, and prosperity) [14]. It also broadens the scope from mostly product-related queries (at the product level) to questions about sectors (at the sector level) or even the economy as a whole (economy level). Furthermore, it broadens current LCA to encompass relationships other than technology ones, such as physical ones (including restrictions in accessible resources and land), economic and behavioural ones, and so on. Furthermore, normative issues like as discounting, weighing, and weak versus strong sustainability might be explicitly incorporated as part of deepening. Unlike LCA, LCSA is a transdisciplinary integration framework of models rather than a model in and of itself [18]. LCSA works with a variety of disciplinary models and guides the selection of the best ones for a given sustainability issue. The key difficulty thus is to structure, choose, and make a multiplicity of models practically available in respect to various types of life cycle sustainability questions.

Although this is perfectly compliant with ISO's criterion "there is no one technique for performing LCA", it represents a considerable departure from previous LCA practice. The expansion to economic and social repercussions also contradicts ISO's clear limitation to environmental considerations. Climate Change is the most significant environmental impact value caused by the usage of wood raw resources. It is advised that the usage of wood as a raw material for production be reduced in order to mitigate the effects of climate change. The LCA can be used to calculate the environmental impact of utilising wood.

## References

1. Moreno AI, Font R, Conesa JA (2017) Combustion of furniture wood waste and solid wood: Kinetic study and evolution of pollutants. Fuel 192:8
2. Cesprini E, Resente G, Causin V, Urso T, Cavalli R, Zanetti M (2020) Energy recovery of glued wood waste—A review. Fuel 262
3. Besserer A, Troilo S, Girods P, Rogaume Y, Brosse N (2021) Cascading recycling of wood waste: a review. Polymers 13(11)
4. Owoyemi JM, Zakariya HO, Elegbede IO (2016) Sustainable wood waste management in Nigeria. Environ Socio-Econ Stud 4(3):9
5. Plis A, Kotyczka-Morańska M, Kopczyński M, Łabojko G (2016) Furniture wood waste as a potential renewable energy source. J Therm Anal Calorim 125(3):15
6. Sahoo K, Bergman R, Alanya-Rosenbaum S, Gu H, Liang S (2019) Life cycle assessment of forest-based products: a review. Sustainability 11(17)
7. Faraca G, Boldrin A, Astrup T (2019) Resource quality of wood waste: The importance of physical and chemical impurities in wood waste for recycling. Waste Manag 87:12
8. Maier D (2021) Building materials made of wood waste a solution to achieve the sustainable development goals. Mater Des 14(24)
9. Cetiner I, Shea AD (2018) Wood waste as an alternative thermal insulation for buildings. Energy Build 168:10
10. Hossain MU, Poon CS (2018) Comparative LCA of wood waste management strategies generated from building construction activities. J Clean Prod 177:10
11. Rosas JG, Gómez N, Cara-Jiménez J, González-Arias J, Olego MÁ, Sánchez ME (2020) Evaluation of joint management of pine wood waste and residual microalgae for agricultural application. Sustainability 13(1)
12. Barjoveanu G, Pătrăuţanu OA, Teodosiu C, Volf I (2020) Life cycle assessment of polyphenols extraction processes from waste biomass. Sci Rep 10:12
13. Rios FC, Grau D, Chong WK (2019) Reusing exterior wall framing systems: a cradle-to-cradle comparative life cycle assessment. Waste Manag 94:15
14. Hossaini N, Reza B, Akhtar S, Sadiq R, Hewage K (2015) AHP based life cycle sustainability assessment (LCSA) framework: a case study of six storey wood frame and concrete frame buildings in Vancouver. J Environ Plann Manag 58(7):24
15. Pacheco-Torgal F, Cabeza LF, Labrincha J, De Magalhaes AG (2014) Eco-efficient construction and building materials: life cycle assessment (LCA), eco-labelling and case studies. Woodhead Publishing
16. Takano A, Hafner A, Linkosalmi L, Ott S, Hughes M, Winter S (2015) Life cycle assessment of wood construction according to the normative standards. Eur J Wood Wood Prod 73(3):13
17. Kayo C, Dente SM, Aoki-Suzuki C, Tanaka D, Murakami S, Hashimoto S (2019) Environmental impact assessment of wood use in Japan through 2050 using material flow analysis and life cycle assessment. J Ind Ecol 23(3):13
18. Pajchrowski G, Noskowiak A, Lewandowska A, Strykowski W (2014) Wood as a building material in the light of environmental assessment of full life cycle of four buildings. Construct Build Mater 52:8

# Concern on Wood Waste Utilization: Environment and Economic Evaluation

**Noorshashillawati Azura Binti Mohammad**

**Abstract** This chapter highlights the concern on wood waste utilization regarding on environment and economic evaluation. Wood waste is the part of the effluents that can comprise discarded wood, whole trees, stumps, or clipped branches. Wood waste is also derived from downstream (sawmills) to furniture, boards, and moldings. Forest waste (waste from deforestation) and residues from wood processing plants are also two types of wood waste associated with sawmill operations. Wood waste can be decreased without negatively impacting the world's forests by improving the productivity of primary wood consumption and using raw wood resources produced from sustainable forest management [1–5]. Due to the defects in the felled trees, the production of sawn timber is considered wasted. Only about 47% of the logs that reach the sawmill are converted into salable timber. The remaining residue containing 33% wood chips, 7% sawdust, 8% shavings, and 5% bark should be discarded or otherwise used [6]. The timber industry is an important industry in Malaysia. At the same time, the timber industry has a significant impact on the environment in general (air, water, and soil) and in particular on land and resource management. So, we must give emphasis to the solution. For example, the adoption of cleaner technology and waste minimization (Krajnc and Domac in Energy Policy 35:6010–6020, 2007). The main factors of environmental degradation are recognized as:

- Inefficient use of timber creates excessive waste and leads to the over-clearing of forests and plantations.
- Burning of branches and treetops in forests and plantations.
- Open burning of wood waste from industry on-site or off-site.
- Inadequate or unlicensed on-site incinerators
- Illegal disposal of waste (especially sawdust) into rivers, wastelands, and so on.

The vast amount of waste generated from wood processing operations in many countries presents challenging opportunities for utilizing wood waste. Consequently, the timber sector is anticipated to see both timber costs and waste disposal costs rise. Subsequently, wood waste is it is anticipated that wood waste will gradually become

N. A. B. Mohammad (✉)
Department of Wood Industries, Faculty of Applied Polymer Sciences, UiTM Pahang Branch, Tun Razak Jengka, 26400 BandarPahang, Malaysia
e-mail: noorshashillawati@uitm.edu.my

 
S. N. Sarmin et al. (eds.), *Wood Waste Management and Products*, Sustainable Materials and Technology, https://doi.org/10.1007/978-981-99-1905-5_3

a valuable resource. Wood waste is a sustainable, inexpensive, and widely accessible source of energy that has the potential to replace fossil fuels in a variety of uses, made up of heat, power, and biofuels. The expanded use of agricultural biomass can aid agriculturally based countries in achieving energy security and providing jobs without contributing to environmental damage [1, 4, 7, 8].

## 1 Wood Waste Resource

A better understanding of the content and element of wood waste is essential. Wood waste is also classified into two types: industrial waste generated from industries and final waste generated after the products have been used [3, 4]. The following categories make up wood waste:

Types of wood waste:

1. Bark
2. Slabs
3. Long offcuts (Trimmings)
4. Short offcuts (Short ends)
5. Peeler cores
6. Sawdust
7. Shavings
8. Sander dust
9. Rejects.

However, the wood waste can be divided into groups:

1. Bulk wastes, which include all larger wastes, are easily segregated.
2. Particle wastes (also known as silo wastes) are a mixture of tiny and fine wood particles resulting from various processes and normally collected in a silo via dust extraction system. These wastes are more difficult and expensive to separate.

Wood waste should be seen as a dynamic and changing material flow rather than a homogenous product. The largest producer of wood waste is the construction and demolition sector. Wood fraction accounts for 20–40% of construction and demolition waste in Europe. Additionally, the furniture business is a significant contributor to waste. Packaging accounts for a lesser part of total production. Wood recycling typically implies the reduction of the material to small particles (chips, fibers, etc.) that can be repurposed in the manufacturing of engineered wood products. Table 1 provides a categorization of wood wastes based on their source [3, 8, 9].

**Table 1** Classification of wood waste according to its origin [9]

| Origin | Type | Class |
|---|---|---|
| Packaging | Pallets and boxes (untreated, no MDF) | 1–2 |
| | Pallets and boxes (with MDF/treated wood) | 3 |
| Construction/demolition | Wood from construction and rebuilding (untreated, no MDF) | 1–2 |
| | Old wood from demolition and rebuilding (with MDF/treated wood) | 3 |
| Furnitures | Furniture (untreated, no MDF) | 1–2 |
| | Furniture (with MDF/treated wood) | 3 |
| | Furniture, upholstered | 3 |
| Others | Impregnated wood (wood treated with CCA, creosote or PSP) | 4 |
| | Composite building materials from demolition | 3 |
| | Miscellaneous (items made from plastic, glass, metal, carboard) | 3 |

## 2 Potential Usability of Wood Waste

Wood waste can be recycled into several products such as wood composites, power production (heat and electricity), composting, and animal manure. These low-cost, unused feedstocks have the potential to boost the viability of wood waste. As a result, the wood pellet sector has grown to capitalize on the potential given by the rising demand for renewable energy sources, creating long-term value for the bioeconomy [3, 6]. It is anticipated that bioenergy in British Columbia could generate over 1600 MW of heat and/or electricity and produce 3.2 million tons of wood pellets. The entire capacity of BC wood pellet generation in 2017 was 2.4 million tons, accounting for 66% of total Canadian wood pellet capacity but falling far short of the county's maximum manufacture [6].

Wood waste is a type of rubbish that comprises waste wood from several sources, like wood packing, destruction, and construction, the wood-based sector, private dwellings, and rail systems [1]. This garbage can be used as a secondary source of raw materials to produce energy and a wide range of new goods, such as chemicals, biofuels, and organic materials [2]. Lipophilic and hydrophilic residues in wood bark can be turned into high-value items like cosmetic chemicals and medications [3]. Yang et al. [4] reported that bio-oil derived from waste wood works well as an extender and modifier for petroleum asphalt binder in asphalt payment. Biofuel production and wood composites might be used as additional wood waste applications with a significant added value. In addition, minimizing wood waste could aid the timber industry in decreasing its environmental effect while simultaneously satisfying the rising wood demand without further deteriorating the world's forests [5]. Consequently, forest-based companies should highlight on reduce, recover, and increasing the usage of wood waste from harvested and processed wood.

A large variety of essential industrial goods can be produced from leftover wood and the leftovers of wood-based industrial operations. Sawmills collect 1 ton of sawdust, shavings, slabs, and edgings for every 1000 board feet of timber produced; almost 75% of this useless material is composed of wood, and 25% is bark [7]. It can be utilized for energy and non-energy purposes. The generation of energy from wood waste includes burning, cogeneration, pellets, and briquettes, while non-energy uses include the production of composite boards, surface products, compost, and cement board [8]. Multiple study has uncovered a few value-added strategies for transforming wood waste into profitable goods. For example, a study conducted in Finland identified several emerging markets for wood products (textiles, chemicals, biofuels, and alternative plastics) [9]. A study conducted in Zimbabwe found that most offcuts and chips from wood-based companies are used as fuel by local people to generate steam for kiln dryers in commercial sawmills [10]. Another Japanese study reported that furniture manufacturers generate 15 million cubic meters of wood waste, over 90% of which is recycled into processed wood and fuel [11]. Virgin wood from Paulownia fortuniei, a fast-growing species, meets the minimum standard EN 300 Type 1 (1997) for General Purpose Oriented Particleboard (OSB) panels for use in dry environments [13].

There is a heated debate about even if burning wood instead of coal can help reduce greenhouse gas emissions (GHG). Proponents argue that the carbon dioxide emitted when trees are felled and burned is captured by new trees that sprout in their stead. So, there are no GHG emissions as carbon is part of the sustainable cycle. Biofuels can also help with energy security and job creation. Wood fuel is more location and climate-independent, easier to store, distribute, and most importantly transport over greater distances than other renewable energy sources like photovoltaic (PV) and wind. It is also less location and climate dependent than other renewable energy sources like wind and solar. Critics contend that encouraging wood fuels may lead to a global logging boom that damages forest biodiversity. For instance, biofuel advancement in Europe may have made a significant contribution to deforestation in the Eastern United States, Western Canada, and Southeast Asia [6, 12].

Forests are a vital component of the global ecosystem. In addition, they contribute to national economies through the production of forest products such as timber and non-timber forest products. According to DFRS [20], 40.36% of Nepal's land area is classed as forest, while 4.38% consists of other woody regions, for a total of 44.74%. Recent years have seen around 3,4 million $m^3$ of stem wood harvested from Nepal's forests, as estimated from forest stumps [1, 13]. A significant amount of this is utilized as lumber and poles, with the remainder as fuelwood. Pandey et al. [13], reported that annually, approximately 4.8 million people might be employed in the sustainable production of 900,000 $m^3$ of timber and 1.2 million $m^3$ of fuelwood.

The particleboard sector is currently the principal user of recycled wood. The inclusion of utilization of wood waste from building and demolition into the inner layer of medium-density fiberboard. In 2019, particleboard usage in Europe was 37.07 million $m^3$ [7]. Particleboard contains varying amounts of recycled wood depending on the country. Italy has a 100% rate, Belgium, the United Kingdom, and Denmark have a 50% rate, Germany, France, and Spain have a 15–30% rate, while Switzerland has a 0% rate. [4]. There looks to be space for improvement in several European nations. The fabrication of waste wood plastic [8–11] and waste wood concrete composites are considerably smaller-scale advances pertaining to the cascade use of waste wood [9].

In industrial countries, pulp and wood-based panel sectors utilize an ample proportion of logging and wood processing leftovers. Alternately, wood processors and power plants chop and burn residues to generate steam and electricity. This development is largely explained by strict environmental restrictions for waste management, the requirement to reduce forest fire risks, and the high recreational value of many northern kinds of wood (for which logging leftovers are a detriment). However, the desire to rationalize and maximize efficiency may be the strongest motivating factor.

In terms of usability, the wood wastes fall within 2 broad uses:

1. As power generation resources such as:
    - Boiler for kilns drying, wood conditioning, lacquer-curing, and so on.
    - Co-generation plant fuel.
    - Commercial firewood (brick baking, noodle production, tobacco curing, and steam generation).
2. As secondary raw material such as for:
   Within the wood-based industry:
    - Medium density fiberboard
    - Particleboard
    - Blockboard
    - Laminated board
    - Charcoal briquettes
    - Parquet
    - Pallet manufacturing.

   Outward wood-based industry:
    - Compost and mushroom cultivation
    - Livestock litter/bedding
    - Low-volume wood goods, e.g., in Cottage Industry
    - Pulp and paper industry.

## 3 Economic Evaluation

Wood biomass is a significant renewable energy source, particularly in nations that have historically relied on forestry resources. Wood biomass can have various good socioeconomic and environmental consequences in these nations. Wood offers numerous advantages regarding the bioeconomy and cyclical economy. It's a naturally occurring, biodegradable material with remarkable mechanical and thermal qualities. In comparison to comparable materials made from inorganic or fossil source elements, wood materials often have substantially less of an impact on nature during the manufacturing and final phases [1]. In addition, unlike agricultural resources, wood is not in competition with food. Consequently, the utilization of wood for advanced purposes (energy production, construction materials, etc.) has increased since the beginning of the twenty-first century [2].

A study indicates that by 2030, Europe's wood supply may not be sufficient to meet demand [1]. The increase in wood use is sustained by a rise in the generation of waste wood from wood-based products that have reached the end of their useful life. Thus, recycling this vast deposit might provide copious and inexpensive raw materials to produce new materials. Wood is a natural substance, but reclaimed wood usually contains additives (adhesives, varnishes, paints), various impurities (wood treatment products and heavy metals), and contaminants (glass, plastic, metals, etc.). This diversity makes the recycling process rather difficult. Therefore, current wood waste management solutions are primarily based on landfill, energy recovery, and materials recovery [3]. Although more difficult, the second salvage option should be pursued as it is based on recycling by creating new materials and involves "cascading" use. According to [9], cascading utilization is "the efficient use of resources through residual material use and material recycling to increase the overall availability of biomass within a given system." As a result, the cascading effect makes it possible to delay the release of carbon in the form of $CO_2$ when the material is used as an energy source at the end of its life [9].

Utilizing woody biomass from forest land could increase rural community economies, boost carbon dioxide discharge mitigation from sustainable wood low-grade wood, attract private investment, and protect the environment. Small-diameter wood and harvest wastes that could be utilized for bioenergy and bioproducts are available for utilization in quantities of 210 million oven-dry tons. Their estimated economic value is $5.97 billion (109). The review of current U.S. laws, regulations, and directives affecting the use of forest biomass, as well as the identification of barriers, challenges, and potential opportunities connected to the use of woody biomass from public lands, are necessary to achieve this utilization target supporting the U.S. Department of Agriculture's implementation. The use of forest biomass for bioenergy and bioproducts might, however, expand with better coordination of public policies (regulatory legislation, public subsidies, and support programs) at different levels of government. Covers the definitions of major biomass terminology used in various initiatives encouraging the utilization of forest biomass for bioenergy and other bioproducts. Standards for renewable fuels may encourage the utilization of forest biomass from lands [10, 12].

There are considerable regional differences in the use of biomass, bioenergy technologies, market dominance, and research concerns in these fields. However, in most countries, the socioeconomic advantages of using bioenergy can be emphasized as a key element in the expansion of bioenergy's share of the world's energy supply. Indicators of the economy like employment and financial gains are used to quantify the socioeconomic effects of biomass production and use, but the research also considers social, cultural, and environmental considerations. Despite their potential importance at the local level, these latter aspects have historically been left out of most impact analyses since they are not necessarily amenable to quantitative study. Types of technology, regional economic structure, social characteristics, fabrication processes, and common resources are only a few examples of the factors that may affect the local socioeconomic effects, which are diverse and will vary [7].

Research conducted by the Slovenian Forest Institute and the Croatian Energy Institute Hrvoje Pozar has shown that the method of socioeconomic analysis that is most likely to provide the optimal combination is the level of local bioenergy/renewable energy turned out to be highly dependent on energy development [7]. There are very few reference plants in Croatia and Slovenia, for example, basic modeling is required to aid project construction including technical and political demands. In contrast, there are countless excellent examples of initiatives in Sweden and Austria that are ready for further evaluation. Therefore, it is improbable that a single model applies to all countries. Another critical issue is the biomass resources used to generate energy. Unlike in any other country, wood fuels and biomass in general are primarily produced from forests in Slovenia and Croatia. Croatia and Slovenia are transformational countries (countries in central, eastern, and southern Europe that transitioned to market-oriented democracies in the 1990s and hence have distinct economic and social characteristics [7].

## 4 Wood Waste Management

In terms of environmental and economic implications, timber harvesting is often a significant portion of forest activities and administration. According to a considerable body of information, forest harvesting operations in Sabah, Malaysia can damage up to 50% of the residual stand, and potentially up to 60% [8]. Damaged and damaged trees significantly contribute to logging residues. As previously stated, two-thirds of logging residues in Terengganu, Malaysia, consist of trees injured or destroyed during road construction, logging, and extraction. Lesser logging rates lessen the damage, and in the preceding section's figures, it was assumed that just half of the logging residues are made up of injured trees. However, it is evident that minimizing the impact of forest harvesting might lead to a considerable reduction in logging waste. When compared to traditional logging, using reduced impact logging (RIL) techniques might likely minimize damage by around half save soil and biodiversity, and help preserve the productive potential of the leftover forest products [8].

Wood waste can be used to make a variety of materials, but the government must implement appropriate waste management rules to maximize the usage of wood waste resources. On the upstream side, the volume of wood waste can be kept to an absolute minimum through proper management. The following measures are advised:

- **Reduce**: By using better design equipment, enhancing the skills of the personnel and/or changing the production process to increase the recovery rate or maximize the raw material usage.

- **Reuse**: By returning the "as received" waste to another process or a downstream process.
- **Recycle**: By resizing the waste to become secondary raw materials or to be resource fuel.
- **Recover**: By returning scrap waste to be used as fuel wood.

Proper wood waste management is crucial to saving the environment. In general, each sector should:

- Minimize wood wastage by proactive planning of wood utilization based on resource.
- Access the possibilities for internal utilization of wood waste.
- Keep track of and evaluate the development of wood waste types and volumes relative to production growth.
- Evaluate the viability of recovering energy from wood waste for personal energy needs.
- Evaluate the viability of selling recovered energy to nearby clients or the national power system.
- Survey the possible uses of wood waste within 50–150 km.
- Evaluate the viability of using wood waste for purposes other than energy recovery.
- Evaluate the possibility of wood waste management collaborative ventures between surrounding wood-based companies.

Currently, the demand for and supply of wood waste varies significantly from region to region.

It is considered, however, that the development of a specialized organization of sufficient size is economically possible.

## 5 Conclusions

Examining and evaluating the employability function of bioenergy and socioeconomic factors is a time-consuming and difficult undertaking. There are various models that explore bioenergy systems, but none of them generate these kinds of output results. They are essentially techno-economic models that provide cost and efficiency data to assist developers in making decisions about the design and technique for project developments. And, while they do give specifics on expenses and profits, they don't include revenue information [7, 12–15].

If the recycling idea is to be developed further, waste wood must be seriously considered as a resource material, as it contains a few metals that can be easily claimed and sold. Existing equipment allows for the processing of both waste wood and bark. However, two further requirements must be fulfilled for this to be profitable. First, waste companies must deliver both materials in a recyclable state. At the time of collection, wood, and bark must be separated from other garbage; if the notion of recycling were more commonly accepted, this separation would occur automatically.

Second, prospective clients must be willing to purchase the process material and, if necessary, modify their equipment to accommodate it. There is reason to be positive about the future of wood waste and bark recycling as both requirements are currently being met in some capacity. Charcoal manufactured from waste wood is an innovative and growing use. Bark, a previously unusable byproduct of the forestry industry, is recognized as a high-quality raw material. Bark can be crushed in the same machinery that processes waste wood to generate higher quality fertilizer and useful organic mulch for horticultural and agricultural applications. The processor's cost accounting is very positive [16–18].

## References


1. Basalp D, Tihminlioglu F, Sofuoglu SC, Inal F, Sofuoglu A (2020) Utilization of municipal plastic and wood waste in industrial manufacturing of wood plastic composites. Waste Biomass Valoriz 11(10):5419–5430
2. Cesprini E, Resente G, Causin V, Urso T, Cavalli R, Zanetti M (2020) Energy recovery of glued wood waste—A review. Fuel 262:116520
3. Faraca G, Boldrin A, Astrup T (2019) Resource quality of wood waste: the importance of physical and chemical impurities in wood waste for recycling. Waste Manag 87:135–147
4. Golonis C, Aikaterini R, Konstantinos C, Konstantinos Z, Polixeni G, Andreas P (2022) Environmental and economic assessment of wood pellet production from trees in Greece. Smart Grid Renew Energy 13(7):137–159
5. Derzu D, Mensah-Brown H, Brew-Hammond A (2005) Wood waste cogeneration in Kumasi, Ghana. In: Bioenergy-realizing the potential. Elsevier, pp 213–219
6. Yun H, Clift R, Bi X (2020) Environmental and economic assessment of torrefied wood pellets from British Columbia. Energy Convers Manag 208:112513
7. Krajnc N, Domac J (2007) How to model different socio-economic and environmental aspects of biomass utilisation: case study in selected regions in Slovenia and Croatia. Energy Policy 35(12):6010–6020
8. Ibrahim Y, Yusof Y (2017) Towards sustainable environmental management through green tourism: case study on Borneo rainforest lodge. Asian J Tour Res 2(3):123–143
9. Wu H, Zuo J, Yuan H, Zillante G, Wang J (2023) Investigation of the social and economic impacts of cross-regional mobility of construction and demolition waste in Australia. Resour Conserv Recycl 190:106814
10. Besserer A, Troilo S, Girods P, Rogaume Y, Brosse N (2021) Cascading recycling of wood waste: a review. Polymers 13(11):1752
11. Girods P, Dufour A, Rogaume Y, Rogaume C, Zoulalian A (2008) Pyrolysis of wood waste containing urea-formaldehyde and melamine-formaldehyde resins. J Anal Appl Pyrol 81(1):113–120
12. Charis G, Danha G, Muzenda E (2019) A review of timber waste utilization: challenges and opportunities in Zimbabwe. Procedia Manuf 35:419–429
13. Sahoo K, Alanya-Rosenbaum S, Bergman R, Abbas D, Bilek EM (2021) Environmental and economic assessment of portable systems: production of wood-briquettes and torrefied-briquettes to generate heat and electricity. Fuels 2(3):345–366
14. Pandey S (2022) Wood waste utilization and associated product development from under-utilized low-quality wood and its prospects in Nepal. SN Appl Sci 4(6):1–8
15. Gwak YR, Kim YB, Gwak IS, Lee SH (2018) Economic evaluation of synthetic ethanol production by using domestic biowastes and coal mixture. Fuel 213:115–122

16. Coelho A, De Brito J (2013) Economic viability analysis of a construction and demolition waste recycling plant in Portugal-part I: location, materials, technology and economic analysis. J Clean Prod 39:338–352
17. Skodras G, Grammelis P, Kakaras E, Sakellaropoulos GP (2004) Evaluation of the environmental impact of waste wood co-utilisation for energy production. Energy 29(12–15):2181–2193
18. Höglmeier K, Weber-Blaschke G, Richter K (2014) Utilization of recovered wood in cascades versus utilization of primary wood—A comparison with life cycle assessment using system expansion. Int J Life Cycle Assess 19(10):1755–1766

# Development and Performance of Wood Waste Briquettes in Pyrolysis Reactor

**Mohammed Nasir, Pawan Kumar Poonia, Kaizar Hossain, and Mohammad Asim**

**Abstract** In the present scenario, fossil fuel-based energy comprising oil, coal and natural gas is the main source of global energy. It is leading to many environmental issues like global warming, acid rain and urban smog. Moreover, this fossil fuel is non-renewable and anticipated to be depleting in the next 4–5 decades. Wood-based energy generation is one of the oldest energy sources, and consist of many advantageous characteristics. In this review, the briquette manufacturing technology from waste wood through different processes is discussed. Huge amount of wood-based biomass produced every year throughout the world, in the form of used furniture, temporary houses and industrial waste is a liability to municipal departments of the cities and generally used as a landfill. Such wood waste can be potentially utilized for briquette manufacturing. The wood waste type, amount and availability are varying in different countries depending on domestic and industrial practices. Since the briquette is a product of wood, the product quality is dependent on the raw material characteristics like density, moisture and calorific properties of wood. Other factors like impurities in waste wood and the addition of binding material in briquette manufacturing determine the economics and market value of the briquette. The quality of the briquette is assessed on the basis of product density and calorific value. Different manufacturing technology is being practised based on the size, the moisture content in raw material and the adhesive used. The briquettes are generally burnet in a pyrolysis reactor that requires lower heating temperature, and equipment

M. Nasir (✉)
Department of Forest Products & Utilization, College of Forestry, Banda University of Agriculture and Technology, Banda, UP 210001, India
e-mail: mnknasir@gmail.com

P. K. Poonia
Department of Forestry, College of Agriculture, CCS Haryana Agricultural University, Hisar, HR 125004, India

K. Hossain
Department of Environmental Science, Asutosh College, University of Calcutta, Bhowanipore, WB, India

M. Asim
Institute of Tropical Forestry and Forest Products (INTROP), Universiti Putra Malaysia (UPM), 43400 Serdang, Selangor, Malaysia

 
S. N. Sarmin et al. (eds.), *Wood Waste Management and Products*, Sustainable Materials and Technology, https://doi.org/10.1007/978-981-99-1905-5_4

investment, however, produce high energy and different by-products like bio-oil, biochar and pyrolysis gas which is used in further heating. Therefore, wood waste briquetting is not only a good substitution for energy sources but a genuine disposal of waste material.

**Keywords** Briquets · Biomass · Biochar · Industrial waste · Municipal waste residue · Renewable energy · Pyrolysis

## 1 Introduction

Energy is a basic need of modern civilization, used for generating heat for domestic as well as industrial purposes. Wood is one of the primitive sources of energy used in various forms such as firewood, charcoal, pellets and briquettes [1]. Wood-based energy is widely used for cooking, especially in developing countries like India where a large population lives in rural areas [2]. However, utilizing wood as fuel stock is not easy due to its bulky nature and low densities, which makes it difficult to use efficiently for various purposes and usually make it difficult to transport economically [3]. Furthermore, the energy generation by direct combustion of wood or its products is a wasteful practice and produces greenhouse gases. With the industrial revolution, wood utilization is replaced to a great extent by various non-renewable energy sources like fossil fuels and more recently, nuclear power on a large scale. However, such sources of energy are non-renewable and hazardous to the environment.

With the growing concerns of climate change and limited availability of fossil fuels, wood is regaining its significance in many responsible countries. This trend is increasing every year from 2010 to 2020 [4]. This pattern of biomass utilization varies in rural population as compared to urban population. The rural population collects twigs and branches of trees for their livelihood whereas the urban population consumes the whole trees which lead to deforestation [1]. In India, a huge amount of biomass is produced every year comprising the residues from the crop, used furniture, industrial waste and temporary houses at construction sites [4]. Most of these wastes are the liability of municipal departments of the cities and are generally used as landfill. These can be alternatively used for briquettes manufacturing for generating energy worldwide. Furthermore, briquettes utilization will reduce forest degradation and illegal exploitation of natural forests around the world. The steps involved in briquette manufacturing process are summarized in Fig. 1.

## 2 Historical Developments in Briquette Manufacturing

Briquette production is an age-old practice being used throughout the world for centuries. The fuel shortages during World War I and World War II intensified the briquette utilization, produced from various raw materials. However, after World War

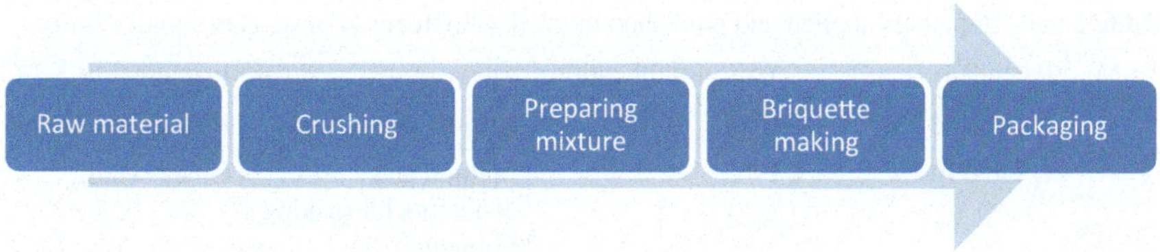


**Fig. 1** Flowchart of briquettes manufacturing

II, the market for briquettes decreased due to large and cheap availability of fossil fuel [5]. However, at the beginning of the 1980s with increasing concern over the environment, briquettes utilization had regained expansion in the whole European countries. It is being used as substitute for firewood in bakeries, and restaurants, as well as factories with fuelwood-burning furnaces, like red brick factories. Some of the key developments in briquette manufacturing in the last century are summarized Table 1. This development is primarily depending on three factors: availability of raw materials, the feasibility of technologies and a growing market for briquettes [6].

Since the raw material for briquettes is cellulosic biomass, energy generation through this method is considered a renewable source of energy. It comprises around 6% of total global energy consumption [4]. A survey conducted in Europe suggested that the production of briquettes was negative during the year 1995–2000 due to a lack of political will and an unorganized briquette market [8]. However, this trend changed dramatically and started increasing during the last ten years [9]. To minimize the stress on natural resource, recycling waste wood for manufacturing valuable products could be an intelligent solution. Considerable attention has been given to recycling materials like glass, iron, aluminium, plastics, etc., however, a little attention is given to recycling wood materials. The post-consumer wood materials are either burned down as a heat source or left to decompose. Over the decades, the waste wood from the wood industries and sawmills were burned directly in the furnace for heating. However this trend is changing in many responsible countries and wood wastes are entirely consumed by the composite industries like particle board and fibre board industries [10]. The other source of wood waste generating millions of tonnes every year from demolished projects, temporary settlements, new construction or land clearing, etc., still has no use and viewed as a burdensome disposal problem. Although these wood wastes are widely available in huge quantities, the main problem is the lack of technical information on methods for efficient and optimum utilization

**Table 1** Development in Briquettes manufacturing [7]

| Sr No | Developments | Year | Invented by |
|---|---|---|---|
| 1 | Equipment used for densified wood products | 1864 | Halsted and Halsted |
| 2 | Fully mechanized method for the production of briquettes from sawdust and shavings | 1920s | Heidenstamm |
| 3 | Production of binderless briquettes | 1982 | Natividad |

**Table 2** List of patents applied and published in the development of briquettes manufacturing

| Sr No | Patent No | Year of publication | Title | Inventor |
|---|---|---|---|---|
| 1 | US2824790A | 1958 | Briquetting of coal | Haig and Anthony |
| 2 | US3091012A | 1963 | Method and apparatus for making briquettes | Gustav et al. |
| 3 | JP4130826B2 | 2008 | Method for producing moulded charcoal for fuel | Mitsuyoshi et al. |
| 4 | US1893417A | 1933 | Smoke less briquettes | Gustav et al. |
| 5 | US1893555A | 1933 | Method of manufacturing briquettes | Gustav et al. |
| 6 | US2164933A | 1939 | Process of baking fuel briquettes | Maurel |
| 7 | EP1502667A1 | 2005 | Municipal waste briquetting system and method of filling land | Grillenzoni |
| 8 | WO2004067686A1 | 2004 | Aromatic wooden briquettes for domestic cooking | Simic |
| 9 | EP2785816A1 | 2014 | Briquette with binder and sodium and/or potassium silicates | Stormanns |

without any adverse effect on the environment. Table 2 includes the list of patents applied and awarded in briquettes manufacturing.

## 3 Source of Wood Waste

The products manufactured from wood require multiple processes, from tree felling to finished goods. It is estimated that approximately only 50% of a whole tree is turned into valuable products, and the rest becomes waste [11]. The waste could be in the form of bark, sawdust, chips, coarse residues, shavings, planer peel and end trimmings generated during the industrial processes. This unused part of wood can potentially pollute the environment in the form of either land, air or water pollution. The average biomass generated from agriculture and forest as waste in European countries was found to be 242 million tonnes (Mt) for 2010–11, and it is expected to increase to 280 Mt by 2030 [9] (Fig. 2). It is assumed roughly half of the biomass of felled trees is wasted as residuals at site, followed by abandoned logs (3.75%), butt trimmings (2.5%), tops and branches (33.75%) and stumps (10%) [12].

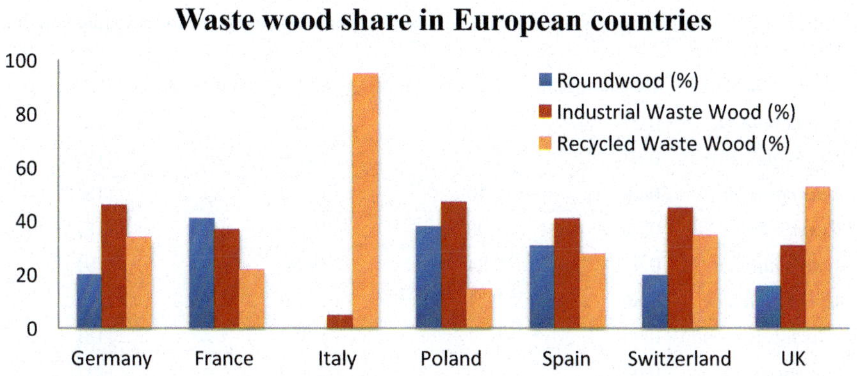


**Fig. 2** Waste wood generated in European countries in the year 2015

The direct burning of this waste wood as a source of energy is considered to be a wasteful practice exhibiting poor burning efficiency and causing indoor air pollution [13]. Converting this wood-based biomass into briquettes can help to meet the energy demands for household cooking and mitigate many issues associated with firewood [14]. The additional advantage of briquettes utilization is, the waste wood obtained from industries or wood waste generated due to rapid industrialization and urbanization can be effectively utilized [15].

## *3.1 Industrial Wood Wastages*

A significant amount of wood waste is produced from the wood processing industries in the form of sawdust, shavings, wood chips and wood cuttings. An estimate from Wood Recyclers Association 2021 reveals that Germany alone produced 11.9 million tonnes of wood waste from wood packaging, demolition and construction, wood processing industry in 2015 [16]. The packaging industries of Finland produce around 207,000 tonne of wood waste every year of which only 15% is used in product recycling and the rest are burnt energy generation.

The wood waste left at the felling site is generally burned in brick kilns to produce heat for cooking. Similarly, the fine or small particles produce in sawmills, while burning in large furnaces heaped up and do not mix well with the oxygen supply, hence resulting in incomplete combustion of biomass. This incomplete combustion generates substantial emissions of greenhouse gas like carbon dioxide and carbon monoxide, which are the primary gases contributing to global warming. However, the briquettes made from such wood particles may have better exposure to air and help in complete combustion and utilization. Ahmed et al. [17] suggested that saw dust including wood powders, wood shavings, wood chips and planer shavings are the ideal materials for briquette production. It has a suitable size, good adhesive force the least abrasion to the briquetting mills. A denser wood species with low moisture

**Table 3** Fuelwood characteristic of commercially important timber species giving huge amount of industrial waste

| Timber species | Density (g/cc) | Ash content (%) | Calorific value (kJ/g) | References |
|---|---|---|---|---|
| *Acacia pinnata* | 0.27 | 1.40 | 18.20 | [18] |
| *Albizia procera* | 0.63 | 2.30 | 18.20 | [18] |
| *Rhus parviflora* | 0.68 | 1.40 | 18.00 | [18] |
| *Toona serrata* | 0.78 | 3.00 | 19.10 | [18] |
| *Acacia nilotica* | 0.9 | 0.84 | 19.50 | [19] |
| *Albizzia lebbeck* | 0.86 | 1.21 | 19.50 | [20] |
| *Salix tetrasperma* | 0.93 | 1.74 | 19.32 | [18] |
| *Tectona grandis* | 0.68 | 1.36 | 20.30 | [21] |

content is always preferred for briquetting due to its high energy content per unit volume and slow burning rates (Table 3).

## 3.2 *Municipal Wood Waste*

The traditional raw material for briquettes manufacturing is a non-commercial wood industrial wood waste. However, this range has been increased to municipal waste and agricultural residue waste which are available in huge quantities throughout the world. Apart from this, large quantities of waste wood are generated in residential, commercial and industrial sources as a container or packaging material. Escamilla-García et al. [22] studied the municipal waste generated every day in Mexico city and observed that it is 65% organic biomass waste including organic matter and cellulosic biomass (Fig. 3). Based on the different sources of the waste wood, it can be classified as (1) Wood packaging and pallets waste; (2) Construction and demolition wood waste and (3) Wood waste and paper waste from industries.

The other raw material for briquette manufacturing could be the used and discarded papers, block board, MDF and HPL generated in huge quantities from residences, offices, classrooms and stores are the products of wood or sawdust. It takes large space and huge effort to manage the dump site. However, they are easy to catch fire and making pellets can increase their density and unify their size, solve pollution and save fuel cost.

| Recycle products | Density (g/cc) | Ash content (%) | Calorific value (kJ/g) | References |
|---|---|---|---|---|
| Block board | 0.81 | 1.12 | 18.4 | [23] |
| Medium density fibreboard (MDF) | 0.80 | 1.10 | 18.6 | [23] |
| High-pressure laminate (HPL) | 1.38 | 0.81 | 20.4 | [2] |

(continued)

(continued)

| Recycle products | Density (g/cc) | Ash content (%) | Calorific value (kJ/g) | References |
|---|---|---|---|---|
| Paper and paper board | 1.12 | 1.5 | 20.08 | [24] |

## 4 Briquette Manufacturing Methods

Briquette manufacturing process is simply a densification process that produces a uniform fuel with high energy density and reduces the transport and handling costs of the bulky biomass. The densification of biomass is generally obtained by the mechanical or hydraulic press of a piston on a die. Further, in order to obtain quality briquettes a binder and heat treatment is also applied depending on the properties of the feedstock [25]. The quality of the final product is evaluated based on the resistance to compression, energy density, compaction rate and equilibrium moisture, etc. However, to produce quality briquettes selection and processing of raw material is a very important factor [26]. Such as dry materials are hard to mould, while damp materials make the wood pellets too loose. To keep the raw materials with suitable moisture, it is necessary to dry them in the range of 12–15%. Further, the size of raw materials should be less than the hole size of the pellet mill die [27]. As the hole size is usually 6, 8 and/or 10 mm, the raw materials should be under 5 mm. The adhesive generally, needed to add as a binder for making pellets. The lignin in wood biomass can also be utilized as in-situ binding material at high temperatures. The quality of the final product is evaluated based on the resistance to compression, energy density, compaction rate and equilibrium moisture, etc.

## 5 Briquettes Manufacturing Technology

In the process of briquette manufacturing a high pressure is applied to the biomass generally mixed with a binder in order to enhance the adhesion between the particles. There are several techniques being used worldwide which can be classified as follows: piston press, screw press, hydraulic press and roller press.

### *5.1 Piston Press*

Piston press is one of the simplest and most common briquetting tools made up of a die and a piston (Fig. 4a). The biomass feedstock mixed with adhesive is supplied into the chamber. The pressing piston is utilized to compress the biomass against the

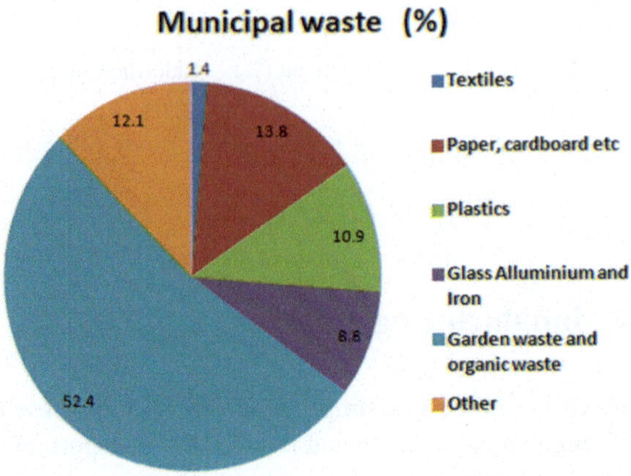


**Fig. 3** A study on municipal waste generated every day (tonne/day) in Mexico City [22]

die at high pressure and give the shape to briquettes as passes through the die. By using this technique, round or square briquettes are often generated. The machine can be used for biomass having a moisture content of up to 22% [28].

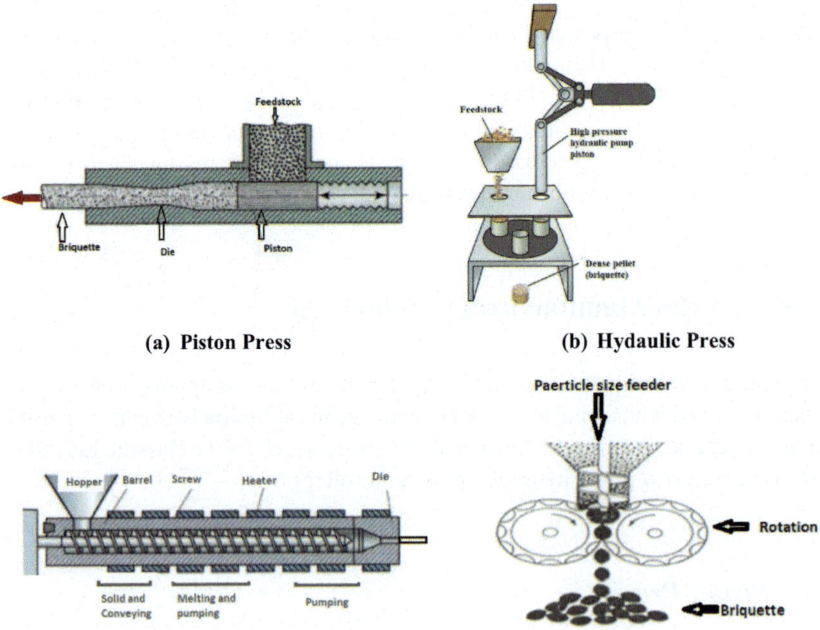


(a) **Piston Press**

(b) **Hydaulic Press**

(c) **Screw Press**

(d) **Roller Press**

**Fig. 4** Techniques of briquettes manufacturing [28]

### 5.2 *Hydraulic Press*

In a hydraulic press briquetting a high-pressure hydraulic fluid system is used to transform the motor's electric energy into mechanical energy (Fig. 4b). The fundamental benefit of this approach is that it may function effectively even in environments with high moisture content. However, it has the disadvantage of a slower manufacturing rate as compared to other techniques and oil leaks.

### 5.3 *Screw Press*

A screw press is made up of a die and a screw extruder (Fig. 4c). It has generally a low diameter in the die region and is conical in shape. In this method of briquette manufacturing, the raw material must be devoid of impurities like metallic fragments, stones, etc. As the material enters the conical die it compresses to the biomass with high pressure. The advantage of this method is that it does not require any binding materials and produces high-quality briquettes. However, it can't give good results in biomass having a particle size of more than 4 mm and moisture content of more than 12%.

### 5.4 *Roller Press*

A roller press is a set of tiny dies with a diameter of around 30 mm to create pellet-sized briquettes (Fig. 4d). In this method of a thick disc with several holes drilled around it is utilized. The biomass material enters the die through a set of rollers, where it takes on the shape of the die. Die types that are used are ring and flat types.

## 6 Pyrolysis Reactor

In contrast to combustion, pyrolysis is the chemical decomposition of organic materials in the absence of oxygen. Therefore, pyrolysis is considered the safest way of getting green energy from waste materials. The major benefit of the pyrolysis process includes lower heating temperature requirements, lower equipment investment, and the production of by-products like liquid phase bio-oil. The criteria for choosing the most favourable pyrolysis reactors are various according to the type and properties of biomass, the energy conversion costs etc. [17]. Due to this fact, the conversion of biomass through these processes provides different types and forms of bioenergy. The bio-oil obtained from the boiler can be used directly as fuel in boilers and after modification into combustion engines. The other products like biochar and pyrolysis

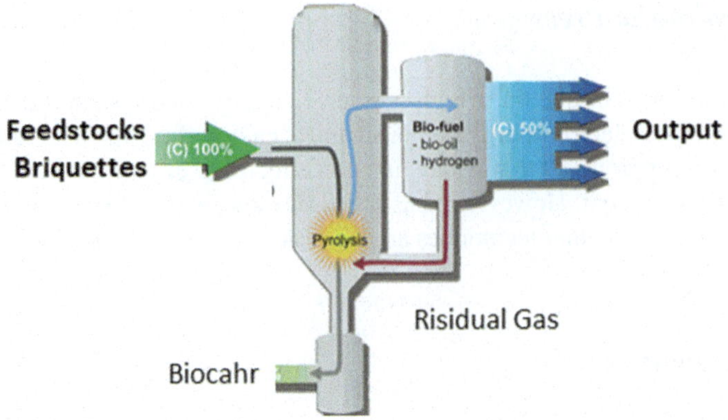


**Fig. 5** Briquette pyrolysis and biochar production sketch. Adopted from Lehmann [29]

gas, can also be used as biofuels to generate heat or power in other processes (Fig. 5). Although the pyrolysis reactor has many advantages, it still has many challenges like the type of reactor, operating parameters and biomass type.

### *6.1 Types of Pyrolysis Reactors*

There are hundreds of types of pyrolysis reactors based on the designs and therefore impossible to include all in this review. In this review classification of pyroysis reactor is classified based on the particle size i.e. fast, Intermediate (small particles >2 mm), and Slow pyrolysis reactor (logs or chips). The "fast pyrolysis reactors" is designed to use powdery biomass (particle diameter <2 mm) as feedstock and generate maximum yields of bio-oil. In the intermediate pyrolysis reactors, small particles (particle diameter >2 mm) are used to operate. The slow pyrolysis reactors are designed to use feedstock in the form of logs or chips. The slow pyrolysis reactors are further classified as kilns, retorts and converters based on the feedstocks used in the form of log, pile-wood, and chips respectively. The converters using small particles operate at conditions comparable to the intermediate pyrolysis reactors.

## 7 Conclusions

Although the forest is one of the renewable resources on planet earth, it is depleting very fast due to unsustainable utilization. Thousands of metric tonnes of wood waste are generated as a result in the form of used furniture, municipal waste and industrial waste. Utilizing this wood waste and transforming it into valuable products is not only conserving our natural resources but a genuine disposal of waste material. The review covered the major developments in briquettes utilization and patents applied and awarded in United State relevant to briquette manufacturing. The factor determining the quality and character of briquettes is addressed in detail. A comparison is made to study the calorific value of wood and found that a densified biomass has always better physical and thermal properties as compared to simple wood. Briquetting parameters were examined and it is observed that temperature, die pressure and moisture content play a major role during briquette preparation. The piston press method is found to be the most widely used technique due to its easiness in construction and functioning. The briquette pyrolysis parameters such as calorific value, moisture content burning rate, and ash content are discussed. It was observed that the calorific value are higher density are the two major factors determining the briquette quality. When the density is higher it burns slowly with high compaction pressure. Further, higher ash content decreases the calorific value of briquettes and increases the risk of metal corrosion.

Promoting a briquette-based energy generation and uses in small, middle, and large-scale combustion plants will save from dangerous gas emission. Utilizing waste wood for briquetting is not only a solution to disposal management but directly reduce the pressure on natural forest. Furthermore, it has numerous economic, social and environmental advantages such as recycling wood, economic gain from waste, and mitigating greenhouse gas emissions.

## References


1. Ravindranath N, Nayak MM, Hiriyur R, Dinesh C (1991) The status and use of tree biomass in a semi-arid village ecosystem. Biomass Bioenerg 1(1):9–16
2. Günther B, Gebauer K, Barkowski R, Rosenthal M, Bues C-T (2012) Calorific value of selected wood species and wood products. Eur J Wood Wood Prod 70(5):755–757
3. Garcia CA, Hora G (2017) State-of-the-art of waste wood supply chain in Germany and selected European countries. Waste Manag 70:189–197
4. Sreevani P (2018) Wood as a renewable source of energy and future fuel 1992:040007
5. Eriksson S, Prior M (1990) The briquetting of agricultural wastes for fuel: food and Agriculture Organization of the United Nations
6. Mendoza Martinez CL, Sermyagina E, de Cassia Oliveira Carneiro A, Vakkilainen E, Cardoso M (2019) Production and characterization of coffee-pine wood residue briquettes as an alternative fuel for local firing systems in Brazil. Biomass and Bioenergy 123:70–77
7. Vinterbäck J (2000) Densification of wood and bark for fuel production—A story of 150 years. In: Wood Pellet Use in Sweden: A systems approach to the residential sector. Silvestria, p 152
8. Hirsmark J (2002) Densified biomass fuels in Sweden: country report for the EU/INDEBIF project
9. Karlhager J (2008) The Swedish market for wood briquettes-production and market development. (Ph.D. regular). Institutionen för skogens produkter. https://core.ac.uk/download/pdf/211579807.pdf
10. Knauf M (2015) Waste hierarchy revisited- An evaluation of waste wood recycling in the context of EU energy policy and the European market. Forest Policy Econ 54:58–60
11. Berger F, Gauvin F, Brouwers HJH (2020) The recycling potential of wood waste into wood-wool/cement composite. Constr Build Mater 260:119786
12. Dionco-Adetayo EA (2001) Utilization of wood wastes in Nigeria: a feasibility overview. Technovation 21(1):55–60

13. Rao S et al (2012) Environmental modeling and methods for estimation of the global health impacts of air pollution. Environ Model Assess 17(6):613–622
14. Purohit P, Chaturvedi V (2018) Biomass pellets for power generation in India: a techno-economic evaluation. Environ Sci Pollut Res 25(29):29614–29632
15. Prvulovic S, Gluvakov Z, Tolmac J, Tolmac D, Matic M, Brkic M (2014) Methods for determination of biomass energy pellet quality. Energy Fuels 28(3):2013–2018
16. Pandey S (2022) Wood waste utilization and associated product development from under-utilized low-quality wood and its prospects in Nepal. SN Appl Sci 4(6):1–8
17. Ahmed A et al (2020) Sawdust pyrolysis from the furniture industry in an auger pyrolysis reactor system for biochar and bio-oil production. Energy Convers Manag 226:113502
18. Bhatt BP, Todaria NP (1990) Biomass
19. Bargali K, Bargali SS (2009) Acacia nilotica: a multipurpose leguminous plant. Nat Sci 7(4)
20. Mishra SS, Gothecha VK, Sharma A (2010) Albizia lebbeck a short review. J Herb Med Toxicol 4(2):9–15
21. Rosamah E, Ferliyanti F, Kuspradini H, Dungani R, Aditiawati P (2020) Chemical content in two teak woods (Tectona grandis Linn.F.) that has been used for 2 years and 60 years. 3BIO: J Biol Sci Technol Manag 2(1):15
22. Escamilla-García PE, Camarillo-López RH, Carrasco-Hernández R, Fernández-Rodríguez E, Legal-Hernández JM (2020) Technical and economic analysis of energy generation from waste incineration in Mexico. Energy Strat Rev 31:100542
23. Haseli M, Layeghi M, Zarea Hosseinabadi H (2018) Characterization of blockboard and battenboard sandwich panels from date palm waste trunks. Measurement 124:329–337
24. Pivnenko K, Olsson ME, Götze R, Eriksson E, Astrup TF (2016) Quantification of chemical contaminants in the paper and board fractions of municipal solid waste. Waste Manag 51:43–54
25. Grover P, Mishra S (1996) Biomass briquetting: technology and practices, vol 46. Food and Agriculture Organization of the United Nations Bangkok, Thailand
26. Kongprasert N, Wangphanich P, Jutilarptavorn A (2019) Charcoal briquettes from Madan wood waste as an alternative energy in Thailand. Procedia Manuf 30:128–135
27. Dinesha P, Kumar S, Rosen MA (2019) Biomass briquettes as an alternative fuel: a comprehensive review. Energy Technol 7(5):1801011
28. Tumuluru JS, Wright CT, Hess JR, Kenney KL (2011) A review of biomass densification systems to develop uniform feedstock commodities for bioenergy application. Biofuels Bioprod Biorefin 5(6):683–707
29. Lehmann J (2007) A handful of carbon. Nature 447(7141):143–144
30. Stormanns F (2014) Briquette with binder and sodium and/or potassium silicates. European Patent 2,785,816,A1
31. Haig GD, Anthony KJC (1958) Briquetting of coal. United State Patent 2,824,790,A
32. Maurel HF (1939) Process of baking fuel briquettes. United State Patent 2,164,933,A
33. Gustav K, George M, Charles C (1933) Smoke less briquettes. United State Patent 1,893,417,A
34. Gustav K, George M, Charles C (1963) Method and apparatus for making briquettes. United State Patent 3,091,012,A
35. Gustav K, George M, Charles C (1933) Method of manufacturing briquettes. United State Patent 1,893,555,A
36. Grillenzoni M (2005) Municipal waste briquetting system and method of filling land European Patent 1502667A1
37. Mitsuyoshi N et al (2008) Method for producing molded charcoal for fuel. Japan Patent, 4130826B2
38. Simic Z (2004) Aromatic wooden briquettes for domestic cooking. WO2004067686A1

# Wood Waste as a Renewable Energy Source: Effect of Pretreatment Technology for Sustainable Bioethanol Production

**Zubaidah Aimi Abdul Hamid and Ahmad Faizal Abdull Razis**

**Abstract** Concerning the availability of further fossil fuel supply, greenhouse gas emissions, global warming, and rising fuel prices, it is necessary to identify renewable, ecologically friendly, and economically feasible new alternatives for raw materials for energy sources. Lignocellulosic biomass (LB) derived from wood waste has great potential as an alternative source to produce second-generation biofuels without affecting global food security. However, the major constraints of LB are the presence of physical and chemical barriers caused by the interconnection of the primary constituents of lignocellulosic biomass (cellulose, hemicellulose, and lignin), which makes this component resistant to hydrolysis into fermentable sugars. Thus, the conversion of LB to bioethanol requires extensive processing, especially at the pretreatment stage. In general, pretreatment procedures for turning wood waste into bioethanol are classified into chemical, physicochemical, and biological. The objective of pretreatment is to improve enzyme accessibility, thereby enhancing the digestibility of cellulose and other components that can increase the production yield. Current findings addressing the application, mechanism, and production yield of different pretreatment strategies, such as chemical, physiochemical, and biological procedures, to produce bioethanol from wood waste have been expansively presented.

**Keyword** Wood waste · Pretreatment technology · Bioethanol · Renewable energy · Sustainable

Z. A. Abdul Hamid (✉)
Faculty of Bioengineering and Technology, Universiti Malaysia Kelantan, Jeli Campus, 17600 Jeli, Kelantan, Malaysia
e-mail: zubaidahaimi.ah@umk.edu.my

A. F. Abdull Razis
Natural Medicines and Products Research Laboratory, Institute of Bioscience, Universiti Putra Malaysia, 43400 UPM Serdang, Selangor, Malaysia

 
S. N. Sarmin et al. (eds.), *Wood Waste Management and Products*, Sustainable Materials and Technology, https://doi.org/10.1007/978-981-99-1905-5_5

# 1 Introduction

A global energy crisis has emerged in recent decades as a result of population growth and technological advancement. Due to the rapid consumption of these non-renewable resources, it is currently anticipated that fossil fuel stocks will run out within 40–50 years [1]. Despite the fact that fossil reserves are limited, they remain a primary energy source, with coal, crude oil, and natural gas accounting for 80% of global energy supply. These fossil reserves make up 81% of the world's energy supply, despite the fact that their supplies are limited [2]. This energy was used to produce power, gasoline to keep the factories and transportation running, and raw materials for chemical products or by-products. Fossil fuel combustion releases greenhouse gases, including carbon dioxide, methane, and nitrogen oxide, that harm the environment and human health by causing acid rain, photochemical smog, and global warming [3].

In 2019, the world's primary energy supply totalled 606 EJ. Renewable energy technologies accounted for 14.1% of total energy supply, a 1.33% increase over the previous five years (Fig. 1). Sustainable, renewable, and environmentally friendly energy sources are becoming more popular in the search for a renewable energy source that can meet future energy demands. Besides that, researchers are continuing to investigate other potential renewable energies, such as solar, wind, hydro, biomass, and geothermal, as a source of biobased products. As shown in Fig. 2, global renewable energy sources have increased in response to current demand from 2015 to 2019.

Demand for biofuels, particularly bioethanol, to replace gasoline is expected to increase in the future. Bioethanol, or ethyl alcohol ($C_2H_5OH$), contains a high-octane number, which results in reducing gas emissions. Higher octane fuels are favoured in spark-ignition internal combustion engines (ICEs), whereas gasoline blended with bioethanol functions as an improver or gasoline substitute in the form of ethyl tertiary butyl ether (ETBE) [4]. Recently, all countries have begun to review their energy

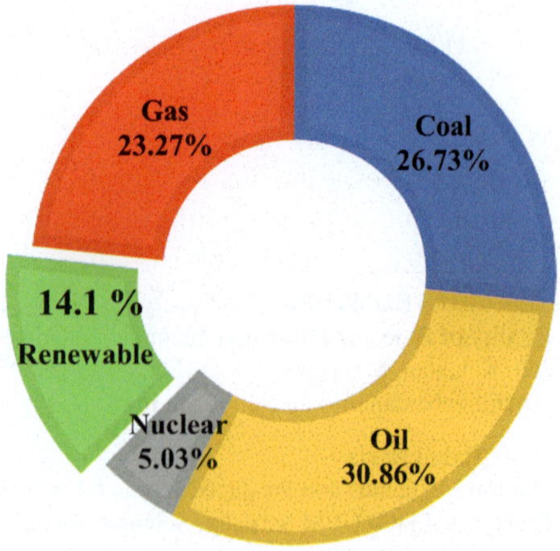


**Fig. 1** Total primary energy source globally for 2019

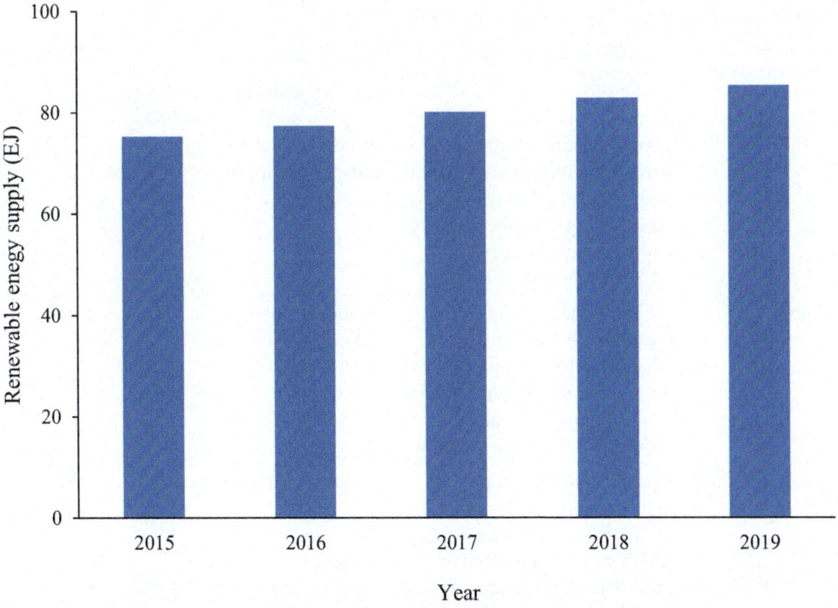


**Fig. 2** Renewable energy source globally from year 2015–2019

policies and implement strategies to reduce their reliance on fossil fuel reserves while reducing greenhouse gas emissions (Table 1). Furthermore, more attention and effort have been focused on making significant progress in sustainable energy studies. Policy mandates for bioethanol-gasoline blends have been implemented in most of the country. The usage of the gasoline blends E10 and E20 (10 and 20% ethanol: gasoline) for transportation use has been widely implemented in the United States and Brazil [5]. An earlier study on bioethanol blending revealed that oxygenation of the fuel mixture improved its combustibility, resulting in fewer greenhouse gas (GHG) emissions to the environment.

## 2 Bioethanol Feedstocks

Bioethanol is one of the most promising alternatives to fossil fuels. Bioethanol is a liquid biofuel made from various types of feedstocks and conversion technologies. Over the last decade (2010–2020), global bioethanol production has grown by 15.22%, while overall consumption increased by 21.51%. Currently, bioethanol production is segmented into three generations based on the feedstock used (Fig. 3) [13].

First-generation biofuels were made from food-based feedstock that contained high sugar and starch content. Previously, the use of food-based feedstock for

**Table 1** Bioethanol mandate in various countries

| Country | Mandate for bioethanol | Policy/program | References |
|---|---|---|---|
| United states | Under this programme, gasoline sold in the United States contains a higher percentage of renewable fuel (e.g., ethanol). Today, most gasoline in US containing E10 and E20 fuels available for Flex-fuel vehicles (FFVs) | Renewable fuel standard (RFS) program under Energy Policy Act of 2005 and Energy Independence and Security Act of 2007 | [6] |
| China | By 2020, the use of E10 fuel (gasoline containing 10% ethanol) will be mandatory throughout the country, moving from 11 experimental provinces to the entire country | (National Energy Administration 2017) | [7] |
| Brazil | Brazil's national policy maintains the rule for blending 18–27.5% ethanol in gasoline, which was implemented in 2015. Currently it set at 27% | National biofuels policy 2017 | [8] |
| European Union | European Union policy calls for a 10% renewable energy (biofuels) achieved in transportation sector by 2020 | Renewable energy directive | [9] |
| Canada | By 2030, Canada aims to increase biofuel consumption from 7 to 15% at the same time increase the annual biofuel production capacity from 3 to 8.5 billion litres | Clean Fuel Standard (CFS) policy | [10] |
| India | After revising its strategy, India aims to reach a 20% ethanol-to-petrol ratio by 2030 | National Policy on Biofuels | [11] |
| Thailand | The government plans to reduce production of octane 91 E10 by 2022 and octane 95 E10 and E85 between 2023 and 2027. By the end of 2037, only premium gasoline (octane 95 fuel) and E20 become predominant in the market | Alternative Energy Development Plan | [12] |

energy production represented more than 96% of global biofuel production in 2020. However, using food as a feedstock for biofuels creates a conflict between food/feed

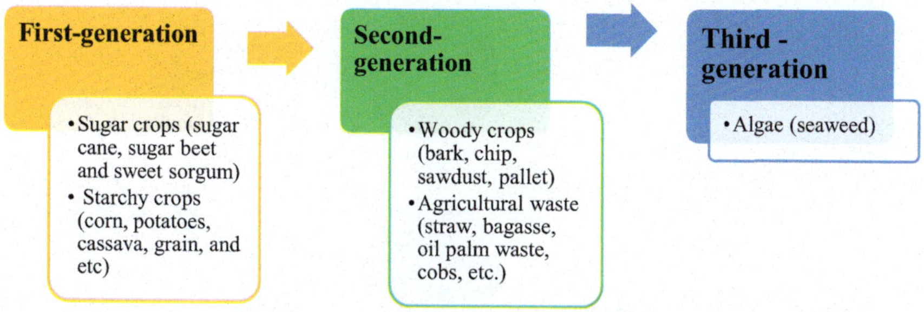


**Fig. 3** Different feedstocks represent the generations in bioethanol production

sources and fuel, resulting in rising prices and insufficient feedstock supplies. Second-generation biofuel technologies are being developed in response to growing criticism of first-generation biofuels and interest in non-food-based biofuels as renewable alternatives to fossil carbon resources. Woody crops and agricultural waste are lignocellulosic biomass that has been used in the second generation of bioethanol production [14]. Even though lignocellulosic biomass is more economically viable than sugar and starch feedstocks, the conversion of sugars from lignocellulosic materials is more challenging in terms of process design and operation. As a result, optimising the bioconversion of lignocellulosic biomass into bioethanol requires process modification, particularly in the pretreatment method [15].

## 3 Wood Waste as a Renewable Energy Source

Wood is a renewable material composed primarily of cellulose (35–50%), hemicellulose (15–35%), aromatic polymers of lignin (27–32%), and extractive (2–5%). Cellulose is made up of long linear chains of D-glucose held together by 1,4-glycosidic bonds. These glucan chains are joined by hydrogen bonds to form cellulose microfibrils with diameters ranging from 3 to 5 nm. Hemicellulose is composed of 1,4-linked glycans with different substituents depending on hardwood and softwood species. Although the significant role of hemicelluloses in imparting cell wall characteristics is still unclear, it has been postulated that hemicelluloses crosslink with cellulose via hydrogen bonds, which may modify the ability of microfibrils to bond with each other. Lignin is also frequently recognised as the cementing agent that imparts stiffness and compressive strength to the cell wall. Furthermore, impregnated lignin can improve the cell wall's water resistance. Small amounts of organic molecules, including resins, lipids, tannins, and stilbenes, can be found in wood as extractives, which are typically linked with antifungal and antimicrobial effects [16].

In the past 30 years, international trade in wood products has grown by 143%, with an estimated value of 244 billion USD. Economic growth and demand from China have been the most significant contributors to this trade expansion, with wood

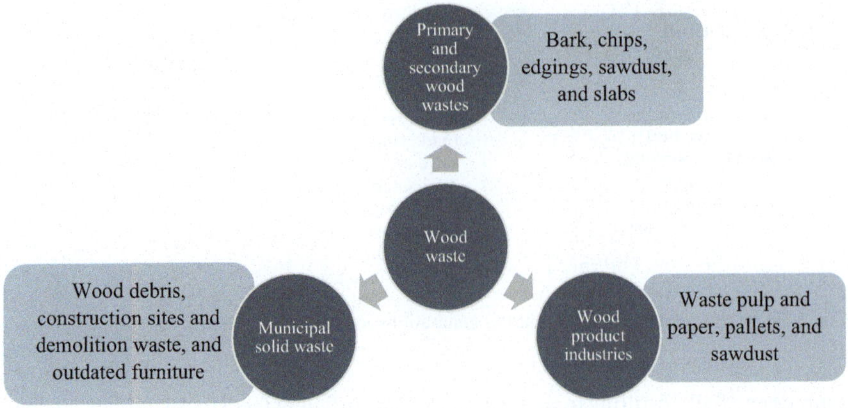


**Fig. 4** Various sources and categories of wood waste

imports to China reaching 760% up to 49 billion USD [17]. Wood is a versatile structural material that has been used for decades in construction, fuel, pulp and paper manufacturing, furniture production, and various other applications. Along with the rise in the usage of wood, the amount of waste wood generated from obsolete wood-based products has also increased. Waste wood is derived from primary and secondary wood wastes from forest waste from land fields [18, 19], wood product industries, and municipal solid waste [20–23]. Ihnat et al. [24] also reported that the major source of wood waste comes from outdated furniture materials such as particle boards (PB), oriented strand boards (OSB), and middle-density fibreboards (MDF). Various sources and categories of wood waste are shown in Fig. 4.

The development of second-generation bioethanol production is intensively explored. Mature technologies from first-generation bioethanol offer a significant advantage in expanding into new forms of lignocellulosic material. Furthermore, enhancing technological processes while lowering the cost of fermentable sugar synthesis and ethanol fermentation are the key goals for second-generation bioethanol from wood waste feedstocks. Pretreatment, saccharification, fermentation, and distillation are the four steps in the manufacturing of ethanol from lignocellulosic materials. The two main methods for converting lignocellulosic biomass into bioethanol are thermochemical and biochemical conversion. Thermochemical conversion is a heat-based method that includes combustion, gasification, hydrothermal processing, torrefaction, pyrolysis, etc. that converts biomass into value-added fuels and chemicals. Previous research has examined the technological, economic, energy, and environmental aspects of biochemical processes versus thermochemical methods. As a result, high productivities and improved energy efficiency have been observed in the synthesis of ethanol via biochemical methods. Biochemical conversion of second-generation of biomass involves the use of chemical, physiochemical, and biological pretreatment to convert biomass into biofuel [25]. Thus, this chapter discussed the use of wood waste as a lignocellulosic material, technology approaches via pretreatment stages on the challenges, and prospects in the bioethanol production process.

## 4 Pretreatment Technology from Wood Waste for Bioethanol Production

Compared to starch and sugar-based feedstocks, the conversion of lignocelluloses into bioethanol is more challenging due to the lignin content and crystalline structure of the wood waste, which makes them inaccessible to enzymatic degradation [26]. Effective pretreatment procedures can enhance overall process efficiency and have a direct effect on bioethanol productivity. Besides, it plays an important role in cellulose bioconversion processes into fermentable sugar by eliminating lignin, separating hemicellulose, and lowering cellulose crystallinity while enhancing porosity for the enzymatic hydrolysis process. Therefore, this process must be effective to prevent the degradation of carbohydrates, dissolve hemicellulose and lignin content, increase surface area and decrease particle size of biomass, and avoid unwanted by-products while increasing the concentration of fermentable sugars after the saccharification and fermentation processes. Physical, physio-chemical, chemical, biological, or a combination of these methods are the most common pretreatments that have been employed on wood waste in past research [27]. However, in this chapter, the effect of the selected pretreatment that has been adopted on the wood waste for bioethanol production will be discussed (Fig. 5).

## 5 Chemical Pretreatment

Chemical pretreatment, including alkaline, acid, ozone, and solvent, is considered one of the viable approaches due to its high efficiency and uncomplicated procedure. Besides, the application of this technique has increased the surface area and porosity of raw material, improving biomass digestibility by de-crystallising cellulose, and removing lignin and hemicelluloses. In acid pretreatment, concentrated or diluted acids are used to break the structure of lignocellulosic materials. Because of its high effectiveness in breaking down the lignocellulose structure and glycosidic bonds, acid pretreatment has received widespread attention for industrial applications. The main downside of this technique is the hazardous and corrosive chemicals used, which require several steps of neutralising, resulting in a high cost [28]. Sofokleous et al. [29] studied the impact of 3% w/v of sulfuric acid ($H_2SO_4$) for an acid pretreatment method on green waste (branches, leaves, and lignocellulosic residues from tree pruning, hedge cuttings, and grass clippings) collected from municipal waste. The use of 10% of solids loading with further hydrolysis using cellulose CellicCtec2 on this green waste resulted in the production of 33.67% of ethanol yield.

Meanwhile, base solutions such as sulphite, sodium hydroxide, ammonium hydroxide, and calcium hydroxide are used in alkaline pretreatment [30]. Bay et al. [31] investigated the impact of Sodium hydroxide (NaOH) pretreatment on pine and poplar wood residue at 93 °C in a water bath for 2 h with the following use of enzyme solution of Cellic®CTec2 (Novozymes, Denmark) for enzymatic hydrolysis and

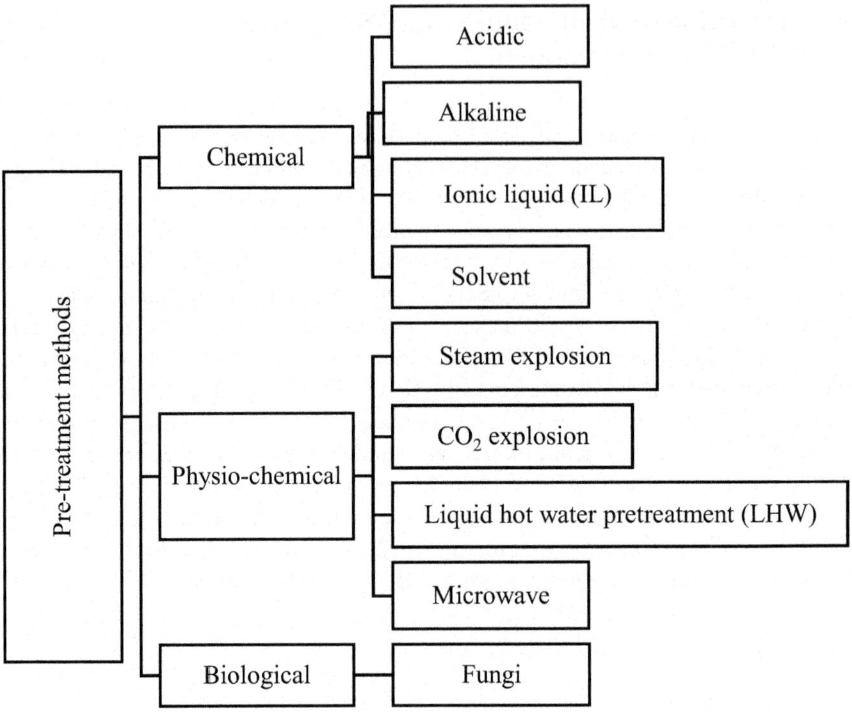


**Fig. 5** Different types of wood waste pretreatment methods

Saccharomyces cerevisiae utilising in fermentation, which resulted in high ethanol yields of 297.5 and 249%. In another work, Amândio et al. [32] converted eucalyptus globulus bark wastes into wood pulp using the kraft pulping method. During this procedure, wood chips are treated in a heated combination of high pH solutions including water, sodium hydroxide (NaOH), and sodium sulphide ($Na_2S$). Commercial cellulases and Ethanol Red® were employed in the hydrolysis and fermentation, respectively. As a result, after 20.5 h of fermentation, the highest ethanol concentration of $50.8 \pm 0.5$ g·L$^{-1}$ was obtained. Milder conditions, low sugar decomposition, and high delignification are the factors that make alkaline pretreatment preferable in bioethanol production. This method requires a larger amount of alkali and is time-consuming, which is the major drawback of this method [33]. For this reason, the combined procedure was occasionally used to fix this issue. Previous studies by Hossain et al. [5] described the use of rubberwood sawdust (RWS) as a raw material and the pretreatment with a mixture of diluted sulfuric acid ($H_2SO_4$) and sodium hydroxide (NaOH). The pretreatment was carried out using 2 M of acid and alkaline concentration, 90 °C, and residence time in acidic conditions for 90 min. As a result, 45% of bioethanol was produced from this process.

Currently, pretreatment of ionic liquids (ILs) has received a lot of attention. 1-butyl-3-methylimidazolium chlorides ([BMIM][Cl], 1-allyl-3-methylimidazolium chlorides [AMIM][Cl]) and 1-ethyl-3-methylimidazolium acetate ([EMIM][OAc]) are the most frequently employed for cellulose dissolution. This method was considered environmentally friendly because ionic liquid is a non-derivatising solvent for cellulose, allowing it to be recycled and reused [34]. Ionic liquid pretreatment is highly stable and can result in the partial or total dissolution of the lignocellulosic material at temperatures ranging from 80 to 160 °C. However, ionic liquid is very costly. Large amounts of hot water and energy are required to remove solvents that penetrate in the wood structure. Shafiei et al. [34] studied the use of 1-ethyl-3-methylimidazolium acetate ([EMIM][OAc]) as an ionic solvent pretreatment and spruce chips and powder as raw materials. The optimum bioethanol yield of treated chip and powder was increased from 2.7 to 66.8% and 9.7 to 81.5% after 15 h at 120 °C. In the other study by Reina et al. [15], it was indicated that 70% of bioethanol was obtained from *Eucalyptus dunnii* bark residues using 1-butyl-3-methylimidazolium chlorides ([BMIM][Cl] solvent at 140 °C for 8 h.

Pretreatment with organic solvents such as ethyl alcohol, methyl alcohol, acetone, ethylene glycol, and tetrahydrofurfuryl alcohol is very effective to break down the internal bonds of lignin and hemicelluloses with or without the use of a catalyst agent. Catalysts can be either organic or inorganic acids (like HCl and $H_2SO_4$) or bases (NaOH, $NH_3$ and $CaCO_3$). Monrroy et al. [35] delignified Pinus radiata wood chips using an ethanol-water mixture and the brown rot fungus Gloephyllum trabeum. The bioethanol yield obtained from this pretreatment was 63.8% after being subjected to 96 h of simultaneous enzymatic saccharification (*Trichoderma reesei* cellulases Celluclast) and fermentation using *Saccharomyces cerevisiae* IR2T9 (SSF) at a 10% concentration.

## 6 Physiochemical Pretreatment

Physical or mechanical pretreatment is a technique that breaks down lignocellulose materials into fine particles, increasing surface area, disrupting cell structure, reducing cellulose crystallinity, and paving the way for subsequent chemical and biological treatments. Milling, high-pressure homogenisation, electron beam irradiation, heat compression, and photocatalysis are examples of physical pretreatments. However, the disadvantages of this approach are high power consumption and low efficiency [36]. Thus, physiochemical pretreatment has been introduced to enhance the effectiveness of physical pretreatment by combining chemical and physical procedures. This method includes steam explosions, $CO_2$ explosions, and liquid hot water pretreatment (LHW).

The steam explosion method is a thermophysical-chemical process that involves the mechanical deconstruction of lignocellulosic materials using saturated steam under high pressure. After the reaction is stopped, the pressure is released, resulting in the disruption of the cell wall structure and the solubilisation of hemicellulose and lignin fractions [27]. Several studies have been conducted to investigate the effects of steam explosion pretreatment on wood waste materials. Recently, Rochón et al. [37] employed this pretreatment on Eucalyptus grandis sawdust at 200 °C for 10 min, followed by hydrolysis using Cellic CTec 2 (Novozymes) and Saccharomyces cerevisiae for fermentation, yielding high ethanol concentrations of 75.6 g/L. Cotana et al. [38] reported that 10.6 g of ethanol was recovered from 100 g of pine tree wood waste following a steam pretreatment at 220 °C for 10 min and enzymatic hydrolysis using cellulases, glucosidases, and hemicellulose, followed by fermentation with *Saccharomyces cerevisiae* yeast.

$CO_2$ pretreatment has gained a lot of interest due to its chemical-free method. To enhance the effectiveness of this procedure, high-pressure $CO_2$ was adopted to encourage the formation of carbonic acids, which are then required to degrade the hemicellulose structure. High-pressure $CO_2$ gas was employed to enhance the surface area of the pre-treated compounds, improving enzyme adsorption, and thereby accelerating the enzymatic hydrolysis step [39]. Wu et al. [40] investigated the impact of high-pressure $CO_2$ on waste aspen branch chips at 190 °C with the addition of low enzyme loading (10 FPU $g^{-1}$ glucan) and fermentation using *Saccharomyces cerevisiae*, which resulted in an increase of 10% of bioethanol yield from the control sample to 8.7 g/L. However, Chen et al. [41] found that using pressurised $CO_2$ in a supercritical state at 180 °C for 1 h resulted in a low bioethanol output of 0.69 g/L. This low bioethanol production may be because 5-hydroxymethyl-2-furaldehyde (HMF) and furfural are making by-products that act as inhibitors and slow down bioethanol synthesis.

Liquid hot water treatment (LHW), also known as autohydrolysis, or hydrothermal pretreatment, is a low-cost, environmentally friendly, and simple-to-use method for selectively removing hemicellulose with low cellulose content and lignin degradation [42]. The LHW technique requires only water as a reactive material, resulting in autoionisation of water towards acid hydronium ions while simultaneously releasing the acetic acids from hemicellulose that work as a catalyst to promote xylan depolymerisation into xylooligosaccharides (XOS) and monosaccharides [43]. However, this process occasionally creates inhibitors, such as 5-hydroxymethyl-2-furaldehyde (HMF), furfural, and phenolics, which function as inhibitors and limit bioethanol production. Romaní et al. [44] report a production of 50 g/L bioethanol from high solid loading of *Eucalyptus globulus* wood after autohydrolysis pretreatment at 210 °C, followed by SSF and Presaccharification and Simultaneous Saccharification and Fermentation, (PSSF) stages. However, the cellulose residues that are generated after LHW treatment still include a significant amount of lignin and need further steps of delignification [45]. This was corroborated by Senila et al. [46] who stated that the inclusion of a delignification process utilising sodium chlorite on local wood waste was conducted after autohydrolysis pretreatment at 190 °C for 10 min with the ethanol content of 17.9 g/L.

Recent studies have focused on microwave irradiation as a promising pretreatment option due to its consistent heating efficiency, simple and low-cost operation, and rapid processing time. Microwaves change the ultrastructure of lignocellulose, which generates uniform heating by inducing dipole rotation, which aligns polar molecules with the rapidly changing electric field, and by ionic conduction, which causes immediate heating of ionic components in the biomass when electromagnetic waves are applied [47]. As a result, polysaccharides are effectively deconstructed in a shorter reaction time, removing lignin and hemicellulose, increasing cellulose accessibility to hydrolytic enzymes. Alio et al. [48] pretreated sawmill mixed biomass from four softwood species (fir, spruce, Scots pine, and Douglas fir) using microwave assisted organosolv pretreatment. The extraction in a microwave reactor (Monowave 450, Anton PaarGmbH, Austria) at 175 °C for one hour with a 60:40 ethanol: water solvent mixture with the addition of 0.25% $H_2SO_4$ as a catalyst with the subsequent fermentation using *Saccharomyces cerevisiae* resulted in the production of 16 g/L of ethanol from 34.5 g/L of sugars (80% of the potential yield).

## 7 Biological Pretreatment

Biological pretreatment seems to be promising since it is an environmentally friendly approach that produces no inhibitors, low energy consumption, and mild treatment conditions. This approach primarily employs fungal and bacterial strains or their respective enzymes to treat the lignocellulosic material for bioethanol production [49]. This technique is gaining more popularity due to its rapid reaction time and minimal nutrient requirements for the next enzymatic hydrolysis process. Numerous microorganisms, including fungus (*Phanerochaete chrysosporium, Trichoderma reesei, Trichoderma viride, Ceriporiopsis subvermispora, Aspergillus niger*, etc.) and bacteria (*Clostridium sp., Cellulomonas sp., Bacillus sp., Thermomonospora sp., Streptomyces sp.*, etc.) have been extensively utilised in the biological pretreatment [50].

In the wood waste pretreatment, *Ceriporiopsis subvermispora* has been widely used to degrade the lignin content and recalcitrant components (cellulose and hemicellulose) in the plant cell wall. This fungus degrades lignin more selectively and causes less cellulose loss. However, the disadvantage of this pretreatment is the prolonged time that is needed to reach the digestibility stage, approximately 4–8 weeks [51]. Nazarpour et al. [52] reported on the effects of biological treatment using the white rot fungus *Ceriporiopsis subvermispora* on the particle size of rubberwood (1 mm) for bioethanol production. As a result, after 90 days, the maximal ethanol concentration and yield were 17.9 g/L and 53.0%, respectively. Mattila et al. [53] used the white rot fungus *Phlebia radiata* to produce bioethanol from recycled wood (recycled construction, municipal wood waste, spruce sawdust, and birch sticks. As a result, 32.4 ± 4.5 g/L of bioethanol was obtained after 30 days of pretreatment followed by hydrolysis and fermentation process.

## 8 Conclusion

Wood waste is gaining favour as a readily accessible and low-cost source of renewable lignocellulosic biomass for liquid fuel generation. However, the pretreatment step must be involved in bioethanol production and reclassified as one of the costliest stages. This chapter provides a comprehensive overview of the chemical, physiochemical, and biological pretreatment technologies currently available for use on wood waste for the generation of bioethanol. According to the findings, each technique has benefits and disadvantages. However, additional research is necessary in terms of the optimisation of the conversion of fermentable sugar and production yield of bioethanol; decreasing inhibitors formed from sugar breakdown during pretreatment; reducing chemical material; energy consumption and production cost; and developing an environmentally friendly process to establish an effective combination of the existing approaches. Thus, understanding the fundamentals of various pretreatment technologies, the diverse compositions of biomass feedstock, and the correlation between biomass feedstock composition and pretreatment methods would then significantly benefit in determining the optimal pretreatment method/combinations for a given biomass feedstock.

## References

1. Zabed H, Sahu JN, Suely A, Boyce AN, Faruq G (2017) Bioethanol production from renewable sources: current perspectives and technological progress. Renew Sustain Energy Rev 71:475–501
2. WBA-World Bioenergy Association (2021) Global bioenergy statistics
3. Nanthakumar C, Sakthikumaran M, Sivashankar G (2018) Carbon dioxide-the frontline greenhouse gas. Carbon 7:4–10
4. Awogbemi O, Kallon DVV, Onuh EI, Aigbodion VS (2021) An overview of the classification, production and utilization of biofuels for internal combustion engine applications. Energies 14:5687
5. Hossain N, Mahlia TMI, Miskat MI, Chowdhury T, Barua P, Chowdhury H, Nizamuddin S, Ahmad NB, Zaharin NAB, Mazari SA, Soudagar MEM (2021) Bioethanol production from forest residues and life cycle cost analysis of bioethanol-gasoline blend on transportation sector. J Environ Chem Eng 9:105542
6. Hoekman SK, Broch A (2018) Environmental implications of higher ethanol production and use in the US: a literature review. Part II-biodiversity, land use change, GHG emissions, and sustainability. Renew Sustain Energy Rev 81:3159–3177
7. He L, Wang B, Xu W, Cui Q, Chen H (2022) Could China's long-term low-carbon energy transformation achieve the double dividend effect for the economy and environment? Environ Sci Pollut Res 29:20128–20144
8. Deshmukh M, Pendse DS, Pande A (2022) Effects of blending bioethanol with gasoline on spark-ignition engine-a review. J Integr Sci Technol 10:87–99
9. Sakthivel P, Subramanian KA, Mathai R (2020) Experimental study on unregulated emission characteristics of a two-wheeler with ethanol-gasoline blends (E0 to E50). Fuel 262:116504
10. Saini R, Osorio-Gonzalez CS, Brar SK, Kwong R (2021) A critical insight into the development, regulation and future prospects of biofuels in Canada. Bioengineered 12:9847–9859
11. Saravanan AP, Pugazhendhi A, Mathimani T (2020) A comprehensive assessment of biofuel policies in the BRICS nations: implementation, blending target and gaps. Fuel 272:117635
12. Bajpai P (2020) Developments in bioethanol. Springer Nature
13. Kumar B, Bhardwaj N, Agrawal K, Verma P (2020) Bioethanol production: generation-based comparative status measurements. Biofuel production technologies: critical analysis for sustainability. Springer, Singapore, pp 155–201

14. Tse TJ, Wiens DJ, Reaney MJ (2021) Production of bioethanol—A review of factors affecting ethanol yield. Ferment 7:4–268
15. Reina L, Botto E, Mantero C, Moyna P, Menéndez P (2016) Production of second-generation ethanol using Eucalyptus dunnii bark residues and ionic liquid pretreatment. Biomass Bioenergy 93:116–121
16. Ramage MH, Burridge H, Busse-Wicher M, Fereday G, Reynolds T, Shah DU, Wu G, Yu L, Fleming P, Densley-Tingley D, Allwood J (2017) The wood from the trees: the use of timber in construction. Renew Sustain Energy Rev 68:333–359
17. Zang D, Hu Z, Yang Y, He S (2022) Research on the relationship between agricultural carbon emission intensity, agricultural economic development and agricultural trade in China. Sustainability 14(18):11694
18. Millati R, Cahyono RB, Ariyanto T, Azzahrani IN, Putri RU, Taherzadeh MJ (2019) Agricultural, industrial, municipal, and forest wastes: an overview. In: Sustainable resource recovery and zero waste approaches, pp 1–22
19. O'Dwyer J, Walshe D, Byrne KA (2018) Wood waste decomposition in landfills: an assessment of current knowledge and implications for emissions reporting. Waste Manag 73:181–188
20. Akhtari S, Malladi KT, Sowlati T, Mirza F (2021) Incorporating risk in multi-criteria decision making: the case study of biofuel production from construction and demolition wood waste. Resour Conserv Recycl 167:105233
21. Azambuja RDR, Castro VGD, Trianoski R, Iwakiri S (2018) Recycling wood waste from construction and demolition to produce particleboards. Cienc Tecnol 20:681–690
22. Cetiner I, Shea AD (2018) Wood waste as an alternative thermal insulation for buildings. Energy Build 168:374–384
23. Wang L, Yu IKM, Tsang DCW, Li S, Li J-S, Sun Poon CS, Wang YS, Dai JG (2017a) Transforming wood waste into water-resistant magnesia-phosphate cement particleboard modified by alumina and red mud. J Clean Prod 168:452–462
24. Ihnat V, Lübke H, Balbercak J, Kuna V (2020) Size reduction downcycling of waste wood. Rev Wood Res 65:205–220
25. Cecchi T, Carolis CD (2021) Biocascading: general strategy for the recovery of valuable substances from food waste. Biobased products from food sector waste. Springer, Cham, pp 109–167
26. Jafari V, Labafzadeh SR, Jeihanipour A, Karimi K, Taherzadeh MJ (2011) Construction and demolition lignocellulosic wastes to bioethanol. Renew Energy 36:2771–2775
27. Rezania S, Oryani B, Cho J, Talaiekhozani A, Sabbagh F, Hashemi B, Rupani PF, Mohammadi AA (2020) Different pretreatment technologies of lignocellulosic biomass for bioethanol production: an overview. Energy 199:117457
28. Solarte-Toro JC, Romero-García JM, Martínez-Patiño JC, Ruiz-Ramos E, Castro-Galiano E, Cardona-Alzate CA (2019) Acid pretreatment of lignocellulosic biomass for energy vectors production: a review focused on operational conditions and techno-economic assessment for bioethanol production. Renew Sustain Energy Rev 107:587–601
29. Sofokleous M, Christofi A, Malamis D, Mai S, Barampouti EM (2022) Bioethanol and biogas production: an alternative valorisation pathway for green waste. Chemosphere 296:133970
30. Awogbemi O, Von Kallon DV (2022) Pretreatment techniques for agricultural waste. Case Stud Chem Environ Eng 100229
31. Bay MS, Karimi K, Esfahany MN, Kumar R (2020) Structural modification of pine and poplar wood by alkali pretreatment to improve ethanol production. Ind Crops Prod 152:112506
32. Amândio MS, Rocha JM, Serafim LS, Xavier AM (2021) Cellulosic bioethanol from industrial eucalyptus globulus bark residues using kraft pulping as a pretreatment. Energies 14:8–2185
33. Muthuvelu KS, Rajarathinam R, Kanagaraj LP, Ranganathan RV, Dhanasekaran K, Manickam NK (2019) Evaluation and characterization of novel sources of sustainable lignocellulosic residues for bioethanol production using ultrasound-assisted alkaline pre-treatment. Waste Manag 87:368–374
34. Shafiei M, Zilouei H, Zamani A, Taherzadeh MJ, Karimi K (2013) Enhancement of ethanol production from spruce wood chips by ionic liquid pretreatment. Appl Energy 102:163–169

35. Monrroy M, Ibanez J, Melin V, Baeza J, Mendonça RT, Contreras D, Freer J (2010) Bioorganosolv pretreatments of P. radiata by a brown rot fungus (Gloephyllum trabeum) and ethanolysis. Enzyme Microb Technol 47:11–16
36. Zhang J, Zhou H, Liu D, Zhao X (2020) Pretreatment of lignocellulosic biomass for efficient enzymatic saccharification of cellulose. In: Lignocellulosic biomass to liquid biofuels. Academic Press, pp 17–65
37. Rochón E, Cabrera MN, Scutari V, Cagno M, Guibaud A, Martínez S, Böthig S, Guchin N, Ferrari MD, Lareo C (2022) Co-production of bioethanol and xylosaccharides from steam-exploded eucalyptus sawdust using high solid loads in enzymatic hydrolysis: effect of alkaline impregnation. Ind Crops Prod 175:114253
38. Cotana F, Cavalaglio G, Gelosia M, Nicolini A, Coccia V, Petrozzi A (2014) Production of bioethanol in a second generation prototype from pine wood chips. Energy Procedia 45:42–51
39. Toscan A, Morais ARC, Paixao SM, Alves L, Andreaus J, Camassola M, Dillon AJP, Lukasik RM (2017) High-pressure carbon dioxide/water pre-treatment of sugarcane bagasse and elephant grass: assessment of the effect of biomass composition on process efficiency. Bioresour Technol 224:639–647
40. Wu Y, Ge S, Xia C, Cai L, Mei C, Sonne C, Park YK, Kim YM, Chen WH, Chang JS, Lam SS (2020) Using low carbon footprint high-pressure carbon dioxide in bioconversion of aspen branch waste for sustainable bioethanol production. Bioresour Technol 313:123675
41. Chen WC, Lin YC, Ciou YL, Chu IM, Tsai SL, Lan JCW, Chang YK, Wei YH (2017) Producing bioethanol from pretreated-wood dust by simultaneous saccharification and co-fermentation process. J Taiwan Inst Chem Eng 79:43–48
42. Carvalheiro F, Duarte LC, Gírio F, Moniz P (2016) Hydrothermal/liquid hot water pretreatment (autohydrolysis): a multipurpose process for biomass upgrading. In: Biomass fractionation technologies for a lignocellulosic feedstock based biorefinery. Elsevier, pp 315–347
43. Rigual V, Santos TM, Domínguez JC, Alonso MV, Oliet M, Rodriguez F (2018) Evaluation of hardwood and softwood fractionation using autohydrolysis and ionic liquid microwave pretreatment. Biomass Bioenergy 117:190–197
44. Romaní A, Ruiz HA, Pereira FB, Teixeira JA, Domingues L (2014) Integrated approach for effective bioethanol production using whole slurry from autohydrolyzed Eucalyptus globulus wood at high-solid loadings. Fuel 135:482–491
45. Zhang W, Zhang X, Lei F, Jiang J (2020) Co-production bioethanol and xylooligosaccharides from sugarcane bagasse via autohydrolysis pretreatment. Renew Energy 162:2297–2305
46. Senila L, Varaticeanu C, Roman M, Miclean M, Roman C (2014) Bioethanol production from wood waste. Studia Universitatis Babes-Bolyai, Ambientum, p 59
47. Pooja NS, Sajeev MS, Jeeva ML, Padmaja G (2018) Bioethanol production from microwave-assisted acid or alkali-pretreated agricultural residues of cassava using separate hydrolysis and fermentation (SHF). Biotech 8:1–12
48. Alio MA, Tugui OC, Rusu L, Pons A, Vial C (2020) Hydrolysis and fermentation steps of a pretreated sawmill mixed feedstock for bioethanol production in a wood biorefinery. Bioresour Technol 310:123412
49. Zabed HM, Akter S, Yun J, Zhang G, Awad FN, Qi X, Sahu JN (2019) Recent advances in biological pretreatment of microalgae and lignocellulosic biomass for biofuel production. Renew Sustain Energy Rev 105:105–128
50. Sharma A, Choudhary J, Singh S, Singh B, Kuhad RC, Kumar A, Nain L (2019) Cellulose as potential feedstock for cellulase enzyme production: versatility and properties of various cellulosic biomasses. In: New and future developments in microbial biotechnology and bioengineering. Elsevier, pp 11–27
51. Awogbemi O, Von Kallon DV (2022) Valorization of agricultural wastes for biofuel applications. Heliyon e11117
52. Nazarpour F, Abdullah DK, Abdullah N, Zamiri R (2013) Evaluation of biological pretreatment of rubberwood with white rot fungi for enzymatic hydrolysis. Materials 6:2059–2073
53. Mattila HK, Kačar D, Mali TLE, Lundell TK (2018) Lignocellulose bioconversion to ethanol by a fungal single-step consolidated method tested with waste substrates and co-culture experiments. AIMS Energy

# Valorization of Wood Waste as Biosorbent for the Removal of Organic and Inorganic Contaminants in Water

**Nurul Syarima Nadia Sazman, Nurul Izzati Izhar, Nur Ramadhan Mohamad Azaludin, Shaari Daud, Hartini Ahmad Rafaie, and Zul Adlan Mohd Hir**

**Abstract** The valorization of wood waste as biosorbent has sparked intense interest from researchers, exclusively for removing organic and inorganic contaminants in aqueous solutions. Highlights on several desirable features including higher porosity, outstanding physicochemical properties, and selectivity offer a new vision towards sustainable chemistry for environmental protection. A decline in water quality poses significant domestic and industrial challenges. However, their performance and effectiveness for removing such contaminants from water depends on how they are fabricated, how they work together, and what mechanisms are at play. The use of this material and approach in the water recovery process suggests that developing an enhanced protocol is necessary for successfully and realistically removing the contaminants from the environment.

**Keywords** Adsorption · Biosorbent · Water recovery · Wood waste

## 1 Introduction

Recently the existence of organic and inorganic pollutants in the water stream due to the rapid developments around the world has aroused massive attention of researchers in the world. This is because it can contribute to poisonousness, destructiveness, and non-degradability in the ecosystem [1–3]. The surface and subterranean water (aqueous environment) will be continuously contaminated by the effluents that result

N. S. N. Sazman · N. I. Izhar · N. R. Mohamad Azaludin · S. Daud · Z. A. Mohd Hir (✉)
Faculty of Applied Sciences, Universiti Teknologi MARA Pahang, 26400 Bandar Tun Abdul Razak Jengka, Pahang, Malaysia
e-mail: zuladlan@uitm.edu.my

H. Ahmad Rafaie
Faculty of Applied Sciences, Centre of Foundation Studies, Universiti Teknologi MARA, Selangor Branch, Dengkil Campus, 43800 Dengkil, Selangor, Malaysia

Z. A. Mohd Hir
Catalysis for Sustainable Water and Energy Nexus Research Group, School of Chemical Engineering, College of Engineering, University Teknologi MARA, 40450 Shah Alam, Selangor, Malaysia

 
S. N. Sarmin et al. (eds.), *Wood Waste Management and Products*, Sustainable Materials and Technology, https://doi.org/10.1007/978-981-99-1905-5_6

from activities like agriculture, industry, and residential that contain these organic and inorganic substances. Thus, it is important to find a solution to overcome this matter. One of the best ways that can be used to solve this problem is by using an adsorbent that can effectively remove the contaminants. Basically, adsorbents can be divided into two; the commercial adsorbent and biosorbents derived from agricultures.

Commercial adsorbent that was produced synthetically such as activated carbon, carbon nanotubes, and zeolite showed good adsorption and high removal rates towards several pharmaceuticals belonging to different therapeutic classes [4–7]. However, the application of large-scale approach will involve higher cost. Thus, the finding of another low-cost substitute and biodegradable adsorbent is needed. This low-cost substitute adsorbent is known as biosorbent. Biosorbents can be defined as biological materials that are being employed to eliminate contaminants submissively from aqueous solution. Agricultural wastes, algae, bacterial, and industrial wastes were an example of biological materials. The scientific community has become interested in these biological materials because they are renewable sources, inexpensive, biodegradable, and after complete usage, do not create secondary pollution [8]. The presence of some functional groups in these materials which act as hydrophobic interaction during the sorption process makes it more attractive to the researchers [9, 10].

Due to their abundance in nature, low cost, superior mechanical and chemical resilience, and biodegradability, wood wastes from forestry and agriculture have gained popularity among the global community in recent years [11]. On another hand, this wood waste can be a good biosorbent because it consists of lignocellulosic materials. Lignocellulosic materials are usually composed of polysaccharides (cellulose and hemicelluloses) and an aromatic polymer (lignin). Furthermore, some parts of wood waste such as barks are rich with tannins. Lignin with the presence of polar functional groups, such as hydroxyl, carbonyl, carboxyl, and phenyl as active sites make it favourable for the removal of organic and inorganic contaminants from water and wastewater. For example, the adsorptive removal of dyes (and other contaminants) from water is mainly accomplished by the interaction of dyes and specific functional groups of lignocellulosic polymers (e.g., −OH and −COOH). The availability in various industrial waste especially in wood waste; the biodegradability, cost-effectiveness, antibacterial, stabilizer, and antioxidant properties make it an environmentally friendly material. Compared to other naturally occurring materials that are manmade, lignin materials provide a wide range of benefits. Nevertheless, compounds such as tannins that are mostly found in barks also are good to remove heavy metals as well.

Sawdust, a low-cost locally available material and solid waste, can be used as a biosorbent for contaminant removal. Sawdust is a common byproduct made up of fine wood particles. It is classified as agricultural waste as well as a byproduct of manufacturing industries. It is widely available in the countryside for free or at a low cost. The materials comprise several organic constituents such as lignin, cellulose, and hemicellulose. These organic constituents will effectively bind harmful compounds by the functional groups that are present in those constituents, respectively. In the studies, sawdust proved to be a promising effective material for eliminating harmful components from aqueous solutions [11–14]. Another wood waste that can be used as biosorbent is tree bark. Tree barks are relatively coming from the exploitation of wood by mankind, and it became an important waste since it is also easily found in abundance. Pine bark is one of the examples of wood species that is found to be good in the sorption of heavy metals [15, 16] and organic contaminants [17, 18] in wastewater treatment processes. This is due to the higher tannins' contents and derivatives of polyphenols in barks that act as active components in the sorption process.

## 2 Wood Waste

In 2012, it has been reported that commercial and institutional (ICI) production generated about 11,500 tonnes from the 33,000 tonnes of solid waste produced daily, with an average of 0.41 kg of ICI garbage created per individual per day. Plastic, paper, food, and wood are the four major types of industrial waste in Malaysia, making up correspondingly 39.1, 35.1, 5.8, and 4.5% in total. For wood waste, the average calorific value ranges from 16.488 to 20.092 MJ/kg [19]. While in Finland, the statistics show that over 3.4 million tonnes of wood waste are produced each year. This represents around 3.4% of the total volume of waste generated in the country. According to the data, the main industries that produce wood waste are the manufacturing of paper, sawn products, electricity, and building. About 224,000 tonnes of other construction waste were generated in 2013, of which 142,000 tonnes (63%), were wood waste. 4.2% of the total amount of wood waste produced is wood waste from construction [20]. Nonetheless in 2015, Germany generated waste of about 401 million tonnes, with 11.9 million tonnes of waste wood coming from the wood packing, building and demolition, and wood processing industries [21].

Forest-based industries could come up with a better management and conservation of natural resources, while also creating opportunities for employment and providing income. The varied wood-based sectors create enormous loads of waste, which must be managed with care, recycled, distributed, or disposed [22]. Wood waste is produced from a variety of sources, including sustainable forest management (SFM) companies, sawmills, furniture producers, panel board manufacturers, as well as building, demolition, and pruning operations. Consequently, in terms of its technological characteristics, it is a very heterogeneous biomass [23]. Wood biomass is composed mainly of three polymers: cellulose (40–45% of the wood's dry weight), hemicellulose (30% of the wood's dry weight), and lignin (20–30%) [24]. This renewable biomass is specifically used in the removal and management of contaminants, intelligent and effective separation of oil and water, the production of activated carbon, the fabrication of polyhydroxyalkanoate bioplastics, the extraction of cellulose nanocrystals, and the source of carbon for the storage of energy (supercapacitors) [25]. Due to the growing number in worldwide population and the resulting rise in energy consumption, $CO_2$ emissions, and changes in climate are driven by the burning of fossil fuels, utilizing wood wastes as a source of energy is also considered an essential option.

According to Pandey et al., about half of the wood gets converted into useful items, while the other half is wasted [22]. As wood is used in major industrial operations, waste products such as sawdust, chips, bark, coarse residues, slabs, peeler log cores, planer shavings, and end trimmings are produced. The byproduct of wood such as sawdust or wood shaving, are usually acquired from woodworking activities including the wood's sawing, planning, sanding, and milling. It is a lignocellulosic compound that is inexpensive, ubiquitous, and reasonably plentiful, but has disposal issues. The majority of it was used in paper mills and cattle ranches in recent decades, while the remainder was simply dumped onto the surrounding area without being treated [25]. The major issues, however, were created by the lack of strategies for managing wood waste and an absence of adequate legislation as well as a lack of economic advantages for using wood waste and weak environmental regulatory enforcement. Hence, the proper utilization of wood waste may effectively minimize environmental effects without endangering the global forest [22].

In 1995, Malaysia generated dried sawdust of about 0.266 million tonnes. As a result, Cipta Briquette Sdn. Bhd., which is based in Bintulu, Sarawak, contributes to the massive amounts of sawdust generated by the timber sector, using around 30–40 different species of tree for the manufacturing and export of charcoal briquette. The finest types of trees in Malaysia include cengal, balau, meranti, jati, keruing, kempas, bitangor, and other types recognized as the best in the local timber industry. The timber industry in Malaysia, as reported by the Forest Research Institute Malaysia (FRIM), generates 3.4 million $m^3$ of wood waste annually, comprising bark, sawdust, slabs, chips, and other unprocessed materials with 55% of normal restoration rate [26]. Agglomerated materials including oriented strand boards (OSB), middle density fiberboards (MDF), and particle boards (PB), which are typically made from old furniture, are also likely the major source of wood waste [27]. From previous study by Huang et al., it should be emphasized that sanding dust of high-pressure laminates (HPL) is a solid waste formed from the cutting and sanding process of HPLs [28]. It has been calculated that domestic fireproof board industries produced around 10,260 tonnes of dust annually, as it has been commonly employed in interior decorating, furniture, countertops, external walls, and other industries, due to its rich surface colour, texture, and unique physical appearance. HPL sanding dust has a tiny particle size and a low specific gravity, making it simple to pollute the atmosphere.

Adsorption is the most frequently used technique to remove dyes employing traditional adsorbents, which include activated carbon. Unfortunately, despite their exceptional adsorption abilities, the expensive cost of traditional adsorbents frequently limits their utilization. Finding affordable adsorbents that might replace the pricey traditional ones is therefore crucial. The inexpensive adsorbent is by definition ubiquitous in nature, a manufacturing byproduct, or waste that needs almost no processing [29]. Waste wood biomass satisfies the criteria for being a cost-effective adsorbent by being relatively inexpensive, widely accessible throughout the year, and requiring little to no processing. Moreover, since it is an organic (biological) compound, it classifies as a biosorbent. The subcategory of adsorption is biosorption which the adsorbent being the biological origin [24]. Sawdust is mostly a captivating material for removing the organic contaminants in wastewater among the waste material of lignocellulosic originating from the wood and agri-food industries [29]. Velić and co-workers [29] have tested several tree species' sawdusts as adsorbents for organic contaminants removal, which are meranti, beech, and chirpine woods for methylene blue, direct brown, and Congo red dyes removal, respectively. The results were found to be remarkably successful for MB removal with percentages removal of MB ranging from 93.25 to 98.50%. It turned out that poplar sawdust was the most effective. While the MB adsorption mechanism onto sawdust's poplar is interpreted in terms of the Langmuir and Freundlich adsorption isotherm models.

## 3 Physicochemical Characterization of Wood Waste as Biosorbent

The physicochemical and characterization analysis of wood waste as biosorbent is of great importance to provide new insight into the properties and predict the adsorption performance to remove the organic and inorganic contaminants in water. The performance of an adsorption system can be controlled and monitored using the properties of wood waste as a biosorbent. The development of an adsorbent is substantially influenced by its constitutional characteristics; as a result, the principal adsorbent mechanism depends heavily on the type and characteristics of the adsorbent. Consequently, it is essential to comprehend how the composite adsorbent feature impacts adsorption capacity.

The characteristics of the particles' shape and homogeneity, as well as the biosorbent's surface characteristics, can be identified and evaluated using scanning electron microscopy (SEM). Ibrahim et al. [30] developed a modified sawdust sorbent from oak wood utilizing chemical modification method by incorporating poly (amido amine) (PAMAM) as a functional polymer containing active amino groups and epichlorohydrin as a cross-linker [30]. The obtained samples have been undergoing post treatment process with $TiO_2$ or ZnO nanoparticles (NPs) for the elimination of some dangerous water pollutants, including heavy metal ions and an anionic dye. The obtained SEM showed variations in sawdust morphology before and after treatment (Fig. 1a, b). They reported that due to the heterogeneous deposition of PAMAM on sawdust surfaces and the attraction force between PAMAM dendritic nanostructures causing agglomeration of some PAMAM dendrimers, as a result, modified sawdust had rougher surfaces (Fig. 1b) than for unaltered sawdust (Fig. 1a). Images of sawdust that has been altered by the addition of $TiO_2$ or ZnO NPs reveal finely scattered nanoparticles on the sawdust surfaces (Fig. 1c, d). $TiO_2$ NPs, in contrast to ZnO NPs, were dispersed more evenly. Furthermore, the EDX spectra of nanomodified materials (Fig. 1c, d) demonstrate the presence of Ti and Zn elements.

An earlier study prepared and examined the characteristics of different inexpensive wood-based biosorbents, such as pineapple, bamboo stems, and banana pith, for an alternative way to be applied in heavy metal removal from wastewater solutions [31]. Based on surface morphology (Fig. 2 (I)), the existence of microporous structure in

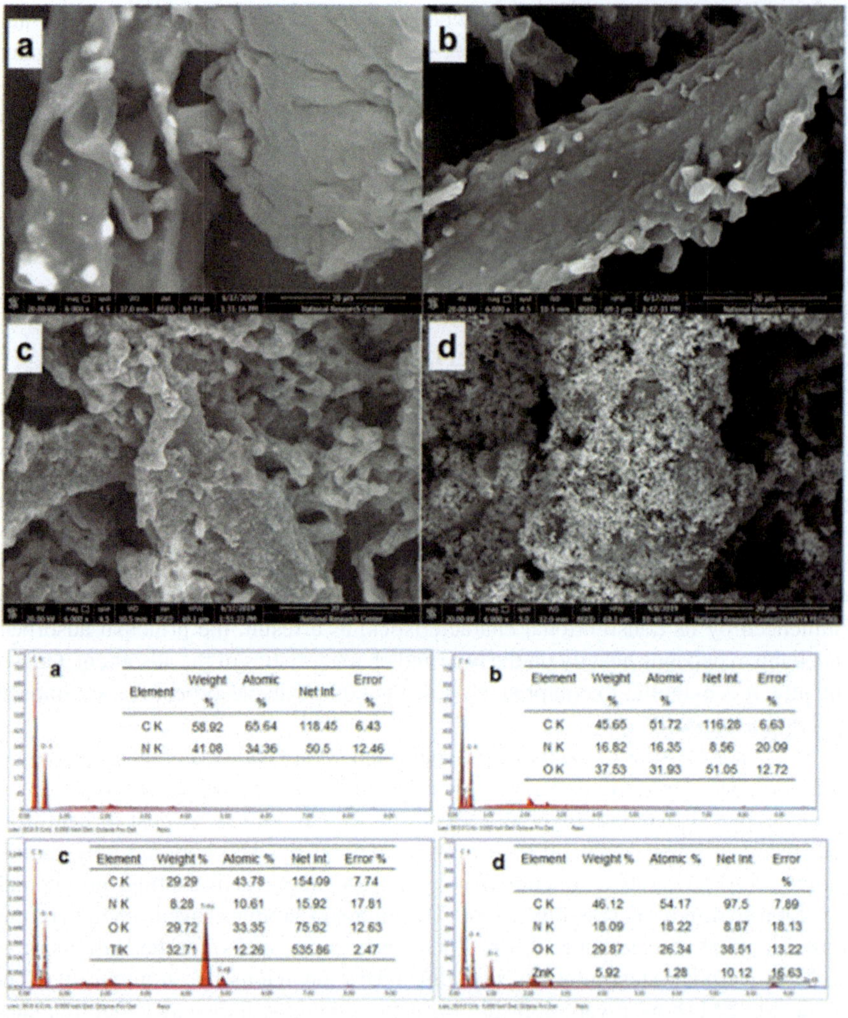


**Fig. 1** Scanning electron microscope of sawdust morphology **a** before and **b** after treatment, modified sawdust post-treated with **c** $TiO_2$ and **d** ZnO NPs [30]

the treated biosorbents has been observed in all samples and this porous structure is crucial for removing heavy metals from wastewater. Additionally, according to BET data, the biosorbents made from banana pith, pineapple stem, and bamboo stem exhibit good combinations of surface area (9.072–116.01 $m^2/g$) and a narrower range of average particle sizes (30–78 mm). According to FTIR data that have been published (Fig. 2 (II)), the sample's pre-treatment spectra have a variety of peaks and bond formation. It has been noted that the enabling surface functional groups (such as C≡N stretching, the stretching vibration of C=O, $-CH_3$ wagging, and C–O stretching vibration) contribute to the adsorption of Ni (II).

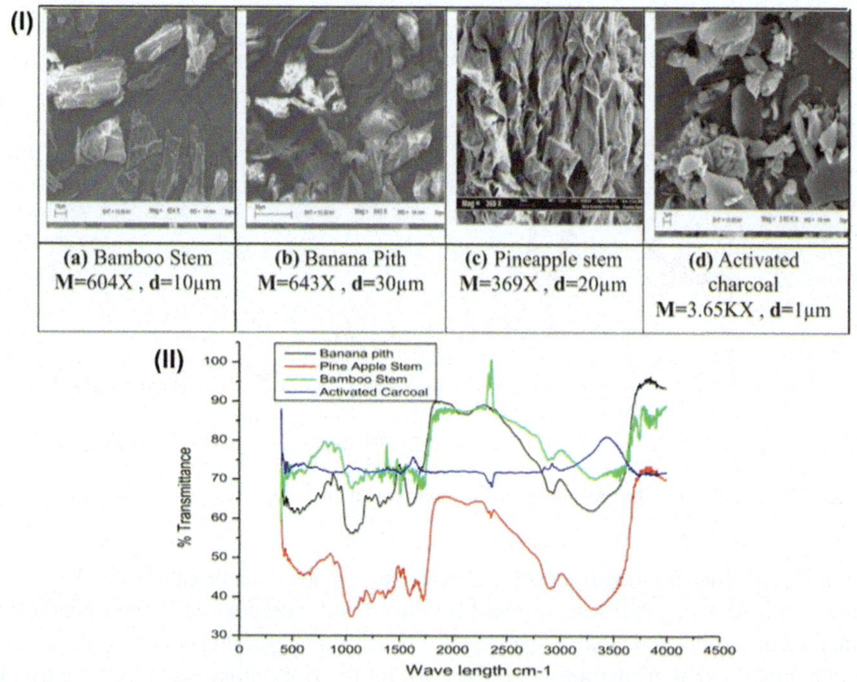


**Fig. 2** (I) SEM images of adsorbents with corresponding magnification (M) and particle diameter (d) and (II) FTIR for all adsorbents [31]

Recently, researchers are widely searching for an inexpensive and non-conventional adsorbent for dye removal in wastewater [32]. Due to their aromatic properties, dye molecules are typically poisonous, carcinogenic, and mutagenic to humans, posing serious environmental and health risks. Thus, it is necessary and essential to remove harmful dyes from industrial effluents before they are released into the environment. In addition to morphological research, XRD analysis is essential for determining the structural and crystallinity characteristics of wood waste-sorbent. Ghaedi et al. [33] was producing an activated carbon (AC) from oak tree that has been used as adsorbent for the removal of noxious anionic dye sunset yellow [33]. The XRD research they performed revealed that the AC is composed of both crystalline and amorphous structures due to the presence of sharp and broad peaks. The crystalline nature is about 26% while the amorphous nature is about 74%showing that the prepared samples are mostly in amorphous structure with low crystallinity. Further analysis revealed that the XRD patterns for the AC sample are identical to the (002), (100), and (101) planes of graphite peaks (Fig. 3).

Another investigation conducted by Che et al. described the creation of a wood-based filter embellished with silver nanoparticles (Ag NPs) and its use in the purification of water [34]. Wood has demonstrated significant potential for filtration and separation applications since it is a naturally occurring porous cellulosic material

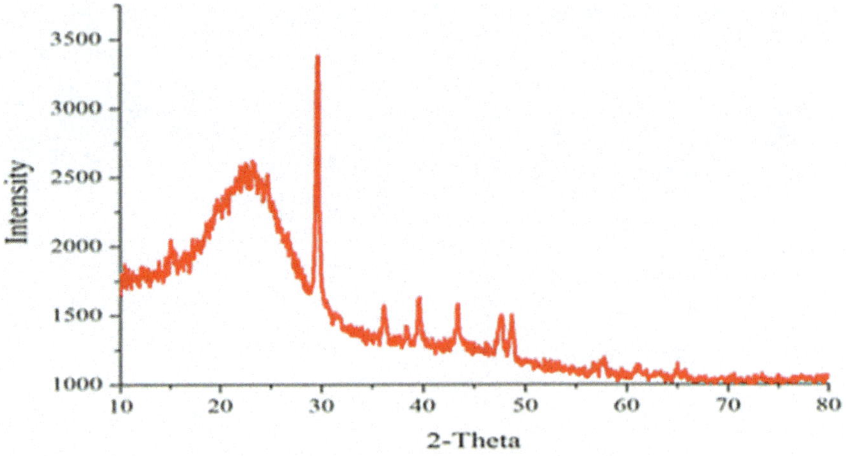


**Fig. 3** XRD pattern of the produced AC derived from oak tree [33]

with a distinctive hollow structure and membranes made of nanoscale holes. To boost the purification capabilities, Ag was introduced and incorporated in their synthesis method because wood alone is inadequate at purifying water that contains small-size contaminants. The formation of Ag NPs within the wood filter was further verified by XRD (Fig. 4) and they observed that the starting natural wood exhibited typical crystalline cellulose patterns with diffraction angles (2θ) at around 16°, 22°, and 35°. Meanwhile, Ag/wood composites XRD spectra revealed the appearance of new peaks at 38.1°, 44.4°, 64.6°, and 77.7°, which were assigned to the (111), (200), (220), and (311) lattice planes of Ag, respectively [34].

**Fig. 4** XRD pattern of natural wood and Ag/wood [34]

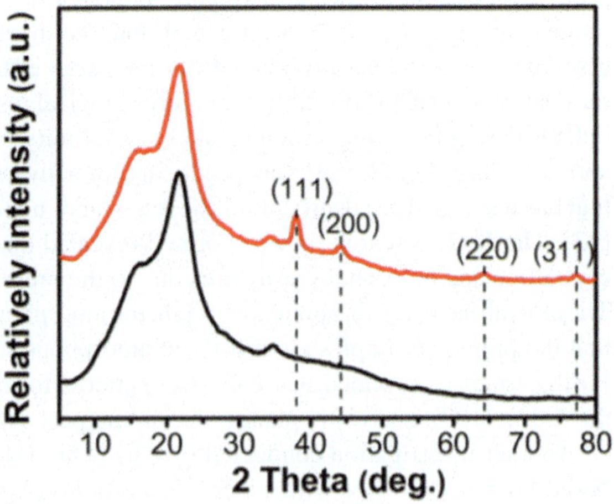

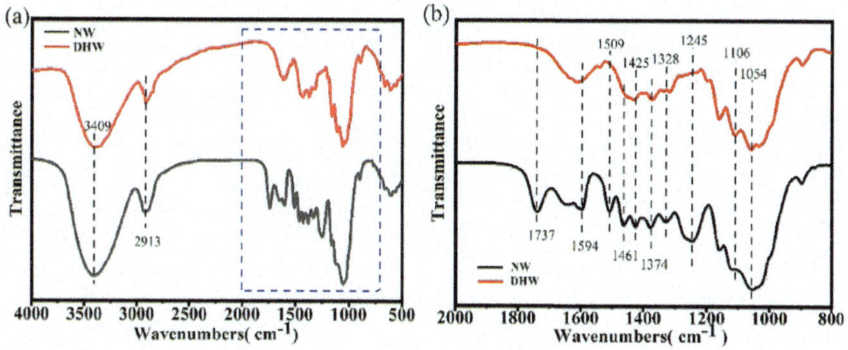


**Fig. 5** FTIR spectra of **a** oven-dried poplar wood (NW) and hemicellulose removed (DHW) blocks, **b** zoomed-in detailed curves [35]

Fourier transform infrared spectroscopy (FTIR) analysis was done to identify the vibrational frequency changes in the functional groups in the adsorbent and this qualitative analysis of specific functional groups on the biosorbent surface is significant which affects the efficiency of the biosorption process. A study done by Wu et al. [3] reported a novel type of poplar sponge with exceptional adsorption capabilities as one of the alternative ways to address the issues of water pollution. They introduced a poplar sponge, wood-based (Populus tomentosa Carr), as base material due to its better mechanical properties, and removes lignin with the aid of microwave, which greatly reduces the soaking time of the wood in the medicine and reduces the loss of wood strength. The removal of lignin and hemicellulose from modified wood cell walls was confirmed by FTIR data (Fig. 5), resulting in a compressible rebound 3D structure that was further confirmed by high-resolution scanning electron microscopic image analyses. They reported that the obtained adsorbent structure has outstanding compressible rebound and water absorption capabilities [35].

Figure 6 displays the FTIR spectra of Euroamerican poplar (EP), one of the best biosorbents for removing Congo Red dye, both before and after dye biosorption [36]. The very wide band dominates both spectra at 3339.7 $cm^{-1}$ (3324.8 for CR loaded EP), which can be attributed to hydroxyl groups (−OH), i.e., vibrations of OH bond stretching, probably connected to inter- and intramolecular formation of hydrogen bonds within cellulose and lignin. The second most prominent band is the 1028.7 $cm^{-1}$ intense band, that could be assigned to the polysaccharides, i.e., C–O, C=C, and C–C–O stretching in lignin, cellulose, and hemicellulose. The band with a maximum at 2914.8 $cm^{-1}$ could be associated with the stretching of aliphatic groups (−CH). The 1729.5 (1736.9 for CR loaded EP) band can be assigned to C=O stretching while the presence of branched-chain aromatic radicals is indicated by bands from 1423.8 to 1505.8 $cm^{-1}$. The slight shifting of the peaks (frequency and intensity change) in the FTIR spectrum of CR loaded EP compared to that of EP could be assigned to the biosorption of CR on the surface of the biosorbent [36].

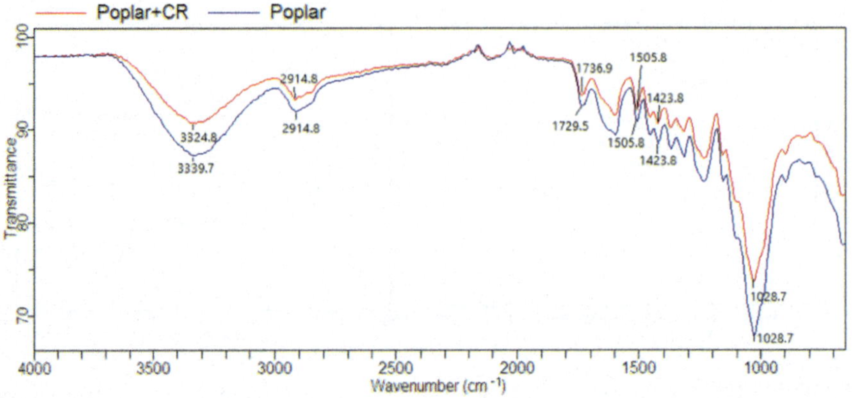


**Fig. 6** The FTIR spectral characteristics of EP before and after CR biosorption [36]

Guan et al. [37] revealed the highly compressible wood sponges with a spring-like lamellar structure as efficient and reusable oil adsorbents for cleaning up organic contaminants and oil spills [37]. The lignin component of wood was initially removed selectively using a straightforward delignification procedure utilizing acidified $NaClO_2$ aqueous solution to create the wood sponges. By comparing the natural and delignified wood using FTIR analysis (Fig. 7), it was determined that the lignin had been removed. After the chemical treatment, the lignin's distinctive peaks at 1593, 1505, and 1462 $cm^{-1}$ (aromatic skeletal vibrations) had vanished. Besides, the hemicellulose-related peaks at 1736 and 1235 $cm^{-1}$ remained during the treatment, indicating the selectivity of the $NaClO_2$ solution for removing the lignin fraction. They also added a chemical process using a NaOH solution to eliminate any remaining hemicellulose in the delignified wood in order to further enhance the structure. The complete disappearance of the hemicelluloses at 1736 and 1235 $cm^{-1}$ in the FTIR spectra indicates the significant elimination of hemicellulose.

## 4 Application of Wood Waste for Adsorptive Removal of Organic Contaminants

Global access to safe and clean drinking water is worsening due to the progress of industrialization and the dramatic growth of population in every country. The tremendous concern and awareness towards the environmental and water contamination has steadily increased with regard to the potential threat it brings [9, 38]. Therein, the drive to acquire safe and clean drinking water, without impacting the existing freshwater supply has resulted in establishing several advancements of pollution-free technologies for wastewater treatment and management. Among feasible strategies, the adsorption process is regarded as one of the most efficient and economical ways to restore contaminated water.

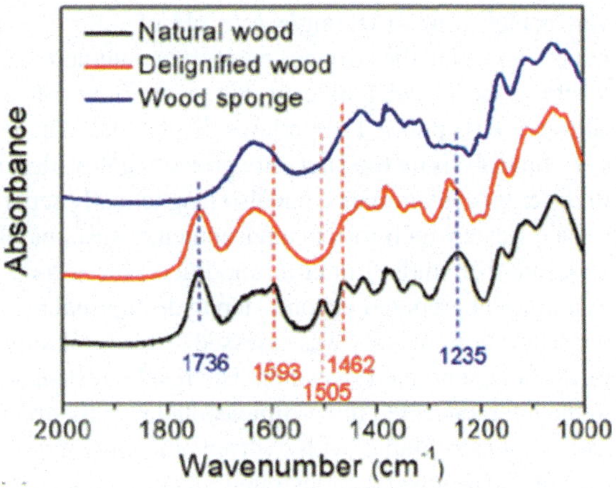


**Fig. 7** FTIR spectra of the different wood samples [37]

Compared to other current water treatment systems, adsorption processes have made significant development because of the safety concerns, ability to remove up to 99.9% of contaminants, energy and cost savings, and ease of operation [9, 39]. The selection of various adsorbents, ranging in price from cheap to high, is crucial when using the adsorption technique. Additionally, the adsorbents must have a significant amount of pore space and a large specific surface area with accessible surface-active sites in order to be effective. High porosity is crucial in determining the adsorbent uptake and selectivity [40, 41]. A suitable adsorbent must therefore be appropriately constructed to ensure that it can effectively and practically remove the contaminants from water and wastewater under a variety of circumstances.

International interest has arisen towards a renewable resource such as biomass and their conversion to value-added adsorbent not only for economical perspective but also greener strategy. Utilizing these waste materials could decrease the amount of waste produced while also producing goods with a high economic value. In order to create affordable and effective adsorbent for the removal of organic pollutants in water, the recovery of Agri-industrial waste, such as wood waste, has been the current focus [29, 42, 43]. Despite being abundant, most wood wastes were disposed of at landfills without any further treatment or recycling purpose and subsequently not only consumed the valuable land resources but also further caused the emissions of greenhouse gases at a large amount. Along with being practical and affordable, the wood's distinctive hierarchical porosity structure makes it a particularly promising adsorbent for high throughput wastewater treatment.

Kang et al. converted pine wood (PW) waste into bio-magnetic adsorbent by co-pyrolysis with red mud (RM) at 500, 650, and 800 °C for the removal of acetaminophen, methylene blue, methyl orange, and ibuprofen in water [44]. The results demonstrated the capability of mixed PW and RM to adsorb the contaminants when pyrolyzed at 800 °C and the mass reduction reached ~56%. Moreover, the result indicates that using real biomass composed of cellulose and lignin further improved the overall adsorption performances. The adsorptive removal of ibuprofen was achieved at 21.01 mg/g at a pH of ~4.5, the highest among other contaminants suggesting the stronger π-stacking interaction of the adsorbent-adsorbate. The leaching study is insignificant due to very low concentration of iron detected in the solution (0.03 ppm). The findings display the effective combination of PW and RM via co-pyrolysis method not only producing flexible adsorbent for adsorptive removal of different contaminants, but also magnetically separable for easy recycling and can reduce waste which fosters environmental sustainability and circular economy.

In another study, Liu et al. modified pine wood sawdust with graphene aerogel as adsorbent prepared via one-step hydrothermal process [45]. The quantification of the adsorption capacity was assessed towards oil removal (chloroform, gasoline, and toluene) in simulated seawater. The results revealed a significant improvement due to the interconnected three-dimensional porous structures, ultra-light, mechanically and chemically stable, with hydrophobic and lipophilic properties. The adsorptivity towards oil removal demonstrated higher Q-values (from 44 to 63 g/g) indicating its strong potential to be exploited in the waste consumption, oil-water separation, and organic solvent adsorption applications.

Synthesizing adsorbent from wood waste-derived biochar is another alternative to remove organic contaminants from water. This study has been conducted by Zhou et al. [46] for the removal of polycyclic aromatic hydrocarbons (PAHs). The surface adsorption capacity of oxygen-rich biochar derived from wood waste significantly affected by the pyrolysis conditions. The presence of multiple functional groups, including carboxyl and hydroxyl, was required for the adsorption of naphthalene ($C_{10}H_8$). The primary functional groups adsorbed for the elimination of PAHs included the benzene ring, −COOH, and $-CH_3$. By taking into account various pyrolysis settings and activation techniques, He et al. examined the production of wood waste-derived biochar with surface functionality and modifiable carbon structures [47]. The study revealed that a temperature of 650 °C was successful in the construction of carbonized structure in biochar. Porous and graphitic carbon skeleton structures can also be obtained critically at temperature of 750 °C. Additionally, activation with steam allowed for the development of mesoporous structure, whereas activation with $CO_2$ and acid treatment resulted in microporous structure.

Ma et al. investigated the efficacy of combining magnetic metal-organic framework (MOFs), specifically ZIF-67, with wood waste for anionic (Congo Red, CR) and cationic (Methylene Blue, MB) dyes removal in water [48]. Based on the findings, it can be concluded that carbonized wood's hierarchical porous structure is advantageous for the rapid dye diffusion and the magnetic core-shell nanoparticles are an efficient way to subsequently adsorb dye molecules. The capacities of CR and MB dyes for adsorption are 1117.03 mg/g and 805.08 mg/g, respectively. The effectiveness is still more than 99% when a very high dye concentration is applied, up to 1200 mg/L. Interestingly, the thermostability of the adsorbent is high since it can be easily burned (few seconds) to remove the adsorbed dyes for better recycling ability.

The conversion of wood waste into commonly versatile adsorbent such as activated carbon (AC) is still attractive due to its facile synthesis process, high contaminants removal, flexibility in adsorbing various contaminants, and high regeneration ability.

In this approach, Firdaus et al. converted teak wood waste into AC for the removal of MB via microwave heat irradiation followed by physiochemical activation (KOH treatment and $CO_2$ gasification). The results revealed high surface area and total pore volume of 1345.25 $m^2/g$ and 0.6140 $cm^3/g$, respectively with MB adsorption up to 66.69 mg/g and 38.23% yield [49]. Meanwhile, Du et al. converted sappan wood waste into AC and underwent chemical treatment using phosphoric acid [32]. The highly porous AC with surface area and total pore volume of 1355.41 $m^2/g$ and 0.710 $cm^3/g$, respectively were obtained at activation temperature of 600 °C for 2 h. The resulting AC adsorbed optimally 50 mg/g of both Brilliant blue RV and Acid blue R dyes. Similarly, Mousavi et al. [50] utilized grape wood waste as the raw materials to be converted into AC for the removal of MB dye [50]. At pH 11, 100 mg/L of MB, and 12.25 g/L of adsorbent, the assessment for MB removal reached 98%. Based on the experimental results, the pseudo second-order fully fitted Langmuir model best described the kinetics ($R^2 = 0.99$), which indicates monolayer adsorption. They suggest that wood waste biomass can be considered as a low-cost lignocellulosic adsorbent with a strong application potential and as a substitute to costly and conventional adsorbents. However, further study should be commenced to explore the possibility of high adsorption capacity through various modification procedures, as well as the removal of organic molecules from all types of water samples.

## 5 Application of Wood Waste for Adsorptive Removal of Inorganic Contaminants

Water nowadays is heavily contaminated with potential disease pathogens as well as cancer-causing organic and inorganic chemicals. The discharge of waste and countless chemical compounds, such as pharmaceuticals and hormones, has increased during the course of the century as a result of rising population and natural phenomena. Despite the development of several analytical techniques, the release of contaminants keeps increasing. The majority of water pollutants are brought on by human activity, including anthropogenic and natural processes like agriculture (fertilizers, pesticides), sewage, mining, industrial chemical wastes, and many more. Additionally, organic and inorganic pollutants like heavy metal ions continue to pollute water, and all of these pollutants are difficult to remove and do not degrade through biological processes [51].

According to Ameta, wastewater treatment is the process that converts wastewater into a beneficial product (with no discernible well-being or environmental problems) that is subsequently returned to the water-cycle or can be repurposed [52]. A wastewater treatment plant is the physical facility used for wastewater treatment (WWTP). These plants are categorized based on different varieties of wastewater that need to be treated. Physical, chemical, and biological processes were used in the conventional wastewater treatment approaches. Although the sequential or simultaneous use of these techniques may offer an effective way to remove or degrade pollutants, their actual application is still severely constrained by issues with efficiency, execution, and cost. While conventional treatment technologies have limitations such as pH restrictions, a broad range of organic-load modifications, the physicochemical behaviour of effluents, etc., and biological wastewater treatment seems promising only for dairy and agricultural wastewater.

Most of these methods, with the exception of adsorption, have intrinsic drawbacks such as high operating and equipment costs, high energy needs, high maintenance costs, incomplete metal ion removal, and secondary waste formation after treatment [53]. Many wastewater treatment techniques and methods have been introduced around the world, but adsorption is still the most effective and popular approach because of many intrinsic advantages, including its simplicity, convenience of use, availability of a variety of adsorbents, and high efficacy. Organic, biological, and inorganic pollutants in wastewater can be treated using adsorption for both their soluble and insoluble forms. Adsorption is also considered as powerful strategy for physical, chemical, and biological process systems and is frequently utilized for the purification of potable water and industrial effluent [51].

An adsorption process involves the buildup of a substance between two phase interfaces, such as a liquid–liquid, gas–solid, or liquid–liquid contact. Adsorbent is the phase in which the substance is collected, while adsorbate is the accumulated substance. Contaminants (adsorbate) adhere to the adsorbed surface in this surface phenomenon. Based on the intermolecular interaction between adsorbate and adsorbent, adsorption is classified as physisorption (physical adsorption) or chemisorption (chemical adsorption) [54].

Copper, nitrates, nitrites, cyanides, selenium, lead, iron, heavy metals arsenic, cadmium, mercury, fluorides, phosphates, aluminium, calcium, titanium and sodium compounds, perchlorates, chromium oxides, nitric acid, chlorides, and sulphates are among the typical inorganic contaminants [55]. To eliminate inorganic pollutants, such as nitrates from wastewater, sugarcane bagasse can be a suitable and affordable adsorbent. Because sugarcane is a significant cash crop grown all over the world, and bagasse is a valuable byproduct that may be used for adsorption, therefore it is widely available. It can be employed directly, chemically altered, or turned into biochar [55]. Sawdust that has been acid-hydrolyzed can be used to treat contaminants such as dyes, heavy metal ions, and inorganic impurities in wastewater [56]. Adsorption of heavy metal ions was achieved with sawdust modified with 0.1 M $HNO_3$, resulting in a complete removal of both copper and lead within 6 h.

The mechanism for adsorptive removal of heavy metal ions can be broken down into two steps: first, adsorption occurs in the porous structure of the adsorbent, and then ion exchange occurs (Fig. 8). Redox interactions can also take place between the active elements of the modified adsorbent and the metal ions, stabilizing metal ion precipitates and detoxifying heavy metals by reducing their valency [57].

In another study, $Ca^{2+}$ and $Mg^{2+}$ cations were observed to be released into solution during adsorption, according to [59]. These liberated cations had strong correlations with the adsorbed Cr(III), proving that cation exchange was the key to elimination. At lower pH of solution, increased concentration of $H^+$ had a negative impact

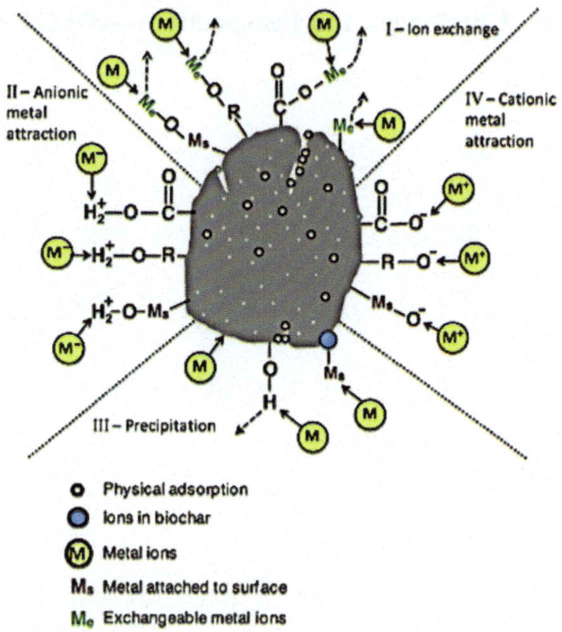


**Fig. 8** Biochar interactions with inorganic contaminants mechanism. Physical adsorption shows as the circles on biochar particle [58]

on cation exchange between Cr(III) and minerals of biochar. The elimination of chromium occurs when the pH of the solution is between 2 and 5, at which point the surface formed from crop straw becomes negatively charged while the trivalent chromium species remain positively charged. Electrostatic attraction between negatively charged Cr(VI) species and positively charged biochar, oxygenated functional group reduction of Cr(VI) to Cr(III), and subsequent complexation of Cr(III) with functional groups of biochar are the main mechanism for recovering hexavalent chromium species. The adsorption processes that control the removal of lead from water include surface complexation, cation exchange, and precipitation [60].

The relative distribution of $Pb^{2+}$ adsorption on sludge biochar was investigated by Lu et al. in a pilot-scale research project for wastewater treatment [61]. The exchange with $K^+$ and $Na^+$ contributed up to 4.8–8.5% of the adsorption in the study, whereas surface complexation made up 42%. Co-precipitation of lead with organic matter and mineral phases of biochar, precipitation as lead phosphate, surface complexation with active carboxyl and hydroxyl functional groups, and surface precipitation as lead phosphate all had an impact on adsorption. The divalent heavy metal such as cadmium has also received high attention due to its propensity to hydrate when in aqueous solution. Because of this, cation exchange and precipitation predominate in the elimination of cadmium via adsorption. Meanwhile, Harvey et al. prepared and classified a variety of plant biochars into two categories: low and high exchange capacity biochars. In their work, cation exchange was discovered to be the most effective method of removing cadmium using high cation-exchange biochars [62].

## 6 Challenges and Perspectives—SWOT Analysis

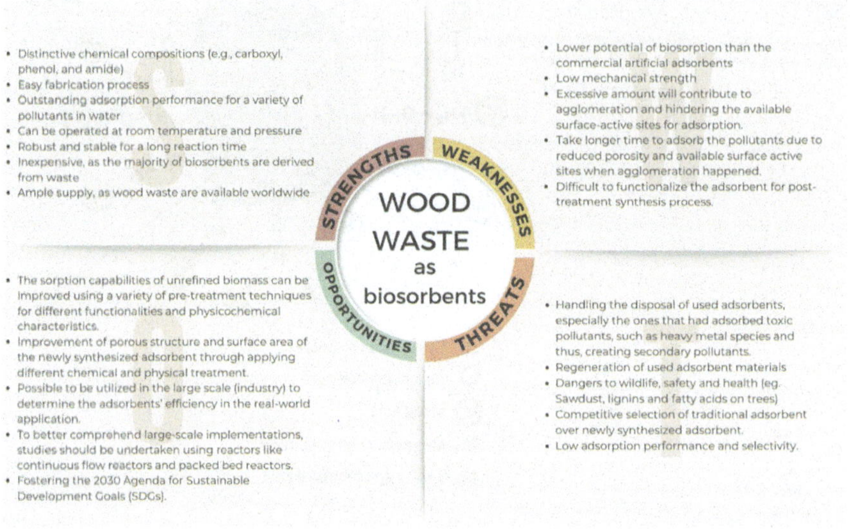


## 7 Conclusion

One of the most creative research initiatives in environmental chemistry has been the construction of wood waste biosorbents using a variety of methods and reaction settings. High porosity, limitless architectural matrices, adaptable functionalities, simple operation procedures, outstanding chemical, and mechanical strength could bestow them with significant advantages for practical applications, specifically for the effective removal of organic and inorganic contaminants in water and the environment. In this scenario, one of the elements that should be focused on to develop better recyclability sorbents is the high stability in sustaining the structures and porosity in diverse environmental circumstances. Therefore, the depth of knowledge regarding the principles underlying this kind of sorbent materials may provide crucial evidence that will inspire scientists, engineers, the government, and the private sector to commit and work together to investigate additional options for the practical application of this technology in addressing a variety of emerging environmental issues.

**Acknowledgements** The authors gratefully acknowledge the financial and management support from Universiti Teknologi MARA under the Lestari SDG Triangle Grant (Project Number: 600-RMC/LESTARI SDG-T 5/3 (019/2021), as well as the services and facilities provided by UiTM Pahang Branch for the technical work.

## References


1. Oyewo OA, Mutesse B, Leswifi TY, Onyango MS (2019) Highly efficient removal of nickel and cadmium from water using sawdust-derived cellulose nanocrystals. J Environ Chem Eng 7. https://doi.org/10.1016/j.jece.2019.103251
2. Rafatullah M, Sulaiman O, Hashim R, Ahmad A (2009) Adsorption of copper (II), chromium (III), nickel (II) and lead (II) ions from aqueous solutions by meranti sawdust. J Hazard Mater 170:969–977. https://doi.org/10.1016/j.jhazmat.2009.05.066
3. Wu Y, Luo H, Wang H, Zhang L, Liu P, Feng L (2014) Fast adsorption of nickel ions by porous graphene oxide/sawdust composite and reuse for phenol degradation from aqueous solutions. J Colloid Interface Sci 436:90–98. https://doi.org/10.1016/j.jcis.2014.08.068
4. Cabrera-Lafaurie WA, Román FR, Hernández-Maldonado AJ (2014) Removal of salicylic acid and carbamazepine from aqueous solution with Y-zeolites modified with extraframework transition metal and surfactant cations: Equilibrium and fixed-bed adsorption. J Environ Chem Eng 2:899–906. https://doi.org/10.1016/j.jece.2014.02.008
5. Martucci A, Pasti L, Marchetti N, Cavazzini A, Dondi F, Alberti A (2012) Adsorption of pharmaceuticals from aqueous solutions on synthetic zeolites. Microporous Mesoporous Mater 148:174–183. https://doi.org/10.1016/j.micromeso.2011.07.009
6. Rakić V, Rajić N, Daković A, Auroux A (2013) The adsorption of salicylic acid, acetylsalicylic acid and atenolol from aqueous solutions onto natural zeolites and clays: clinoptilolite, bentonite and kaolin. Microporous Mesoporous Mater 166:185–194. https://doi.org/10.1016/j.micromeso.2012.04.049
7. Hir ZAM, Abdullah AH (2022) Hybrid polymer-based photocatalytic materials for the removal of selected endocrine disrupting chemicals (EDCs) from aqueous media: a review. J Mol Liq 361:119632
8. Tadkaew N, Hai FI, McDonald JA, Khan SJ, Nghiem LD (2011) Removal of trace organics by MBR treatment: the role of molecular properties. Water Res 45:2439–2451. https://doi.org/10.1016/j.watres.2011.01.023
9. Adewuyi A (2020) Chemically modified biosorbents and their role in the removal of emerging pharmaceuticalwaste in the water system. Water (Switzerland) 12:1–31. https://doi.org/10.3390/W12061551
10. Ngah WSW, Hanafiah MAKM (2008) Removal of heavy metal ions from wastewater by chemically modified plant wastes as adsorbents: a review. Bioresour Technol 99:3935–3948. https://doi.org/10.1016/j.biortech.2007.06.011
11. Rosales E, Meijide J, Tavares T, Pazos M, Sanromán MA (2016) Grapefruit peelings as a promising biosorbent for the removal of leather dyes and hexavalent chromium. Process Saf Environ Prot 101:61–71. https://doi.org/10.1016/j.psep.2016.03.006
12. Albadarin AB, Mangwandi C, Walker GM, Allen SJ, Ahmad MN (2011) Biosorption characteristics of sawdust for the removal of Cd(II) Ions: mechanism and thermodynamic studies. Chem Eng Trans 24:1297–1302. https://doi.org/10.3303/CET1124217
13. George M, Hamed T, Mosavian MTH, Khazaei I, Aliabadi M (2012) Use of Sawdust of aspen tree for the removal of chromium(VI) from aqueous solution
14. Kim N, Park M, Park D (2015) A new efficient forest biowaste as biosorbent for removal of cationic heavy metals. Bioresour Technol 175:629–632. https://doi.org/10.1016/j.biortech.2014.10.092
15. Cutillas-Barreiro L, Ansias-Manso L, Fernández-Calviño D, Arias-Estévez M, Nóvoa-Muñoz JC, Fernández-Sanjurjo MJ, Álvarez-Rodríguez E, Núñez-Delgado A (2014) Pine bark as bioadsorbent for Cd, Cu, Ni, Pb and Zn: Batch-type and stirred flow chamber experiments. J Environ Manag 144:258–264. https://doi.org/10.1016/j.jenvman.2014.06.008
16. Paradelo R, Cutillas-Barreiro L, Soto-Gómez D, Nóvoa-Muñoz JC, Arias-Estévez M, Fernández-Sanjurjo MJ, Álvarez-Rodríguez E, Núñez-Delgado A (2016) Study of metal transport through pine bark for reutilization as a biosorbent. Chemosphere 149:146–153. https://doi.org/10.1016/j.chemosphere.2016.01.087

17. Sousa S, Jimenez-Guerrero P, Ruiz A, Ratola N, Alves A (2011) Organochlorine pesticides removal from wastewater by pine bark adsorption after activated sludge treatment. Environ Technol 32:673–683. https://doi.org/10.1080/09593330.2010.510535
18. Nehrenheim E, Odlare M, Allard B (2011) Retention of 2,4,6-trinitrotoluene and heavy metals from industrial waste water by using the low cost adsorbent pine bark in a batch experiment. Water Sci Technol 64:2052–2058. https://doi.org/10.2166/wst.2011.603
19. Choong JE, Onn CC, Yusoff S, Mohd NS (2019) Life cycle assessment of waste-to-energy: energy recovery from wood waste in Malaysia 28:2593–2602. https://doi.org/10.15244/pjoes/93925
20. Manninen K, Judl J, Myllymaa T (2016) Life cycle environmental impacts of different construction wood waste and wood packaging waste processing methods
21. Garcia CA, Hora G (2017) State-of-the-art of waste wood supply chain in Germany and selected European countries. Waste Manag 70:189–197. https://doi.org/10.1016/j.wasman.2017.09.025
22. Pandey S (2022) Wood waste utilization and associated product development from under-utilized low-quality wood and its prospects in Nepal. SN Appl Sci 4. https://doi.org/10.1007/s42452-022-05061-5
23. Lima MDR, Ramalho FMG, Trugilho PF, Bufalino L, Dias Júnior AF, de Protásio T, Hein PRG (2022) Classifying waste wood from Amazonian species by near-infrared spectroscopy (NIRS) to improve charcoal production. Renew Energy 193:584–594. https://doi.org/10.1016/j.renene.2022.05.048
24. Stjepanovi M, Gali A, Kosovi I, Jakovljevi T, Habuda-Stani M (2021) From waste to biosorbent: removal of Congo red from water by waste wood biomass
25. Mallakpour S, Sirous F, Mustansar C (2021) Sawdust, a versatile, inexpensive, readily available bio-waste: from mother earth to valuable materials for sustainable remediation technologies. Adv Colloid Interface Sci 295:102492. https://doi.org/10.1016/j.cis.2021.102492
26. Shafie SM, Othman Z, Hami N (2017) Potential utilisation of wood residue in Kedah: a preliminary study 60–69
27. Inhat V, Lubke H, BalBercak J, Kuna V (2020) Size reduction downcycling of waste wood—A review. https://doi.org/10.37763/wr.1336-4561/65.2.205220
28. Huang Y, Lu L, Ding C, Pan M (2022) Eco-friendly wood-plastic composites from laminate sanding dust and waste poly (propylene) food pails. Waste Manag 149:96–104. https://doi.org/10.1016/j.wasman.2022.06.012
29. Velić N, Stjepanović M, Begović L, Habuda-Stanić M, Velić D, Jakovljević T (2018) Valorisation of waste wood biomass as biosorbent for the removal of synthetic dye methylene blue from aqueous solutions. South-East Eur For 9:115–122. https://doi.org/10.15177/seefor.18-13
30. Ibrahim NA, Abdellatif FHH, Hasanin MS, Abdellatif MM (2022) Fabrication, characterization, and potential application of modified sawdust sorbents for efficient removal of heavy metal ions and anionic dye from aqueous solutions. J Clean Prod 332:130021. https://doi.org/10.1016/j.jclepro.2021.130021
31. Rajesh Y, Rao L (2022) Materials today: proceedings synthesis and characterization of low-cost wood based biosorbent. Mater Today Proc 57:34–37. https://doi.org/10.1016/j.matpr.2022.01.251
32. Du H, Cheng J, Wang M, Tian M, Yang X, Wang Q (2020) Red dye extracted sappan wood waste derived activated carbons characterization and dye adsorption properties. Diam Relat Mater 102:107646. https://doi.org/10.1016/j.diamond.2019.107646
33. Ghaedi AM, Baneshi MM, Vafaei A, Nejad ARS, Tyagi I, Kumar N, Galunin E, Tkachev AG, Agarwal S, Gupta VK (2018) Comparison of multiple linear regression and group method of data handling models for predicting sunset yellow dye removal onto activated carbon from oak tree wood. Environ Technol Innov 11:262–275. https://doi.org/10.1016/j.eti.2018.06.006
34. Che W, Xiao Z, Wang Z, Li J, Wang H, Wang Y, Xie Y (2019) Wood-based mesoporous filter decorated with silver nanoparticles for water purification. ACS Sustain Chem Eng 7:5134–5141. https://doi.org/10.1021/acssuschemeng.8b06001
35. Wu Y, Cai Y, Yang F, Gan J, Zhang J (2022) Chemical modification of poplar wood featuring compressible rebound 3D structure as water treatment absorbents. J Clean Prod 331:129952. https://doi.org/10.1016/j.jclepro.2021.129952

36. Stjepanović M, Velić N, Galić A, Kosović I, Jakovljević T, Habuda-Stanić M (2021) From waste to biosorbent: removal of congo red from water by waste wood biomass. Water (Switzerland) 13. https://doi.org/10.3390/w13030279
37. Guan H, Cheng Z, Wang X (2018) Highly compressible wood sponges with a spring-like lamellar structure as effective and reusable oil absorbents. ACS Nano 12:10365–10373. https://doi.org/10.1021/acsnano.8b05763
38. Pam AA, Mohd Hir ZA, Abdullah AH, Tan YP (2021) Pb(II) removal in water via adsorption onto deep eutectic solvent fabricated activated carbon. Appl Water Sci 11:1–15. https://doi.org/10.1007/s13201-021-01420-6
39. Subramaniam MN, Goh PS, Lau WJ, Ismail AF (2019) The roles of nanomaterials in conventional and emerging technologies for heavy metal removal: a state-of-the-art review. Nanomaterials 9. https://doi.org/10.3390/nano9040625
40. Tran HN, You SJ, Hosseini-Bandegharaei A, Chao HP (2017) Mistakes and inconsistencies regarding adsorption of contaminants from aqueous solutions: a critical review. Water Res 120:88–116. https://doi.org/10.1016/j.watres.2017.04.014
41. Aris AZ, Mohd Hir ZA, Razak MR (2020) Metal-organic frameworks (MOFs) for the adsorptive removal of selected endocrine disrupting compounds (EDCs) from aqueous solution: a review. Appl Mater Today 21:100796. https://doi.org/10.1016/j.apmt.2020.100796
42. Karić N, Maia AS, Teodorović A, Atanasova N, Langergraber G, Crini G, Ribeiro ARL, Đolić M (2022) Bio-waste valorisation: agricultural wastes as biosorbents for removal of (in)organic pollutants in wastewater treatment. Chem Eng J Adv 9. https://doi.org/10.1016/j.ceja.2021.100239
43. Rohaizad A, Shahabuddin S, Shahid MM, Rashid NM, Hir ZAM, Ramly MM, Awang K, Siong CW, Aspanut Z (2020) Green synthesis of silver nanoparticles from Catharanthus roseus dried bark extract deposited on graphene oxide for effective adsorption of methylene blue dye. J Environ Chem Eng 8:103955. https://doi.org/10.1016/j.jece.2020.103955
44. Kang K, Loebsack G, Sarchami T, Klinghoffer NB, Papari S, Yeung KK-C, Berruti F (2022) Production of a bio-magnetic adsorbent via co-pyrolysis of pine wood waste and red mud. Waste Manag 149:124–133. https://doi.org/10.1016/j.wasman.2022.06.009
45. Liu Z, Gao B, Fu H, Hu L, Xing P (2021) One-step synthesis of modified graphene aerogels for the adsorption of chloroform, toluene, and gasoline using pine wood chips as raw material. Diam Relat Mater 119:108545. https://doi.org/10.1016/j.diamond.2021.108545
46. Zhou X, Shi L, Moghaddam TB, Chen M, Wu S, Yuan X (2022) Adsorption mechanism of polycyclic aromatic hydrocarbons using wood waste-derived biochar. J Hazard Mater 425:128003. https://doi.org/10.1016/j.jhazmat.2021.128003
47. He M, Xu Z, Sun Y, Chan PS, Lui I, Tsang DCW (2021) Critical impacts of pyrolysis conditions and activation methods on application-oriented production of wood waste-derived biochar. Bioresour Technol 341:125811. https://doi.org/10.1016/j.biortech.2021.125811
48. Ma X, Zhao S, Tian Z, Duan G, Pan H, Yue Y, Li S, Jian S, Yang W, Liu K, He S, Jiang S (2022) MOFs meet wood: reusable magnetic hydrophilic composites toward efficient water treatment with super-high dye adsorption capacity at high dye concentration. Chem Eng J 446:136851. https://doi.org/10.1016/j.cej.2022.136851
49. Firdaus M, Yusop M, Aziz A, Ahmad MA (2022) Conversion of teak wood waste into microwave-irradiated activated carbon for cationic methylene blue dye removal: optimization and batch studies. Arab J Chem 15:104081. https://doi.org/10.1016/j.arabjc.2022.104081
50. Mousavi SA, Shahbazi D, Mahmoudi A, Darvishi P (2022) Methylene blue removal using prepared activated carbon from grape wood wastes: adsorption process analysis and modeling. Water Qual Res J 57:1–19. https://doi.org/10.2166/wqrj.2021.015
51. Borah P, Kumar M, Devi P (2020) Types of inorganic pollutants: metals/metalloids, acids, and organic forms. INC. https://doi.org/10.1016/B978-0-12-818965-8.00002-0
52. Ameta SC (2018) Introduction 1–12. https://doi.org/10.1016/B978-0-12-810499-6.00001-2
53. Bahadur V, Gadi R, Kalra S (2019) Clay based nanocomposites for removal of heavy metals from water: a review. J Environ Manag 232:803–817. https://doi.org/10.1016/j.jenvman.2018.11.120

54. Yagub MT, Sen TK, Afroze S, Ang HM (2014) Dye and its removal from aqueous solution by adsorption: a review. Adv Colloid Interface Sci 209:172–184. https://doi.org/10.1016/j.cis.2014.04.002
55. Singh S, Wasewar KL, Kansal SK (2020) Low-cost adsorbents for removal of inorganic impurities from wastewater. INC. https://doi.org/10.1016/B978-0-12-818965-8.00010-X
56. Shukla A, Zhang YH, Dubey P, Margrave JL, Shukla SS (2002) The role of sawdust in the removal of unwanted materials from water. Glas Math J 46:239–257. https://doi.org/10.1017/S0017089504001740
57. Yang X, Zhang S, Ju M (2019) applied sciences preparation and modification of biochar materials and their application in soil remediation. https://doi.org/10.3390/app9071365
58. Ahmad M, Upamali A, Eun J, Zhang M, Bolan N, Mohan D, Vithanage M, Soo S, Sik Y (2014) Chemosphere Biochar as a sorbent for contaminant management in soil and water: a review. Chemosphere 99:19–33. https://doi.org/10.1016/j.chemosphere.2013.10.071
59. Chen T, Zhou Z, Xu S, Wang H, Lu W (2015) Adsorption behavior comparison of trivalent and hexavalent chromium on biochar derived from municipal sludge. Bioresour Technol 190:388–394. https://doi.org/10.1016/j.biortech.2015.04.115
60. Tareq R, Akter N, Azam S (2019) Biochars and biochar composites: low-cost adsorbents for environmental remediation, Elsevier Inc. https://doi.org/10.1016/B978-0-12-811729-3.00010-8
61. Lu H, Zhang W, Yang Y, Huang X, Wang S, Qiu R (2011) Relative distribution of Pb 2D sorption mechanisms by sludge-derived biochar. Water Res 46:854–862. https://doi.org/10.1016/j.watres.2011.11.058
62. Harvey OR, Herbert BE, Rhue RD, Kuo L (2011) Metal interactions at the biochar-water interface: energetics and structure-sorption relationships elucidated by flow adsorption microcalorimetry 5550–5556

# Present Scenario and Future Scope of the Use of Wood Waste in Wood Plastic Composites

**Alcides Lopes Leao, Ivana Cesarino, Milena Chanes de Souza, Ivan Moroz, Otavio Titton Dias, and Mohamad Jawaid**

**Abstract** Wood plastic composites (WPC), a class of biocomposites, represent important tools in the production history of sustainable materials and substantial breakthroughs in the area have been realized to enhance their physical and mechanical properties (Das et al. in Sci Total Environ 512–513:326–336, 2015a). However, disadvantages still exist, such as lower mechanical resistance and consequent necessity to increase part size, inferior dimensional stability, higher density, discoloration, flammability, and rottenness. These factors limit the adequate application of biocomposites in wider markets, and although studies have been conducted to alleviate these drawbacks, further investigation is necessary to solve these problems and relieve as many deficiencies as possible (Nagarajan et al. in ACS Omega 1:636–647, 2016). Current literature on the theme shows that these WPCs can be fabricated from a wide variety of polymer matrices, including LDPE, HDPE, PP, ABS, PVC, and PLA. However, there are disadvantages which limit the applicability of WPCs: processing temperature is limited by wood degradation at high temperatures; the incompatibility between hydrophobic matrices and hydrophilic wood fibres; reduction of mechanical properties if compared to parent materials; and finally, flammability issues derived from the flammable nature of wood. Two routes stand out to overcome these difficulties: the use of additives or other reinforcements in association with wood particulates, and the transformation of the wood particles itself before addition via physical and/or chemical processing. From the environmental point of view, the pyrolysis process emerges as a possible solution to recycle several types of residues (including wood)

---

A. L. Leao (✉) · I. Cesarino · M. C. de Souza · I. Moroz
Department of Bioprocesses and Biotechnology, School of Agriculture, Sao Paulo State University (UNESP), Botucatu, Brazil
e-mail: alcideslopesleao@gmail.com; alcides.leao@unesp.br

O. T. Dias
Faculty of Forestry, Centre for Biocomposites and Biomaterials Processing, University of Toronto, Ontario, Canada

M. Jawaid
Institute of Tropical Forestry and Forest Products (INTROP), Universiti Putra Malaysia, Serdang, Selangor, Malaysia

M. C. de Souza
FIT—Flextronics Institute of Technology, Sorocaba, Brazil

 
S. N. Sarmin et al. (eds.), *Wood Waste Management and Products*, Sustainable Materials and Technology, https://doi.org/10.1007/978-981-99-1905-5_7

into a higher value material known as biochar. Considering the recent publication trends, it seems that biochar production may be the future target for wood residues since biochar can be used for a myriad of applications, including reinforcement of a new generation of biocomposites, called second generation WPCs. Other advantages are the reduction of carbon footprint and the use of low-cost raw materials. This class of products may be called WPBC—Wood Plastic Biochar Composite. Given the significant number of production variables, this material must be better studied and tested in composites of biological basis.

## 1 Introduction

Wood waste is abundant worldwide and has a high potential to be recycled/reprocessed. The term "wood waste" involves all wood and wood-based products that have reached the end of their useful life. Reuse and recycling have been driven by the rising cost of waste disposal. In addition, environmental awareness and more restrictive legislation also contribute to the importance of recycling wood waste. It is also worth mentioning that the reuse and recycling of wood waste reduces the need to cut down trees, benefiting the environment. In addition, they transform this waste into products with greater added value. Products developed from recycled wood waste are used in paper production, panel production, wood pellets, biocomposites, biochar, energy production, compost, etc. [46]. Thus, promoting the circular economy of wood, as depicted in Fig. 1.

Polymeric composites are part of this market and are important in several sectors, from aerospace and automotive applications to everyday life [21]. However, two aspects must be properly considered: the economic point of view and the environmental impact [1]. The former is crucial, reducing the composite price and

**Fig. 1** Schematics of the circular economy of wood. Elaborated by the authors

enhancing its properties (for example, mechanical, electrical, etc.). The latter can be solved by using recycled materials or biobased. One recent trend is to use biochar to replace other carbon fillers/reinforcements (i.e., carbon nanotubes, graphene), with the advantage that is based on green carbon [5, 20, 24, 38]. ADDIN Mendeley Bibliography CSL_BIBLIOGRAPHY

Wood residues (in different forms, such as powder or flakes) can be embedded in polymer matrices for the fabrication of composites named WPCs, protagonists of a growing market valued at 5.76 billion dollars in 2021 with a forecast of 6.42 billion dollars in 2022 [37]. According to this report, the annual growth rate (CAGR) of this market is expected to reach 11.5% from 2022 to 2030 [46], in this scenario, this market will reach more than 15 billion in 2030. To date, polyethylene appears as the most important material used in these applications, accounting for more than 66% of the global revenue in 2021. Factors that drive this demand and growth include the need for sustainable materials in civil construction, including residential, commercial, and industrial sectors; the growth of population and derived housing demands; and a consumer trend of valorization of sustainable (or green) buildings in scope with environmental protection.

Besides direct addition, wood residues can also be transformed to generate higher aggregated value materials with promising applications in replacing traditional reinforcements used in the industries, such as carbon fibre-reinforced plastics (CFRP). The global market of CFRPs represents one of the largest economic platforms in the area. This segment is controlled by carbon black produced by the traditional petrol industry [21]. Within the sustainability concept, the driving force for replacing traditional synthetic composites with biological basis polymers with lesser fossil footprints has increased significantly in the last decade.

The attempt to diminish petrol dependence associated with the low cost versus density relation supported the acceptance of bio reinforcements in many composite applications [1]. Polypropylene (PP) is one of the most studied matrices regarding thermoplastic composites reinforced with natural fibres, totalling 1,769 manuscripts [38]. High-density polyethylene (HDPE) and polyvinyl chloride (PVC), with 330 and 76 manuscripts respectively, are also of importance in these applications. However, there is a growing interest in using engineering polymers in the same manner, specifically acrylonitrile butadiene styrene (ABS) which is the theme for 42 manuscripts. All these data searches were done in the Scopus Database until September 2022.

### *1.1 Wood Plastic Composites*

WPCs make use of industrial by-products made of any kind of wood, timber or similar composition, such as sawdust, wooden chips, and others, associated with either thermoset or thermoplastic polymeric matrices including polyethylene (PE), HDPE, PP, PVC, ABS, polylactic acid (PLA), among others [29, 39]. Due to their interesting properties such as recyclability, less maintenance costs, and easiness of fabrication via different processes, these composites have been explored in several areas, including automotive, construction, optical, and electronical devices [34].

Other wood components present in WPCs include wood pallets, milling residues, pulping sludge, pruning materials, wood panel residues, as well as agricultural fibres such as sugarcane bagasse, rice husks, corn stalks, etc. [42].

The main disadvantages of WPCs are their high flammability and incompatibility between hydrophobic matrices and hydrophilic fibres. Flame retardants act to retard flame propagation in WPCs but at the cost of mechanical properties reduction. There are several types of flame retardants used in WPCs, such as metal hydroxides; materials based on phosphorous; boron; carbon; and nanofillers such as organoclays; in this aspect, the use of biochar as filler aids in mitigating this disadvantage and promoting an adequate balance between flammability and mechanical properties [29]. Generally, ammonium polyphosphate (APP) and magnesium hydroxide ($Mg(OH)_2$) have been utilized as flame retardants for the fabrication of wood PP biochar composite [24]. Higher biochar and lower wood contents was beneficial for reducing flammability. Due to its thermal stability, biochar contributes to an efficient coal formation to restrict $O_2$ transference to PP. In general, biochar addition in association with biomass simultaneously decreased flammability, flexural resistance, and tensile/flexural moduli [12].

Also, the environmental benefit from using materials from renewable resources represents an important step towards restraining the irreversible anthropogenic climate alteration in opposition to the use of fossil-derived materials [26]. In this scenario, biochar also represents an interesting possibility.

The disadvantages of biologic derived fillers are their low resistance and stiffness when compared to inorganic polymeric fillers [15]. Thus, the mechanical properties of composites obtained using these materials can be enhanced by adjusting the filler content and its particle size [32], as well as by using additives designed to raise the polymer performance [29].

The thermal stability of materials is considered an important feature for polymeric composite systems [5]. The stability of the components is a critical factor to determine the processing temperature range of a composite system. There is an inherent incompatibility of natural fibres and engineering polymers, such as polyamides, since natural fibres decompose thermally in the high processing temperatures when associated with engineering polymers [20]. On the other hand, thermal stability of biochar can promote the compounding processing at higher temperatures than natural fibres [36].

Moreover, wood waste destined for use as fillers in composite fabrication may also be converted to biochar before addition to the matrix, thus generating a higher aggregated value filler which also yields better performance in the final composites. High aspect ratio carbon fillers such as carbon fibre, carbon nanotubes, and graphene have been employed to explore its potential, but still without economic success to be implemented. In this context, there is a perspective of studying the use of biochar for this end.

## *1.2 Biochar*

Biochar produced through the pyrolysis process can be a sustainable solid substitute for expensive synthetic fillers [37]. According to literature, single layer graphene costs approximately 200 €/cm$^2$, while graphene oxide costs 100,000 €/kg [43]. On the other hand, carbon black costs around 1 €/kg. Until recently, engineered carbonaceous materials such as CNT, graphene, and graphene oxide have not fulfilled the promise of a new carbon era. Therefore, new materials have been considered. In the last years, biochar appeared as the most promising for integrating carbon into composites obtained from residue management [4]. Biochar is a carbonaceous material mostly used for remediation of environmental contaminants in agriculture [40]. However, due to its stability and high carbon content, it is a good candidate as a substitute for more expensive and environmentally damaging carbon materials [20].

The availability of potential raw material for producing biochar is vast, including agricultural, forestry, and wood processing (sandpaper dust) residues, as well as the organic portion of municipal solid waste, etc. [18]. Until now, biochar has been used only for agricultural uses and in removal of contaminants [40]. Due to its significant surface area, porosity, and hydrophobic nature, biochar can be potentially employed as filler or reinforcement in thermoplastic composites based on wood and polymeric resins [25]. The biochar's high specific surface area allows the thermoplastic matrix to have a higher flux creating a mechanical interaction, called interlocking, enhancing these composites' mechanical properties. This use of biochar in biocomposites may expand its applicability and produce advanced composites for circular bioeconomy [47]. However, to date, a few studies have been reported involving the use of biochar in composites such as WPC, resulting in a vast number of variables to be studied. According to the Scopus Database, until September 2022, circa 5,283 manuscripts about the WPC theme were indexed. Approximately twelve manuscripts related to WPC and biochar, eleven to WPC and thermoplastic resins as matrices, and none to biochar and thermoplastic resins as matrices, and to WPC, biochar, and thermoplastic resins as matrices.

The use of biochar as reinforcement/filler is a current developing trend [49]. When applied to soil, biochar improves its chemical, physical, and microbiological properties, and contributes to carbon sequestration—an alternative for disposal of organic residues with concomitant energy production [7]. However, the chemical-physical properties of biochars may vary significantly due to the diversity of raw materials and production conditions. The search for green fillers to be used in polymeric composites brought new attention to biochar as a potential alternative to conventional inorganic mineral fillers [8].

In line with more sustainable materials, the incorporation of residue-based pyrolyzed biochar into wood- and polyolefin-based biocomposites represented an important step towards sustainability. The number of scientific publications regarding the employment of biochar as filler in polymeric composites is considered low. For example, biochar has been used as filler in epoxy composites [5]. The mechanical properties of composites can be tailored using different contents. For epoxy matrices, superior tensile strength and stress at failure have been observed in low thermally treated biochar concentrations (2–4% in weight). In addition, the dielectric properties comparable to nanotube fillers can be achieved using a 20% in weight content [24].

The yield and physical properties of biochar depend on the biomass source, temperature, and duration of pyrolysis. The ability to tune the surface of biochar opens the door for its use for industrial applications such as adsorption of gases and pollutants and adhesion of polymeric composites [23].

Thus, biochar's use has been described in literature in association with WPCs in a selection of manuscripts. Qingfa et al. [48] reported the application of biochar of filler in injected thermoplastics, and that the extrusion method can be employed in the preparation of biochar/wood/plastic composite materials. The authors also affirm that the proper biochar content may increase tensile and impact mechanical resistances; in this study, the optimal mechanical properties were obtained for a 10% (wt.) biochar content.

The increasing desire on tackling simultaneously the issues related to gas emission in landfills and organic residues management, an environmentally benign idea was presented to develop biocomposites in which biochar derived from wood residue pyrolysis was incorporated into the same wood type, PP and maleic anhydride grafted polypropylene (MAPP). Research was conducted via fabrication of WPCs mixed with 6, 12, 18, 24, and 30% of biochar in weight. The 24% sample displayed increased tensile and flexural strengths and moduli. In this sample, the very fine nature of biochar particles resulted in a good filler performance and, consequently, resulted in a denser composite. As a result, the 24% WPBC sample excelled over the other samples in mechanical performance. The high surface area derived from its fine particles and carbonaceous nature of biochar makes it a potential candidate as reinforcing material in the biocomposites. Composites with 12 and 18% biochar exhibited the best ductility and thermal stability, respectively. This study revealed that WPCs fabricated with biochar addition have an interesting potential in the production of adequate biocomposites for several final applications [16].

In 2016, the same research group [13] produced polypropylene/wood biocomposites with activated biochar derived from residues, which did not have active functional groups, but showed high surface area (335 $m^2/g$). The biocomposites were mechanically characterized by tensile, flexural, impact, and hardness tests. Flammability and thermal stability were determined using calorimetry and thermogravimetry, respectively. Infrared spectroscopic techniques and x-ray diffraction were also employed to understand the chemical alterations which occur in biocomposites. The authors concluded that the compatibilizer content can be reduced from 3 to 1% wt. without affecting the mechanical performance and flammability of the biocomposites. SEM images showed that the fine dispersion of biochar particles in polymer-filled pores was achieved with the absence of compatibilizer, resulting in mechanical anchorage. The reduction in compatibilizer content (from 3 to 1% wt.), together with the use of biochar, resulted in an approximate 18% reduction in the biocomposite production costs. In another manuscript, the fillers were based on previous studies with the following properties: 42% PP, 30% wood, 4% MAPP, and 24% biochar. The obtained results were discussed in detail and reinforced the idea of using this filler material in a PP matrix [15].

Another study of importance was published [31], in which biochar was used as material for partial and total substitution of wood flour in combination with PP to produce composite materials. The evaluation consisted of diverse mechanical and physical properties. The highest flexural strength and flexural modulus were obtained for the composite containing 25% wt. biochar. However, the highest tensile properties for the 5% biochar composite. Finally, the 40% biochar composites displayed the lowest average water absorption in 24 and 48 h. These results indicate that biochar is a potential substitute for conventional wood in a wide range of applications in composite fabrication.

A recent study by Wang et al. [45] involved fabrication of four WPC samples via extrusion. Characterization included testing on mechanical properties, water absorption, thermal, and rheological properties. The improved melt strength in decolorizing carbon (NA) composite melts was achieved because of the higher modulus and viscosity. A strong interfacial adhesion between particles and the matrix and high tensile properties were observed in NA composites. The presence of biochar decreased the degree of crystallinity of Douglas-fir (DF) composites. On the other hand, the thermal properties were improved for biochar-based composites. In terms of water resistance, the DF composites exhibited the highest water absorption (3.7%) and thickness swell (2.9%). During artificial weathering test, colour change and lightness changes were observed for longer weathering exposure for DF composite. In addition, NA and biochar composites resulted in enhanced photostability. The authors suggested that this study could pave the way to utilize the renewable biochar as reinforcing filler in WPC in outdoor applications.

The effect of biochar addition can also influence the crystallization dynamics of the polymeric matrix. In fact, Qingfa et al. [50] studied the effect of biochar from rice husk on the properties of WBPC. The reinforcement material was obtained via rapid pyrolysis of rice husk powder at 500 °C using a fluidized bed reactor under inert atmosphere. To study the effect of biochar content, five different ratios were selected: 30, 40, 50, 60, and 70% in weight. The materials were mixed in a high-speed mixer for 10 min to obtain a homogeneous mixture. In order to improve the experimental efficiency, the mixture was processed in a twin screw extruder with different temperature zones (from 135 to 185 °C) at a feeding and extrusion rate of 30 rpm. The biochar from rice husk was revealed to be viable to be employed as filler material, and also to improve the mechanical performance of the composites. The SEM analysis showed that HDPE was incorporated into the biochar's pores. In relation to crystallization rate, as evidenced by DSC, biochar could reduce this property with lower values as biochar content is increased. Furthermore, there are WPC properties that depend heavily on the crystallization behaviour, thus being of utmost importance to study this behaviour if optimal properties are to be achieved. The flexural, tensile, and impact resistances of WPBC surpassed those of WPC. As the rice husk powder content increased, flexural and tensile strengths initially increased reaching a maximum and then decreased. Although the impact resistance of both WBPC and WPC diminished with fibre addition, WBCP still presented better results than WPC. In summary, biochar produced from rice husk emerged as a feasible option to reinforce PEAD.

The use of biochar as filler in polymers has been demonstrated to be a sustainable approach incorporating aggregated value material based on pyrolyzed residues, and simultaneously mitigating possible problems related to biological residues in an intelligent manner [14]. A study was carried out using date palm residues as reinforcement of PP matrix. These residues were prepared at 700 and 900 °C. The contents were added in different ratios (5–15% in weight) to PP and the obtained composites were characterized using diverse techniques. In addition, the thermal, mechanical, electrical, and rheological properties of the composites were investigated. The mechanical properties of biochar and PP composites were enhanced, while the electric characterization revealed a higher electrical conductivity with increased biochar content. Although the addition of biochar in PP matrix significantly reduced the total crystallinity of the resultant composites, a beneficial effect in the crystallization temperature was identified. In terms of rheological properties, biochar addition caused minimal variation in the storage modulus ($G'$) in comparison to pure PP. Also, the authors concluded that the composite properties can be further improved from the inherent properties of biochar such as porosity, surface area, and surface functionalization, which will result in a better matrix/reinforcement interaction and, in consequence, will yield superior properties [35]. In another study about PP and wood biocomposites, the large amount of free radicals in the composite increased its thermal conductivity. In addition, biochar particles had neglected effect in terms of the fusion behaviour of the material in the thermal system. However, both wood and biochar acted as nucleation agents thus raising the crystallization temperature. The added biochar did not cause an interruption of the polypropylene crystalline structure. Nuclear magnetic resonance studies identified the aromatic nature of biochar and also revealed the enlargement of the composite peak intensities with an increase in the biochar content, mostly because of its inherent amorphous nature and the presence of free radicals; this leads to the idea that biochar addition may yield properties optimization while generating a destination for residues [17].

Other manuscripts were published regarding the influence of residual biomass from Arhar (pigeon pea) stalks and Bael (stone apple) husks, obtained via pyrolysis process, in an epoxy resin matrix (2, 4, and 6% in weight) [28]. The 4% Bael husk biochar composite presented an 183% increase in tensile strength when in relation to pure epoxy. Giorcelli et al. [20] found similar results for epoxy matrix composites with biochar produced from maple tree residues, they reported that the brittle behaviour of the resin changed towards ductility since the elongation at fracture increased from 0.02 to 0.12. The tenacity in tensile behaviour demonstrated an enormous impact with an 11-fold increase in comparison to the pure resin.

According to [44], mechanical performance and properties represent crucial quality measurements of composites. The specific moduli of biochar-reinforced PP matrix composites were inferior to those of talc-reinforced composites, but a noteworthy enhancement of the specific elasticity modulus was observed. In relation to mechanical strength, both talc- and biochar-based composites showed similar tensile strength, while biochar composites exceeded talc composites in flexural strength performance. Biochar can help the automotive industry to achieve its goal of weight (and consequent fuel consumption) reduction concerning vehicle parts, as well as increase the use of sustainable materials and environmental benefits.

Another study about the use of organic residues was published in 2016 using biochar derived from several raw materials (pine sawdust, sewer sludge, and chicken manure) in different pyrolysis conditions. These biochars were utilized to fabricate wood and polypropylene-based biocomposites with a 24% load in weight. The composites were evaluated via mechanical, chemical, thermal, morphological, and flammability analyses. Among all samples, the biocomposite fabricated with chicken manure derived biochar presented superior mechanical properties even with its higher ash contents. Overall, biochar incorporation increased the tensile and flexural moduli of the composites. The biochar addition had neglected effect on the crystalline structure of PP. In addition, wood and biochar particles played a role as nucleating agents for PP recrystallization in the composite. SEM revealed the infusion of PP in the biochar pores and a fine dispersion in most of the composites [13].

Injection moulding has been employed as well in the fabrication of biocomposites from PP and biochar with five different contents (0, 15, 25, 30, and 35% in wt.). The obtained biocomposites were tested via tensile and flexural tests, in addition to cone calorimetry, thermogravimetry, DSC, XRD, and infrared spectroscopy. The addition of high content of biochar in pure PP continuously enhanced its tensile modulus and flexural strength. Also, the incorporation of biochar significantly reduced the heat release rate peak and smoke production of the biocomposites. The biochar's high surface area allowed for PP to flow and create a mechanical networking; thus, improving the mechanical properties. Due to its thermal stability, biochar led to a dense coal structure during combustion which inhibited heat and mass transfer between PP and oxygen from the environment. PP had its thermal stability raised because of biochar insertion [9].

In a similar study, [6] suggested that biochar prepared at 900 °C significantly improved the modulus of composites based on PP, while composites produced with biochar obtained at 500 °C showed a slight enhancement. The higher level of graphitization obtained by high temperature pyrolysis (900 °C) yields a higher modulus thus granting a more pronounced stiffness raise in the composite. The authors noted that high temperature processed biochar had smaller pores than the low temperature counterpart. Moreover, these characteristics can vary depending on the targeted application and the polymeric matrix nature. At 500 °C, lignin was the only lignocellulosic component which did not completely decompose. Stability is attained with dTGA around zero before 900 °C in all samples, suggesting that no unstable component remains in biochars produced at 900 °C. This would be an interesting aspect in the use of this filler in high performance composites.

Tadele et al. [41] conducted a comparative study about the life cycle evaluation of biochar for automotive industry. Tadele et al. [41] and Das et al. [17] pointed out the feasibility of biochar use instead of traditional carbonaceous fillers. The authors demonstrated a substantial cost reduction of a composite containing biochar, while attaining mechanical and thermal properties as the composite based on carbon black as a result of the reduction of amount of compatibilizer. The evaluation of the biochar cost was the main goal of the research proposed by [6]. In this study, biochar derived from perennial sugar cane was employed as filler in composites based on PP/poly(octene-ethylene). The fabricated composite with 10–20 wt.% of filler and exhibited strong interactions between copolymer and biochar. A more detailed and in-depth set of studies about PP and biochar interaction was conducted by Das et al. [10, 12, 13]. The authors measured the magnitude of interactions between PP and filler. They concluded that the addition of biochar different sources leads to improvement in terms of flame-retardant, mechanical, and thermal properties of PP composites Moreover, Elnour et al. [19] studied the relation between biochar properties and PP composites, demonstrating a stiffness increase associated with tensile strength not being affected. In the same period, Poulose et al. [35] used biochar derived from date palm mixed with PP showing the insignificant reduction on the storage modulus in a concentration of up to 15% in weight.

Ogunsona et al. [33] revealed that biochar produced at 500 °C has significantly more active sites than that produced at 900 °C, resulting in an excellent interaction with a polyamide 6 (PA6) matrix. The excellent interfacial adhesion between matrix and reinforcement for the lower temperature composite resulted in increased tensile and flexural strengths. Based on this study, it can be concluded that for a polar polymer such as PA6, lower pyrolysis temperature should be employed to maintain surface functional properties of hydroxyl and carbonyl in biochar.

In another use of biochar in composites, Ho et al. [22] studied the use of bamboo coal (BC) derived from bamboo plants. In this study, the BC particle was used as reinforcing agent in polylactic acid (PLA) composite to enhance its mechanical, thermal, and optical properties. The samples varied in content up to 10% in weight of BC particles. The results indicated that the tensile strength, flexural strength, and ductility index (DI) of the BC/PLA composites increased 43%, 99%, and 52%, respectively, in comparison to pure PLA. This phenomenon was credited to the fine uniform distribution of BC particles in addition to their high surface area. SEM images revealed a fine dispersion of BC particles in melt compounding processes when the particle content was lower than 7.5%.

Nan et al. [31] studied the production of polyvinyl alcohol and biochar, produced via a casting method in solution to investigate its electrical conductivity, thermal and mechanical properties. Interestingly, composites with content up to 10% of biochar exhibited an electrical conductivity like most carbon nanotubes and graphene. The mechanical test results indicate that biochar incorporation decreased the tensile strength of the composites. However, the tensile and storage modulus upon the glass transition temperature has positive effect with biochar addition. The thermogravimetry results revealed increased thermal decomposition temperature as well after reinforcement. In conclusion, biochar is shown in this study to possess elevated potential as an alternative to carbon nanotubes and graphene as filler in composite for electronics.

Arrigo et al. [2] published a study on Polyethylene (PE) containing biochar derived from coffee produced at 700 °C. The authors investigated the rheological and thermal properties of composites in proportions of up to 7.5% in weight. The restriction of the dynamics of polymer chains is related to the confinement of the chains on the biochar's surface. In addition, the homogeneous incorporation of biochar particles improved the thermo-oxidative stability of the composites. A distinct behaviour of composites in terms of mechanical properties was considerably different when comparing the composites with pure PE. In fact, flexural resistance did improve while impact resistance was reduced for the composites. Bajwa et al. [3] described the use of biochar to produce a mixture composed of HDPE, PLA, and wood flour resulting in a product with superior thermal stability.

Das et al. [11] reported production of composites fabricated with 6% wt. biochars with 94% gluten. Charcoals consisted of carbon black, biochar from pine bark, and gluten biochar. The control sample was made of neat wheat gluten. Biochars were fabricated via pyrolysis in a reactor under nitrogen environment. A considerable amount of the biomass was heated up to 800 °C at a rate of 10 °C/min and then maintained for 1 h. Both biochars were shredded and sieved to guarantee that particle size was kept lower than 500 μm. The water absorption of the composites increased with biochar application, in particular the gluten biochar. This is an important consideration since pure gluten tends to absorb water; therefore, limiting its application. Among all samples, the added gluten biochar exhibited the highest modulus and lowest water absorption.

In a study comprising diverse resins—polymethyl methacrylate (PMMA), thermoplastic polyurethane (TPU), ethylene vinyl acetate (EVA), PLA, and ABS, results evidenced biochar as an optimal substitute to traditionally used carbon nanomaterials. Also, it is important to point out the importance of low material cost and low environmental impact during the biochar fabrication. Among the considered new applications are a substitute for electrical wires, smart fabrics, highly flexible wires that do not require critical electrical demands, and thermoplastic polymers traditionally employed in 3D printing technology. An exceptional potential is expected for producing 3D electronics and printed electronics [27].

A patent search in the USA Patent Office for the terms WPC, biochar, and biocarbon results in no patents found, demonstrating the potential of this material. In summary, it can be assumed that the use of biochar as filler or reinforcement in thermoplastic or thermoset composites is under-explored and future developments are highly expected in this research field [28].

Overall, the future scope of wood waste utilization should be inclined to the generation of an aggregated value material which does not significantly harm the environment. This concept—currently best represented by biochar fabrication—is in tandem with the ongoing trend of renewable raw materials and energy in substitution of classic fossil-derived compounds, aiming towards a circular economy.

## References

1. Abbate E, Mirpourian M, Brondi C, Ballarino A, Copani G (2022) Environmental and economic assessment of repairable carbon-fiber-reinforced polymers in circular economy perspective. Materials 15. https://doi.org/10.3390/ma15092986

2. Arrigo R, Jagdale P, Bartoli M, Tagliaferro A, Malucelli G (2019) Structure-property relationships in polyethylene-based composites filled with biochar derived from waste coffee grounds. Polymers (Basel) 11. https://doi.org/10.3390/polym11081336
3. Bajwa DS, Adhikari S, Shojaeiarani J, Bajwa SG, Pandey P, Shanmugam SR (2019) Characterization of bio-carbon and ligno-cellulosic fiber reinforced bio-composites with compatibilizer. Constr Build Mater 204:193–202. https://doi.org/10.1016/j.conbuildmat.2019.01.068
4. Bartoli M, Giorcelli M, Jagdale P, Rovere M (2020) Towards traditional carbon fillers: biochar-based reinforced plastic. In: Huicochea EF (ed) Fillers. IntechOpen, London
5. Bartoli M, Giorcelli M, Rosso C, Rovere M, Jagdale P, Tagliaferro A (2019) Influence of commercial biochar fillers on brittleness/ductility of epoxy resin composites. Appl Sci (Switzerland) 9. https://doi.org/10.3390/app9153109
6. Behazin E, Misra M, Mohanty AK (2017) Sustainable biocarbon from pyrolyzed perennial grasses and their effects on impact modified polypropylene biocomposites. Compos B Eng 118:116–124. https://doi.org/10.1016/j.compositesb.2017.03.003
7. Conz RF (2015) Universidade de São Paulo Escola Superior de Agricultura "Luiz de Queiroz" Caracterização de matérias-primas e biochars para aplicação na agricultura Rafaela Feola Conz Piracicaba Rafaela Feola Conz
8. Das C, Tamrakar S, Kiziltas A, Xie X (2021) Incorporation of biochar to improve mechanical, thermal and electrical properties of polymer composites. Polymers (Basel) 13:1–32. https://doi.org/10.3390/polym13162663
9. Das O, Bhattacharyya D, Hui D, Lau KT (2016a) Mechanical and flammability characterisations of biochar/polypropylene biocomposites. Compos B Eng 106:120–128. https://doi.org/10.1016/j.compositesb.2016.09.020
10. Das O, Bhattacharyya D, Sarmah AK (2016b) Sustainable eco-composites obtained from waste derived biochar: a consideration in performance properties, production costs, and environmental impact. J Clean Prod 129:159–168. https://doi.org/10.1016/j.jclepro.2016.04.088
11. Das O, Hedenqvist MS, Johansson E, Olsson RT, Loho TA, Capezza AJ, Singh Raman RK, Holder S (2019) An all-gluten biocomposite: comparisons with carbon black and pine char composites. Compos Part A Appl Sci Manuf 120:42–48. https://doi.org/10.1016/j.compositesa.2019.02.015
12. Das O, Kim NK, Kalamkarov AL, Sarmah AK, Bhattacharyya D (2017) Biochar to the rescue: balancing the fire performance and mechanical properties of polypropylene composites. Polym Degrad Stab 144:485–496. https://doi.org/10.1016/j.polymdegradstab.2017.09.006
13. Das O, Sarmah AK, Bhattacharyya D (2016c) Biocomposites from waste derived biochars: mechanical, thermal, chemical, and morphological properties. Waste Manag 49:560–570. https://doi.org/10.1016/j.wasman.2015.12.007
14. Das O, Sarmah AK, Bhattacharyya D (2015a) A sustainable and resilient approach through biochar addition in wood polymer composites. Sci Total Environ 512–513:326–336. https://doi.org/10.1016/j.scitotenv.2015.01.063
15. Das O, Sarmah AK, Bhattacharyya D (2015b) Structure-mechanics property relationship of waste derived biochars. Sci Total Environ 538:611–620. https://doi.org/10.1016/j.scitotenv.2015.08.073
16. Das O, Sarmah AK, Bhattacharyya D (2015c) A novel approach in organic waste utilization through biochar addition in wood/polypropylene composites. Waste Manag 38:132–140. https://doi.org/10.1016/j.wasman.2015.01.015
17. Das O, Sarmah AK, Zujovic Z, Bhattacharyya D (2016d) Characterisation of waste derived biochar added biocomposites: Chemical and thermal modifications. Sci Total Environ 550:133–142. https://doi.org/10.1016/j.scitotenv.2016.01.062
18. Duku MH, Gu S, Hagan EB (2011) Biochar production potential in Ghana—A review. Renew Sustain Energy Rev 15:3539–3551. https://doi.org/10.1016/j.rser.2011.05.010
19. Elnour AY, Alghyamah AA, Shaikh HM, Poulose AM, Al-Zahrani SM, Anis A, Al-Wabel MI (2019) Effect of pyrolysis temperature on biochar microstructural evolution, physicochemical characteristics, and its influence on biochar/polypropylene composites. Appl Sci (Switzerland) 9:7–9. https://doi.org/10.3390/app9061149

20. Giorcelli M, Khan A, Pugno NM, Rosso C, Tagliaferro A (2019) Biochar as a cheap and environmental friendly filler able to improve polymer mechanical properties. Biomass Bioenergy 120:219–223. https://doi.org/10.1016/j.biombioe.2018.11.036
21. Greene JP (2021) Polymer composites. In: Automotive plastics and composites. William Andrew Publishing, pp 191–222
22. Ho MP, Lau KT, Wang H, Hui D (2015) Improvement on the properties of polylactic acid (PLA) using bamboo charcoal particles. Compos B Eng 81:14–25. https://doi.org/10.1016/j.compositesb.2015.05.048
23. Isaeva VI, Vedenyapina MD, Kurmysheva AY, Weichgrebe D, Nair RR, Nguyen NPT, Kustov LM (2021) Modern carbon-based materials for adsorptive removal of organic and inorganic pollutants from water and wastewater. Molecules 26:1–95. https://doi.org/10.3390/molecules26216628
24. Khan A, Savi P, Quaranta S, Rovere M, Giorcelli M, Tagliaferro A, Rosso C, Jia CQ (2017) Low-cost carbon fillers to improve mechanical properties and conductivity of epoxy composites. Polymers (Basel) 9. https://doi.org/10.3390/polym9120642
25. Leng L, Xiong Q, Yang L, Li H, Zhou Y, Zhang W, Jiang S, Li H, Huang H (2021) An overview on engineering the surface area and porosity of biochar. Sci Total Environ 763:144204. https://doi.org/10.1016/j.scitotenv.2020.144204
26. Lenzen M, Moran D, Kanemoto K, Foran B, Lobefaro L, Geschke A (2012) International trade drives biodiversity threats in developing nations. Nature 486:109–112. https://doi.org/10.1038/nature11145
27. Lepak-Kuc S, Kiciński M, Michalski PP, Pavlov K, Giorcelli M, Bartoli M, Jakubowska M (2021) Innovative biochar-based composite fibres from recycled material. Materials 14:1–11. https://doi.org/10.3390/ma14185304
28. Minugu OP, Gujjala R, Shakuntala O, Manoj P, Chowdary MS (2021) Effect of biomass derived biochar materials on mechanical properties of biochar epoxy composites. Proc Inst Mech Eng C J Mech Eng Sci 235:5626–5638. https://doi.org/10.1177/0954406221990705
29. Mokhena TC, Sadiku ER, Mochane MJ, Ray SS (2021) Mechanical properties of fire retardant wood-plastic composites: a review. Express Polym Lett 15:744–780. https://doi.org/10.3144/expresspolymlett.2021.61
30. Nagarajan V, Mohanty AK, Misra M (2016) Biocomposites with size-fractionated biocarbon: influence of the microstructure on macroscopic properties. ACS Omega 1:636–647. https://doi.org/10.1021/acsomega.6b00175
31. Nan N, DeVallance DB, Xie X, Wang J (2016) The effect of bio-carbon addition on the electrical, mechanical, and thermal properties of polyvinyl alcohol/biochar composites. J Compos Mater 50:1161–1168. https://doi.org/10.1177/0021998315589770
32. Nwanonenyi SC, Obidiegwu MU, Onuegbu GC (2013) Effects of particle sizes, filler contents and compatibilization on the properties of linear low density polyethylene filled periwinkle shell powder. Int J Eng Sci (Ghaziabad) 2:1–8
33. Ogunsona EO, Misra M, Mohanty AK (2017) Impact of interfacial adhesion on the microstructure and property variations of biocarbons reinforced nylon 6 biocomposites. Compos Part A Appl Sci Manuf 98:32–44. https://doi.org/10.1016/j.compositesa.2017.03.011
34. Partanen A, Carus M (2016) Wood and natural fiber composites current trend in consumer goods and automotive parts. Reinf Plast 60:170–173. https://doi.org/10.1016/j.repl.2016.01.004
35. Poulose AM, Elnour AY, Anis A, Shaikh H, Al-Zahrani SM, George J, Al-Wabel MI, Usman AR, Ok YS, Tsang DCW, Sarmah AK (2018a) Date palm biochar-polymer composites: an investigation of electrical, mechanical, thermal and rheological characteristics. Sci Total Environ 619–620:311–318. https://doi.org/10.1016/j.scitotenv.2017.11.076
36. Quosai P, Anstey A, Mohanty AK, Misra M (2018) Characterization of biocarbon generated by high- and low-temperature pyrolysis of soy hulls and coffee chaff: for polymer composite applications. R Soc Open Sci 5. https://doi.org/10.1098/rsos.171970
37. Research G (2022) Wood plastic composites market size, share & trends analysis report by product (polyethylene, polypropylene), by application (automotive components), by region, and segment forecasts, 2022–2030

38. Savi P, Jose SP, Khan AA, Giorcelli M, Tagliaferro A (2017) IMWS-AMP 2017 : IEEE MTT-S international microwave workshop series on advanced materials and processes, vol 8, Pavia, Italy, pp 20–22
39. Scapolio LGB, Moroz I, Jr, Cesarino AR, Leão I, Scuracchio AL, C.H., 2022. Eucalyptus bark residue application fr Poly(Vinyl Chloride) composite production: influence of fiber size and content. Curr Appl Polym Sci 05. https://doi.org/10.2174/2452271605666220524105918
40. Srinivasan P, Sarmah AK, Smernik R, Das O, Farid M, Gao W (2015) A feasibility study of agricultural and sewage biomass as biochar, bioenergy and biocomposite feedstock: production, characterization and potential applications. Sci Total Environ 512–513:495–505. https://doi.org/10.1016/j.scitotenv.2015.01.068
41. Tadele D, Roy P, Defersha F, Misra M, Mohanty AK (2020) A comparative life-cycle assessment of talc-and biochar-reinforced composites for lightweight automotive parts. Clean Technol Environ Policy 22:639–649. https://doi.org/10.1007/s10098-019-01807-9
42. Teuber L, Osburg VS, Toporowski W, Militz H, Krause A (2016) Wood polymer composites and their contribution to cascading utilisation. J Clean Prod 110:9–15. https://doi.org/10.1016/j.jclepro.2015.04.009
43. Tripathi M et al (2020) Free-standing graphene oxide and carbon nanotube hybrid papers with enhanced electrical and mechanical performance and their synergy in polymer laminates. Int J Mol Sci
44. Wang T, Rodriguez-Uribe A, Misra M, Mohanty AK (2018) Sustainable carbonaceous biofiller from miscanthus: size reduction, characterization, and potential bio-composites applications. Bioresources 13:3720–3739. https://doi.org/10.15376/BIORES.13.2.3720-3739
45. Wang X, Yu Z, McDonald AG (2019) Effect of different reinforcing fillers on properties, interfacial compatibility and weatherability of wood-plastic composites. J Bionic Eng 16:337–353. https://doi.org/10.1007/s42235-019-0029-0
46. Wood Recycling [WWW Document] (n.d.). https://recyclinginside.com/wood-recycling/, Recycling Inside
47. Yaashikaa PR, Senthil Kumar P, Varjani S, Saravanan A (2020) A critical review on the biochar production techniques, characterization, stability and applications for circular bioeconomy. Biotechnol Rep 28:e00570. https://doi.org/10.1016/j.btre.2020.e00570
48. Zhang Q, Cai H, Yang K, Yi W (2017) Effect of biochar on mechanical and flame retardant properties of wood—Plastic composites. Results Phys 7:2391–2395. https://doi.org/10.1016/j.rinp.2017.04.025
49. Zhang Q, Khan MU, Lin X, Cai H, Lei H (2019) Temperature varied biochar as a reinforcing filler for high-density polyethylene composites. Compos B Eng 175:107151. https://doi.org/10.1016/j.compositesb.2019.107151
50. Zhang Q, Yi W, Li Z, Wang L, Cai H (2018) Mechanical properties of rice husk biochar reinforced high density polyethylene composites. Polymers (Basel) 10:1–10. https://doi.org/10.3390/polym10030286

# Viability of Building Materials Made of Wood Waste: Sustainability and Its Performances

**Krishna Manjari Sahu, Swapnita Patra, and Sarat K. Swain**

**Abstract** Where wood waste (WW) management is a big challenge for proper utilization towards sustainable applications, numerous numbers of attempts are made to solve the problem. Different forms of WW have been utilized for various important purposes like sound resistance, fire retardant, insulating material, and strength enhancing material including various fancy and decorative applications. The present chapter summarizes the possibilities of designing viable building materials in modern society. Out of sundry kinds of applications of WW, the performance in making building material has taken special interest by researchers. Various types of WW and their sustainable applications are schematically established for better understanding of readers. WW is used as a special additive in the preparation of concrete to enable construction. It is noticed that WW containing concrete has been used as a potential material in wall panelling, making of stoneware tiles, designing of $CO_2$ sink, etc. Thermal insulators and electrical insulating materials prepared from WW are useful in making of sustainable building materials. The importance and advantage of WW are analysed for manufacturing of sustainable and green building materials. Further, importance of WW in designing of acoustic platform for resistance of echo of sound in different halls, studios, and other gathering in close rooms has also been discussed. The present chapter reveals new ideas regarding making of various sustainable building materials from WW, for which the present article is a solution to the challenges in WW management.

**Keywords** Wood waste · Sustainable material · Building material · Insulating material · Acoustic

## 1 Introduction

The vulnerability in the condition of earth has become the reason behind the utilization of environment friendly as well as sustainable materials and techniques in

K. M. Sahu · S. Patra · S. K. Swain (✉)
Department of Chemistry, Veer Surendra Sai University of Technology, Burla, Sambalpur 768018, Odisha, India
e-mail: skswain_chem@vssut.ac.in

 
S. N. Sarmin et al. (eds.), *Wood Waste Management and Products*, Sustainable Materials and Technology, https://doi.org/10.1007/978-981-99-1905-5_8

various areas like construction [1], energy production [2], packing applications [3]. to diminish the aftermath of using traditional raw materials [4]. Owing to the accelerated consumption of natural resources and for the balanced management of remaining reserves and renewable raw materials, the recovery and utilization of waste generated during production and consumption processes are the absolute necessity of the moment. This coexists with the demand for environmental protection, which implies the eradication of threats and reduction of losses that might occur due to the possible accumulation and elimination of waste [5]. The management of natural resources is the biggest challenge to reduce anthropogenic environmental pressure in the next few decades [6]. A number of natural sustainable resources are exploited as substitute elements for cement and aggregates in the construction field to make eco-friendly material [7]. Out of which, wood has emerged as a prominent, sustainable, environmental friendly substitute material in the construction business for the preparation of building materials [8]. It is a natural and renewable resource that can be feasibly utilized in the manufacturing of wood products. Recycled wood waste (WW) products are widely exploited as replacement of pristine raw materials. Production of these materials are beneficial in reduction of expenditure for transportation and logging process and also the incineration and landfilling costs are minimized for the disposure of WW. Further, the recycling and reusing of WW are favourable in reducing utilization of water and energy in manufacturing process and consequently there is decrease in environmental depletion. The recycling process is necessary to manage current drastic condition of earth due to rapid consumption of fossil fuel, and the continuous climate change [9]. Various forms of WW i.e., sawdust, wood fibre, wood shaving, wood chip, wood flitches, bark, and wood offcuts are depicted in Fig. 1.

Wood is the oldest known construction material and has been used to establish thousands of building structures. Some of the wood-based fabricated products have higher strength-to-weight ratio than steel-based products. The entire substitution of wood in place of conventional building material increases the risk of catching fire and because of this it has been replaced by various non-flammable materials. Due to the above reasons, partial replacement of building material with wood is recommended. Environmentalists and researchers are keen to discover environment friendly substitute products and implement them practically in construction business [10]. There are a variety of materials that can be produced from the natural resource but we have to consider the expense and viability of using these materials in the engineering of sustainable products. Not only wood but also the WW is the main attraction for researchers to fabricate more eco-friendly building materials. WWs are commonly generated from the process of harvesting of wood, processing, and during the production of material from wood and after the use of wood-based products, i.e. post-consumer waste. The post-consumer WW is originated from wood-based products whose life span has been completed and whose economic and technical value determined them as waste materials. These waste materials should be exploited to conserve the raw materials which come under the list of sustainable development programme derived directly from forests. The aforementioned idea is essential for waste management and the management should be done in proper and appropriate

**Fig. 1** Different types of wood waste

way. Reuse of WW should be implemented in the period of economic boom, especially when there is deficiency in raw wood material. Wood recovered from various used and worn-out wood product has appeared as prominent alternative materials for raw forest resources to be utilized in different applications. Post-consumer WW generated in Europe has been used in place of raw wood material of around 22% which includes more than 9% and 12% for industrial and energy purposes, respectively. The construction sector along with industry and demolition sites are largest part of wood market, where the wood and wood-based products are used and greater amount of post-consumer waste is generated. The amount and generation pathways of WW are related to the geographical area and the type of construction procedure followed in that particular area. The percentage of WW in countries like Spain and Norway is generally more than 10% of total waste produced. Waste management strategies should be carried out to eliminate disposal problem and the most concerning one is to use in the recycling process. Although WW is the most common waste generated from industry and demolition sites, reuse and recycling of these waste products can also be successfully accomplished in construction field as secondary raw material. The quality of WW products need to hold balance with the stability and life expectancy along with the safety measures. For such concern, systematic selection of recycling process of WW used for the fabrication of different materials should

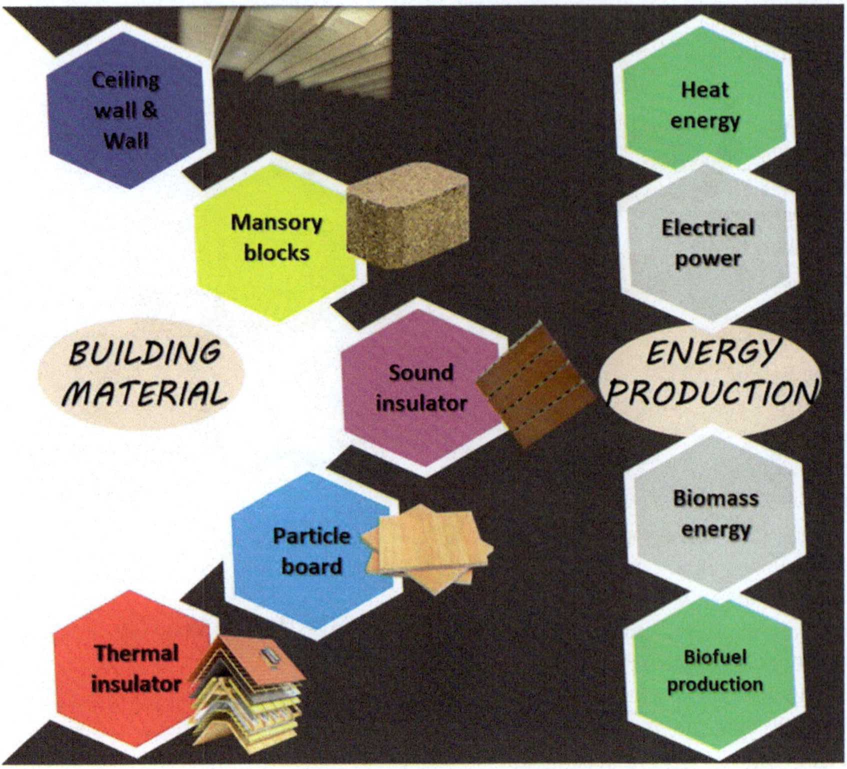


**Fig. 2** Applications of wood waste in energy production and building materials

be done [5, 11]. WW generally exists in different forms and depending upon the applications, it is used as an alternative material in place of raw material to enhance its characteristics and outcomes and most importantly to reduce the negative impact on environment through sustainable development. Multivarious applications of WW in energy production and building materials are represented in Fig. 2.

## 2 Wood Waste as Sustainable Building Material

United Nations (UN) has agreed on Sustainable Development Goals with 17 goals and 169 target matters related to sustainability. Sustainability aims at the three dimensional concept, i.e. environmental, economic, and social welfare of the country to achieve the goals set by UN. Sustainable development has the objective to meet the standard of human development goals while maintaining the quality of natural resources. Based on the available resources, situation, and its impact on the surrounding, the extent to which Sustainable Development Goals can be achieved are determined by the researchers [12]. As a consequence of urbanization and industrialization, cities are developing rapidly and due to this construction design, implementation, and maintenance of building requires huge amount of energy and natural resources. The materials used for the construction purpose are called building materials. Over the time period, there is substantial change in selection of building materials. For instance, building materials such as wood used in ancient times are biodegradable but do not have enough strength to hold the building while in modern days building materials such as insulator, brick, and plastic are non-renewable and have adverse impacts on the environment. The dependency over synthetic products made from metal and non-metal raw materials greatly enhances the consumption of energy and carbon dioxide emission which are also responsible for climate change and global warming. Due to the above reasons, selection of materials to construct sustainable buildings is one of the important discussions among environmental scientists. Eco-friendly and green building materials have already established their dominance over conventional materials since these materials provide recommended indoor environment and are most importantly able to meet the sustainable development goals adopted by the UN [13]. Sustainable building materials should maintain positive interrelationship with surrounding ecosystem. Sustainable building encourages using of less amount of energy and production of less amount of pollutants. It all starts with planning, designing, and choice of materials to construct the building. Sustainability in case of building construction starts with ground levels such as choice of raw material, method to prepare construction material, and energy use in these process. All the above methods must satisfy the goals of sustainability to meet sustainable development. For example, bricks use in the construction require more energy to dry and to give maximum strength. Energy consumption is high during this process so the firing of bricks can be done by using agricultural waste or solar energy through mirror and reflecting construction instead of conventional energy source [14]. Sustainable development needs sustainable materials. WW is a notable sustainable material and can largely be used as a sustainable material to produce different building materials. The improper disposal of WW into the surroundings affects both economic growth and the environment. The burning of WW releases vast amount of greenhouse gases which leads to increase in global warming. Reuse or recycling of WW to develop new products decreases the rate of deforestation. The sustainability of environment and economy is dependent upon the country's mindset. There are some countries like Nigeria and Brazil that have maintained the quality product as well as the surroundings by using WW in different applications [15, 16]. Some sustainable and unsustainable building materials recently being used are compiled in Fig. 3.

## 3 Applications of Wood Waste in Building Materials

WW is used in the fabrication of a variety of building materials such as concrete, thermal insulator, and sound insulator. In this chapter, required physical, mechanical,

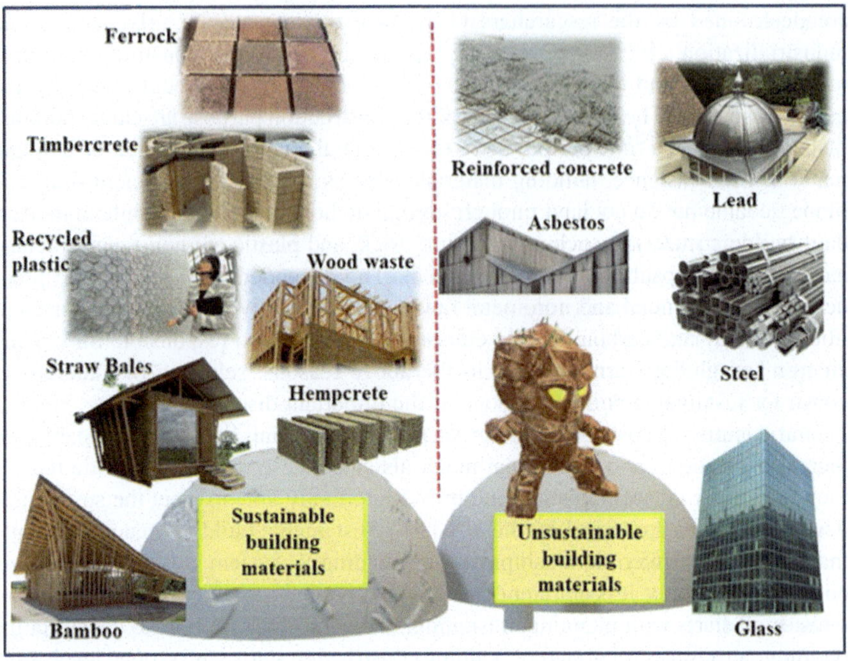


**Fig. 3** Sustainable and unsustainable building materials

and thermal properties such as heat of hydration, flowability, compressive strength, flexural strength, thermal conductivity, and sound pressure level are analysed systematically in accordance with their respective applications. The sustainability and viability of these materials for the successful use as building material have also been discussed in the following section.

## *3.1 Concrete*

Demand of cement in construction purposes is increasing due to rapid urbanization and so is the production of cement. It causes extreme emission of carbon dioxide which ultimately put its negative influence on the environment [17]. It is also reported that the production of 600 kg of cement bags is responsible for the emission of 400 kg of carbon dioxide [18]. Therefore, to development and adoption of new technologies and approaches are the need of the hour to control the negative effects of cement on environment. Varieties of waste and byproducts have already been taken into account to utilize as supplementary cementitious material to lessen the effect [19]. Research has already been effectively carried out to produce green construction material by substituting cement with WW ash [20]. Disposal and environmental impact can be reduced with the use of WW ash as a replacement agent in concrete. The replacement helps in reduction in utilization of cement, production cost of cement, and this also leads to lowering the release of greenhouse gases to the surroundings. Since WW ash is a fine material as compared to cement, it helps to fill the voids present within the matrix of hardened concrete and makes the penetration of NaCl difficult. Due to this, concrete made from WW ash as partial replacement of cement is suitable for building material in coastal region to protect the strctures from salty climatic conditions [21]. Other factors are also considered during the partial substitution of cement. Pozzolanic characteristic is the major feature of a material that is used as the replacement of cement. WW ash exhibits pozzolanic characteristics same as the cementitious materials. For the preparation of concrete, cement is the key component but its utilization can be reduced by adding WW ash to prepare sustainable concrete and mortar without affecting the quality of the resultant product. Addition of WW ash in concrete preparation process has helped in enhancing the microstructure property and strength. However, the replacement of cement and addition of WW ash has certain limitations, as it subsides the strength of concrete. The recommended amount for the addition of WW ash is not more than 20%. According to Ramosh et al., the efficacious use of WW ash in high quality cement can be feasible and up to 10% of cement can be saved [22].

Wood wool is the thin fine strand having 0.5–3 mm thickness. It can be prepared from WW instead wood and can be utilized in making of advanced wood cement composite. Regular binders such as Portland and white cement are mixed with wood wool to fabricate composite. WW must be treated before using it in manufacturing of composite material but positive impact of treated WW is still obscure. WW derives from pallets is a best substitution for spruce to fabricate wood wool cement board. Berger et al. [23] have done comparative studies of composites' characteristics by scrutinized outcomes of chemical and mechanical properties. The wood wool used for the fabrication of composite is prepared from natural spruce wood and WW which is obtained from construction and demolition sites. Since WW is very much compatible with white cement, they have replaced the traditional spruce wood in the manufacturing of wood-based cement boards. Influence of WW on cement, environment assessment of WW and analysis of thermal properties of board are carried out to determine the ability of WW to be use in preparation of building materials. It is seen that the influence of WW on cement is minimal and this effect is measured by isothermal calorimetric analyser. The assessment of environmental impact of WW also shows that there is no significant effect of contaminates and is safe to use in preparation of composites. The thermal conductivity is found to be below acceptable value, i.e. 0.08 W $m^1$ $K^{-1}$. Like other constitutes, WW also has its replacement limit up to 50% to show successful outcome. Above this limit, the heterogeneity of the board increases which is not a good characteristic of a good reinforcement. The performance of material also drastically decreases due to the lowering of flexural strength and density above this limit [23]. Different materials those are fabricated to modify compositions of concrete with the help of WW along with characterization techniques and their applicability in different applications are listed in Table 1.

Recently, cementitious materials reinforced with cellulosic fibre are vastly exploited for construction of residence and for the manufacturing of exterior

**Table 1** Different materials fabricated in modifying different compositions of concretes with wood wastes and their characterization techniques and applications

| Sl. No | Fabricated material | Type of wood waste used | Characterization | Applications | References |
|---|---|---|---|---|---|
| 1 | Wood-gypsum composites | Wood shavings and Sawdust | Compressive strength, SEM, flexural strength, density, hardness, thermal conductivity | Partition panels | [24] |
| 2 | Wood-crete | Sawdust | Thermal conductivity, compressive strength, density | Wall panelling or other non-and semi-structural applications | [25] |
| 3 | Mortars modified with wood waste ash | Wood waste ash | Compressive strength, dynamic modulus of elasticity, SEM, XRD, EDX, XRF, density | Mortar | [26] |
| 4 | Sawdust incorporated porcelain stoneware ceramic tile | Sawdust | Thermal conductivity, Bending strength, Young's modulus, Optical microscopy, EDS, TG-DTA | Porcelain stoneware tiles | [27] |
| 5 | Wood bio-concretes | Wood shavings | Life cycle emissions calculation | $CO_2$ sink | [28] |
| 6 | Concrete containing wood chipping | Wood chipping | Compressive strength, splitting tensile strength, flexural strength, modulus of elasticity, rebound, hammer, ultrasonic pulse velocity | Lightweight concrete | [29] |
| 7 | Wood-magnesium oxychloride cement (MOC) board | Wood fibre | Thermal conductivity, flexural strength | Particleboard | [30] |
| 8 | Concrete containing wood waste powder | Wood waste powder | Workability, density, air content, compressive strength, ultrasonic pulse velocity | Walls or floors construction | [31] |
| 9 | Self-compacting cementitious systems | Sawdust | XRD, TGA, SEM, Particle size distribution, compressive strength, shrinkage stains, air content measurement | Pastes, mortars and concrete | [32] |

(continued)

**Table 1** (continued)

| Sl. No | Fabricated material | Type of wood waste used | Characterization | Applications | References |
|---|---|---|---|---|---|
| 10 | Wood chip concrete | Waste wood chip | Compressive strength, ultimate strain, density, specific strength, SEM, optical microscope, flexural strength | Panel | [33] |

designing products namely siding and roofing materials. Asbestos fibres were the desirable fibre due to high strength-to-weight ratio, to engineer construction composites which are used for the above-mentioned applications. However, the carcinogenic nature of asbestos forced the researcher to find the alternatives and in this regard, the researchers have found cellulosic fibres as a viable alternative [34]. Autoclaved aerated concrete (ACC) has been largely utilized to prepare masonry because it is considered as an insulation material with outstanding thermal characteristic and construction performance [35]. It is prepared by the combination of lime, cement, gypsum, quartz powder, water and small amount of aluminium powder [36]. Addition of fibres to prepare AAC reduces cracks and breakages generated in AAC and helps in enhancement of tensile strength and flexural strength [37]. Owing to the low material and energy consumption value, AAC can be used an eco-friendly material. In addition to this, its negative influence on environment is reduced by adding WW fibre instead of other fibres. Rongsheng et al. have produced wood fibre reinforced AAC and carried out the comparative study of performance of wood fibre reinforced AAC with traditional ACC and polyester fibre reinforced AAC. The physical and mechanical properties influenced by the content of fibre are investigated by SEM analysis. Mostly, the addition of wood fibre to fabricate AAC shows better outcomes than that of polyester fibre. With the increase in the incorporation of wood fibre, thermal conductivity and volume density increase while the opposite result is found for fluidity, swollen height, and porosity. Among other mechanical properties, flexural strength is improved by the incorporation of wood fibre. The microstructure of developed material was studied by SEM analysis and results showed the physical interaction of wood fibre with the AAC matrix. From, the experimental data it is clear that wood fibre generated from WW can be used in AAC products [38].

Cement is not the only component of concrete that can be replaced by WW, it can also be utilized as sand replacement to prepare concrete masonry blocks. However, due to the differences in chemical and physical properties of sand and WW, the addition of WW in place of sand leads to have some impacts on characteristics like heat of hydration, flowability, and compressive and flexural strength of concrete masonry blocks. So the amount of WW addition in the preparation of concrete plays a vital role in determining the performance of material. Antoun et al. have created the concrete masonry by taking different replacement ratio of olive WW (OWW), i.e. 25, 50, and

100%. From the study, it is clear that the requirements set by ASTM for non-load bearing concrete masonry blocks is fulfilled by local OWW. The masonry blocks prepared from OWW appear to have effective strength after 7 days of preparation. Flowability of the prepared blocks was tested immediately after the mixing and heat of hydration was analysed after 48 h of casting while the flexural and compressive strength were evaluated within the intervals of 3, 7, 14, 18, and 40 days. The materials with 25 and 50% replacement of sand by OWW showed better results with higher flexural strength and lower heat of hydration value than the reference material. Decrease in heat of hydration and the delay in setting time of block are due to the presence of zinc and sucrose in OWW. The compressive strength remain unchanged as an acceptable value with respect to the reference value. But the addition of super plasticizers and accelerators to mixture showed enhancement in the performance of compressive and flexural strength. Owing to the above properties of WW-based material, it can be used in non-structural applications namely sidewalks, borders, ditches, filler blocks, and masonry block [39].

Crushed stones and mineral admixtures have shown increase in consumption by the concrete industries, hence consequently the scarcity in supply chain leads to illegal mining to fulfill the demand. For the purpose of sustainable construction, WW has largely been used as recycling waste due to ease of availability and cost efficiency. Lumber and timber industries are producing large number of products made from WW. These WWs are generated in every step starting from cutting trees to sawdust formation. The houses for poor people in West Indies are the combination of WW and clay. Logs obtained from trees are used for the construction of window, door, and trusses and also for the furnishing of the house [40]. In 1993, in one of the earthquake damaged zone of India, the fast-tract houses were made from cement particle board based on wood particle. Also the viability of this material for construction purpose has been examined. Therefore, the partial replacement of coarse aggregate is possible through the use of wood particles in the preparation of concrete. The utilization of unconventional material to produce concrete and to decrease the load on natural resources in construction business is encouraged by the researchers. Thandavamoorthy [41] has developed a concrete reinforced with WW collected from carpentry work to replace coarse aggregates and investigated the mechanical and durability properties to check the feasibility of using such concrete in building material. The compressive strength of prepared material with 15% replacement shows higher value, i.e. 32.36 MPa than general concrete. Acid, alkaline and fire resistance test are also carried out for the evaluation of durability of WW based concrete. The WW concrete has low resistance to acid and alkali as compared to general concrete. However, the 15% WW contained concrete has higher resistance value than other compositional replacement. Along with the structural application of this WW concrete, it also has the capability to use as sound insulation panels in hospitals and sound barriers on highways.

### 3.2 Insulating Building Material

Twentieth century is the starting point of employment of insulating material as one of the crucial building material. Commonly used components for the fabrication of insulators are generally synthetic materials such as polyisocyanurate, expanded polystyrene, and extruded polystyrene. Thermal and sound resistance capacities of synthetic materials are very high but its impact on environment and human health is not acceptable. Owing to the above facts, the use of natural materials is mandatory for the creation of healthy and sustainable environment. Recently, numerous investigations have been carried out for the replacement of synthetic materials by the eco-friendly material to generate insulator [42, 43]. There are varieties of thermal and soundproof materials that have been fabricated with the help of WW, out of which some of them are discussed in this chapter.

### 3.3 Thermal Insulator

Consumption of operational energy can be reduced by the use of insulating material to envelop the building, since these materials require less energy in process of cooling and heating. At present, thermal insulators are largely made up of synthetic materials. The effective outcomes of these thermal insulators have high resistance to heat transfer but these are less eco-friendly material. Development of thermal insulators by utilizing renewable natural resources lowers the production value and helps in designing of sustainable environment. The mechanical and thermal property of natural resources-based thermal insulator has shown better results as compared to conventional thermal insulator [43]. To analyse thermal properties, the evaluation of thermal conductivity is necessary. Thermal conductivity is the amount of heat that passes in a body in presence of thermal gradient. For specific kind of material the interval variation of conductivity is very high and depends on various factors like temperature and water content. Evaluation of these properties is useful while computing thermal resistance of building materials [44]. It is also reported that the international standard for the material to be used as thermal insulator material, thermal conductivity coefficient value must be less than 0.1 W/mK [45].

To decrease the energy consumption and to increase the efficiency of thermal insulator, WW products are utilized in preparation of thermal insulator. Recently, researchers are interested in fabricating composite material for thermal insulator. One of the green composite can be formed by the combination of WW and polymer. The most two important characteristics that are required to use thermal insulating material for construction purpose are thermal stability and compressive strength. Other characteristics such as low thermal conductivity value, and low water absorption capacity and high mechanical strength are needed. For this purpose, Abu-Jdayil et al. [46] have targeted to engineer a green polymeric composite material for construction purpose with thermal stability and insulating property. The reinforcing material is date palm wood powder (DPWP) which is added into the matrix of polylactic acid. The composite is made by taking various weight percentages, i.e. from 0 to 50% of DPWP. The synthesized composite material showed that the addition of WW powder decrease the thermal conductivity up to 0.0692. Polylactic acid-DPWP composite has the potential to be used as a building material owing to its high compression strength and lower thermal conductivity, thermal diffusivity and water absorption value. The 50-DPWP composite has the thermal conductivity and compressive strength of 0.0757 W/(mK) and 65.5 MPa, respectively while the water absorption is about 1.2% in a duration of 24 h. Along with thermal property, evaluation of physical and mechanical properties has been carried out to check the viability of this green composite and to use as thermal insulator for domestic and industrial applications as an alternative traditional insulator [46]. Agoua et al. [44] have targeted to prepare a composite from different sizes of sawdust and glue derived from polystyrene. Both the materials are reused as part of waste management to produce building material which is considered an environmental and economic alternative. The formulation of this composite is done to achieve physical and mechanical properties to use in desirable applications. They have evaluated thermal conductivity of various prepared samples with different granulation in variety of glue content and concluded that it can be used as insulation material to imitate ceiling and dividing walls [44]. Important advantages of WW to be implemented in thermal insulation for building materials are shown in Fig. 4 for better understanding.

Wood chip; a type of WW is generated from wood cutting. United States, Canada, China, Brazil, and Germany are some of the major countries to produce 61.9 million cubic metres of timber. Timber wood is mainly obtained from different timber production facilities and factories and are this wood is the main resources of wood chips. Hanifi et al. have engineered a thermal insulator by taking wood chips obtained from carpenters, olive seed, epoxy resin and polyvinyl chloride (PVC). From the obtained findings, it is seen that all these materials are suitable for the production of thermal insulator. Since olive seeds and wood chips contribute as fibres in the composite, the resultant flexure strength of synthesized olive oil seed, PVC, and wood chips doped material is 44% higher than reference sample. Since, the important phase of the composite is made from PVC and wood chip, increase in their content increases the compressive strength of the composite. Reference sample has 2.4 times lower compressive strength value as compared to the developed sample with 12 additives. Also it is seen that the sample with high water absorption capacity is of lower unit weight. Both the flexure and compressive strength value of developed composite is above the standards requirement. Thermal conductivity coefficient decrease with increase in addition of olive seed. So, the addition of olive seed should be maintained to get the required value of thermal conductivity [45]. Lakrafil et al. have taken two WWs i.e., wood shavings and sawdust obtained from carpenters and two leather wastes to produce insulating materials. Comparative studies have been done for the insulation ability of these waste materials with the component used as building material. The measurement of thermal conductivity shows that these materials have the capability to compete with conventional insulating materials [47].

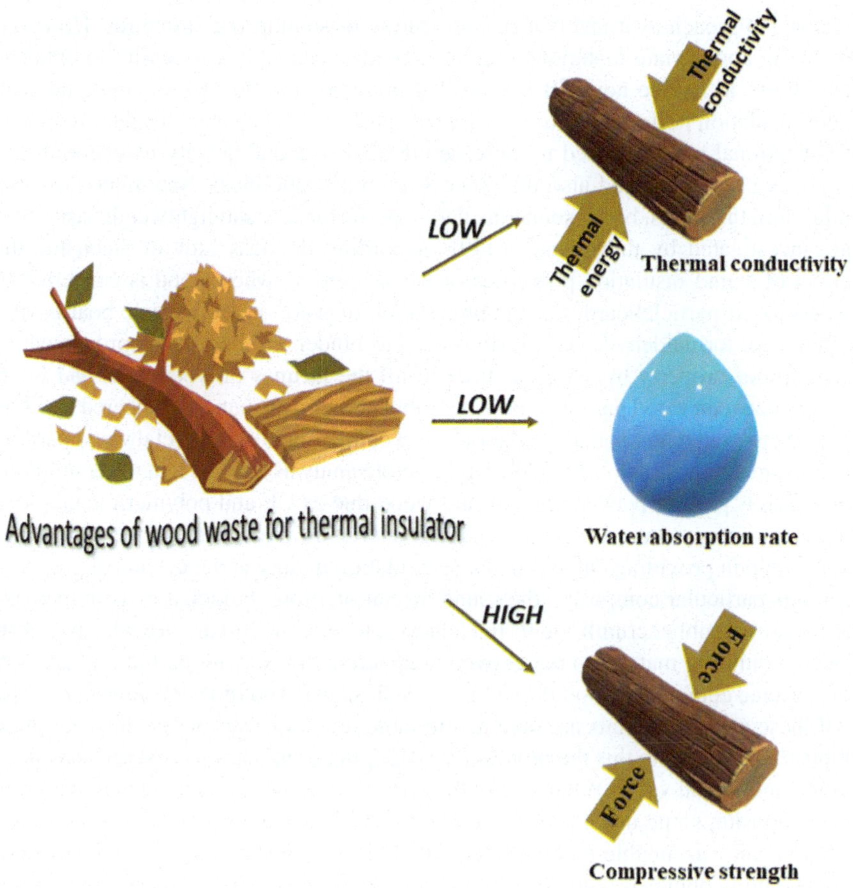


**Fig. 4** Advantages of wood waste to be implemented in thermal insulator applications

## 3.4 *Sound Insulator*

In recent years, noise pollution due to burgeoning of modern industry, construction, advanced traffic system, and transportation have affected human health and environmental condition. To manage, prevent, and reduce noise damage several investigations have been done by the researchers [48]. The three distinct approaches to control the noise pollution are cessation of the noise generation sources, avoidance of sound from entering ears and alternation of noise propagation path. Effect of noise can be minimized by controlling the source, but sometimes it is difficult to control with the available technologies. Therefore, to overcome these difficulties, sound insulation or soundproof materials have been employed to eliminate and to obstruct sound waves during the way of its transmission. There are various cited reports that have used sound insulation techniques to diminish effects of noise and this process is the most realistic approach as it involves manufacturing of soundproof structures. However, selecting appropriate materials to create sound insulator is a necessity to enhance the ability to reduce noise. It is seen that homogenous single layer material with poor insulation performance are not able to achieve desired sound insulation effects. Conventional methods used to increase the thickness and density of materials are not very convenient and also the process are not economical. New materials with minimum thickness, better sound insulation performance, and lightweight structures are investigated by the scientists [49]. According to Mass Law of Acoustic, the effect of sound insulation properties of wood increase when wood is converted to wood-based particleboard. For the preparation of wood-based particle boards very often urea–formaldehyde (UF) resin is used as binder. UF helps in improvement of soundproof property by acting as filler to fill the lacunas of wood cells and interspaces between wood particles. WW tire rubber composite panel engineered by Zhao et al. possessed better sound insulation effect compared to commercial wood particle board and composite floorboard. Microstructure analysis showed that the continuous interface is present between rubber and wood due to UF and polymeric methylene diphenyl diisocyanate (PMDI). These characteristics of WW tire rubber composite panel help in providing enhanced sound insulation quality and mechanical property. For this particular composite, the sound insulation property increases with increase in content of rubber crumb and PMDI adhesive level [50]. Torkaman et al. have fabricated a building material, i.e. lightweight concrete blocks with partial replacement of Portland cement by wood fibre waste, rice husk ash, and limestone powder waste. All these waste materials are obtained in large amounts from forest and limestone industries and due to this the process of developing of material is cost effective with high commercial value. When one of these three materials is used in the preparation of composite, some unwanted characteristics arise which are not appropriate for a suitable and sustainable building material. The most advantageous performance of composite is obtained with the combination of all three waste materials with 25% replacement of Portland cement. Recycling of wood fibre waste, rice husk ash, and limestone powder waste not only helps in production of lighter concrete blocks with good physico-mechanical properties but also a viable solution to reduce environmental problems. It also provides economical designing of building materials such as ceiling panels, absorption materials, and sound barrier panels [51]. Use of WW in sound insulation is portrayed in Fig. 5.

The recycling of timber waste to produce cement bonded particleboard is an appealing process. WW reinforced with cement provides structural durability and enhances the strength of wood particle board by reducing the vulnerability of WW caused by biological degradation and environmental weathering. Due to the granular skeleton, WW particleboards are light weight and may be used as thermal and sound insulating material. Conventional particleboard compromises indoor air quality due to the phenol formaldehyde resin whereas cement bonded particleboard is formaldehyde free and has no impact on human health and environment [52]. Eco-friendly thermal and sound insulating material can also be formed with the help of WW. The sustainable and advanced composite materials are developed by Fornés et al. based on either WW fibre or natural wood fibre reinforced phosphogypsum composite

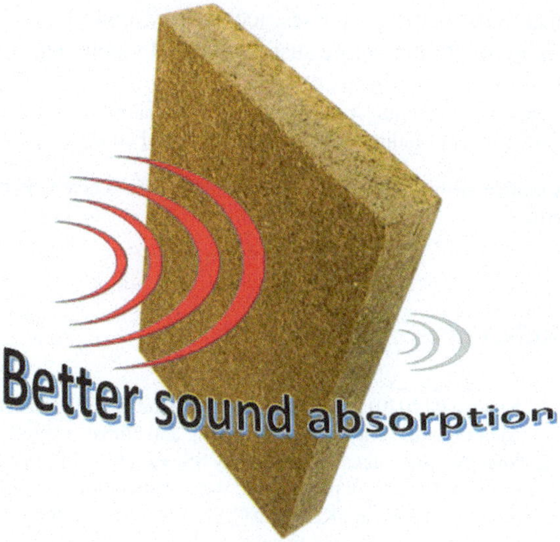


**Fig. 5** Use of wood waste in sound insulation

material for the comparative analysis. The reinforcing agent is added with a variation from 0 to 5 wt% in the matrix. They concluded that small addition of WW fibre constructively reinforced the matrix. Better mechanical outcomes of the developed material are seen with the addition of WW fibre than the natural wood fibre. Compressive strength with 0.5 wt% of WW fibre is found to be 25.1 MPa while the composite with 1 wt% of natural wood fibre provides compressive strength of about 21.9 MPa. Due to the decrease in density, the further addition of fibre reduces the compressive strength of the material. The optimal characteristic of the advanced environment friendly composite is obtained as the result of 3 wt% addition of WW fibre with 13.5 MPa, 0.39 W/mK, and 64.5 dBA of compressive strength, thermal conductivity, and sound pressure level value, respectively. The fabricated composite material is recommended to be used in wall bricks and blocks preparation [53].

## 4 Conclusion

In view of the raise in global warming and green house emission, numerous natural and sustainable materials have been incorporated into the construction business. Considering the sustainable development concept spread at the end of the twentieth century, new composite materials are introduced in modern countries. The utility, performance, and viability are the main concern to determine the successful employment of such materials. WW-based building material has shown its dominance in construction application due to the sustainability, viable mechanical and thermal properties, and low environmental effect on earth. Apart from WW, other eco-friendly materials, better processes, technology, and greener binder should be evaluated in future to produce more sustainable and viable building materials.

**Acknowledgements** Authors convey their thanks to the Department of Science and Technology, Government of India, for providing INSPIRE Fellowship to S. Patra for pursuing Ph.D. degree.

**Conflicts of Interest** The authors declare there is no conflict of interest in publishing this review article.

## References

1. Khadka B (2020) Rammed earth, as a sustainable and structurally safe green building: a housing solution in the era of global warming and climate change. Asian J Civil Eng 21(1):119–136
2. Shah AV, Srivastava VK, Mohanty SS, Varjani S (2021) Municipal solid waste as a sustainable resource for energy production: state-of-the-art review. J Environ Chem Eng 9(4):105717
3. Sivaprasad S, Byju SK, Prajith C, Shaju J, Rejeesh CR (2021) Development of a novel mycelium bio-composite material to substitute for polystyrene in packaging applications. Mater Today: Proc 47:5038–5044
4. Stolz J, Boluk Y, Bindiganavile V (2019) Wood ash as a supplementary cementing material in foams for thermal and acoustic insulation. Constr Build Mater 215:104–113
5. Ratajczak E, Bidzińska G, Szostak A, Herbeć M (2015) Resources of post-consumer wood waste originating from the construction sector in Poland. Resour Conserv Recycl 97:93–99
6. Habert G, Bouzidi Y, Chen C, Jullien A (2010) Development of a depletion indicator for natural resources used in concrete. Resour Conserv Recycl 54(6):364–376
7. Carević I, Serdar M, Štirmer N, Ukrainczyk N (2019) Preliminary screening of wood biomass ashes for partial resources replacements in cementitious materials. J Clean Prod 229:1045–1064
8. Harb P, Locoge N, Thevenet EF (2018) Emissions and treatment of VOCs emitted from wood-based construction materials: impact on indoor air quality. Chem Eng J 354:641–652
9. Kim MH, Song HB (2014) Analysis of the global warming potential for wood waste recycling systems. J Clean Prod 69:199–207
10. Wimmers G (2017) Wood: a construction material for tall buildings. Nat Rev Mater 2(12):1–2
11. Faraca G, Boldrin A, Astrup T (2019) Resource quality of wood waste: the importance of physical and chemical impurities in wood waste for recycling. Waste Manag 87:135–147
12. Salvia AL, Leal Filho W, Brandli LL, Griebeler JS (2019) Assessing research trends related to sustainable development goals: local and global issues. J Clean Prod 208:841–849
13. Omer MA, Noguchi T (2020) A conceptual framework for understanding the contribution of building materials in the achievement of sustainable development goals (SDGs). Sustain Cities Soc 52:101869
14. Petrovic EK, Vale B, Zari MP (2017) Materials for a healthy, ecological and sustainable built environment: principles for evaluation. Woodhead Publishing
15. Owoyemi JM, Zakariya HO, Elegbede IO (2016) Sustainable wood waste management in Nigeria. Environ Socio-Econ Stud 4(3):1–9
16. Kern AP, Amor LV, Angulo SC, Montelongo A (2018) Factors influencing temporary wood waste generation in high-rise building construction. Waste Manag 78:446–455
17. Zeidabadi ZA, Bakhtiari S, Abbaslou H, Ghanizadeh AR (2018) Synthesis, characterization and evaluation of biochar from agricultural waste biomass for use in building materials. Constr Build Mater 181:301–308
18. Cheah CB, Ramli M (2011) The implementation of wood waste ash as a partial cement replacement material in the production of structural grade concrete and mortar: an overview. Resour Conserv Recycl 55(7):669–685

19. Ayobami AB (2021) Performance of wood bottom ash in cement-based applications and comparison with other selected ashes: overview. Resour Conserv Recycl 166:105351
20. Ramos T, Matos AM, Sousa-Coutinho J (2013) Mortar with wood waste ash: mechanical strength carbonation resistance and ASR expansion. Constr Build Mater 49:343–351
21. Manikanta BT, Rao VR, Achyutha M, Reddy K (2018) Performance of wood ash blended reinforced concrete beams under acid (HCl), base (NaOH) and salt (NaCl) curing conditions. Int J Eng Technol 7:1045–1048
22. Tamanna K, Raman SN, Jamil M, Hamid R (2020) Utilization of wood waste ash in construction technology: a review. Constr Build Mater 237:117654
23. Berger F, Gauvin F, Brouwers HJH (2020) The recycling potential of wood waste into wood-wool/cement composite. Constr Build Mater 260:119786
24. Morales-Conde MJ, Rodríguez-Liñán C, Pedreño-Rojas MA (2016) Physical and mechanical properties of wood-gypsum composites from demolition material in rehabilitation works. Constr Build Mater 114:6–14
25. Aigbomian EP, Fan M (2013) Development of wood-crete building materials from sawdust and waste paper. Constr Build Mater 40:361–366
26. Castrillón JJ, Gil H (2020) Mechanical properties of mortars modified with wood waste ash. J Indian Acad Wood Sci 17(1):90–99
27. Novais RM, Seabra MP, Labrincha JA (2015) Wood waste incorporation for lightweight porcelain stoneware tiles with tailored thermal conductivity. J Clean Prod 90:66–72. https://doi.org/10.1016/j.jclepro.2014.11.045
28. Caldas LR, Saraiva AB, Lucena AF, Da Gloria MHY, Santos AS, Toledo Filho RD (2021) Building materials in a circular economy: the case of wood waste as $CO_2$-sink in bio concrete. Resour Conserv Recycl 166:105346
29. Mohammed BS, Abdullahi M, Hoong CK (2014) Statistical models for concrete containing wood chipping as partial replacement to fine aggregate. Constr Build Mater 55:13–19
30. He P, Hossain MU, Poon CS, Tsang DC (2019) Mechanical, durability and environmental aspects of magnesium oxychloride cement boards incorporating waste wood. J Clean Prod 207:391–399. https://doi.org/10.1016/j.jclepro.2018.10.015
31. Bourzik O, Akkouri N, Baba K, Nounah A (2022) Study of the effect of wood waste powder on the properties of concrete. Mater Today: Proc 58:1459–1463
32. Usman M, Khan AY, Farooq SH, Hanif A, Tang S, Khushnood RA, Rizwan SA (2018) Eco-friendly self-compacting cement pastes incorporating wood waste as cement replacement: a feasibility study. J Clean Prod 190:679–688
33. Fu Q, Yan L, Ning T, Wang B, Kasal B (2020) Interfacial bond behavior between wood chip concrete and engineered timber glued by various adhesives. Constr Build Mater 238:117743
34. Khorami M, Ganjian E (2011) Comparing flexural behaviour of fibre–cement composites reinforced bagasse: wheat and eucalyptus. Constr Build Mater 25(9):3661–3667
35. Qu X, Zhao X (2017) Previous and present investigations on the components, microstructure and main properties of autoclaved aerated concrete—A review. Constr Build Mater 135:505–516
36. Huang XY, Ni W, Cui WH, Wang ZJ, Zhu LP (2012) Preparation of autoclaved aerated concrete using copper tailings and blast furnace slag. Constr Build Mater 27(1):1–5
37. Bonakdar A, Babbitt F, Mobasher B (2013) Physical and mechanical characterization of fiber-reinforced aerated concrete (FRAC). Cement Concr Compos 38:82–91
38. Xu R, He T, Da Y, Liu Y, Li J, Chen C (2019) Utilizing wood fiber produced with wood waste to reinforce autoclaved aerated concrete. Constr Build Mater 208:242–249
39. Antoun M, Issa CA, Aouad G, Gerges N (2021) Sustainable masonry blocks: Olive wood waste as substitute for fine aggregates. Case Stud Constr Mater 15:e00590
40. Dasarathy AK, Thandavamoorthy TS (2013) Atteunation of noise using barrier in the form of enclosures. J Appl Res 83–89
41. Thandavamoorthy TS (2016) Wood waste as coarse aggregate in the production of concrete. Eur J Environ Civil Eng 20(2):125–141

42. Mawardi I, Aprilia S, Faisal M, Rizal S (2022) An investigation of thermal conductivity and sound absorption from binderless panels made of oil palm wood as bio-insulation materials. Results Eng 13:100319
43. Binici H, Gemci R, Kucukonder A, Solak HH (2012) Investigating sound insulation, thermal conductivity and radioactivity of chipboards produced with cotton waste, fly ash and barite. Constr Build Mater 30:826–832
44. Agoua E, Allognon-Houessou E, Adjovi E, Togbedji B (2013) Thermal conductivity of composites made of wastes of wood and expanded polystyrene. Constr Build Mater 41:557–562
45. Binici H, Aksogan O (2016) Eco-friendly insulation material production with waste olive seeds, ground PVC and wood chips. J Build Eng 5:260–266
46. Abu-Jdayil B, Barkhad MS, Mourad AHI, Iqbal MZ (2021) Date palm wood waste-based composites for green thermal insulation boards. J Build Eng 43:103224
47. Lakrafli H, Tahiri S, Albizane A, Bouhria M, El Otmani ME (2013) Experimental study of thermal conductivity of leather and carpentry wastes. Constr Build Mater 48:566–574
48. Farooqi ZUR, Sabir M, Zeeshan N, Murtaza G, Hussain MM, Ghani MU (2020) Vehicular noise pollution: its environmental implications and strategic control. In: Autonomous vehicle and smart traffic. IntechOpen
49. Liu M, Peng L, Fan Z, Wang D (2019) Sound insulation and mechanical properties of wood damping composites. Wood Res 64(4):743–758
50. Zhao J, Wang XM, Chang JM, Yao Y, Cui Q (2010) Sound insulation property of wood-waste tire rubber composite. Compos Sci Technol 70(14):2033–2038
51. Torkaman J, Ashori A, Momtazi AS (2014) Using wood fiber waste, rice husk ash, and limestone powder waste as cement replacement materials for lightweight concrete blocks. Constr Build Mater 50:432–436
52. Wang L, Chen SS, Tsang DC, Poon CS, Shih K (2016) Value-added recycling of construction waste wood into noise and thermal insulating cement-bonded particleboards. Constr Build Mater 125:316–325
53. Fornés IV, Vaičiukynienė D, Nizevičienė D, Tamošaitis G, Pupeikis D (2022) The improvement of the thermal and acoustic insulation properties of phosphogypsum specimens by adding waste wood fibre. Constr Build Mater 331:127341

# Building Material in Circular Economy: The Suitability of Wood Waste in Bio-concrete Development

**Messaouda Boumaaza, Ahmed Belaadi, Hassan Alshahrani, Mostefa Bourchak, and Mohammad Jawaid**

**Abstract** The production of Ordinary Portland Cement (OPC) and concrete, as well as the production of aggregates, significantly increases carbon dioxide ($CO_2$) emissions. However, bio-concretes can act as eco-friendly substitutes for conventional concretes making the demand for them on the rise for the many uses they are put to. Construction materials are usually recycled or turned into waste following demolition. As a result, a minor fraction of the economic value and sustainability inherent in them gets exploited by the construction industry. Consequently, the necessity for improving material effectiveness within the resource base will likely rise with the increase in human demand, as it would also be necessary to secure resources for the future. Circular economy (CE) principles may help mitigate the aforementioned problems within the construction industry if they are applied to recirculating construction materials. This chapter presents an approach toward using advanced technologies of implementing CE in the management and Life Cycle Assessment (LCA) of bio-concretes, which are produced by combining wood waste ash (WWA) in place of cement, wood waste (WW) as fine aggregates and wood aggregates (WA) as coarse waste. Thereafter, the chemical and physical properties, the microstructural characteristics, and the strength of wood waste–based bio-concrete (WWBC) are examined. Additionally, the ecological consequences and perspectives of WWBC

M. Boumaaza
Laboratory of Civil and Engineering Hydraulic (LGCH), University 8 Mai 1945 Guelma, BP 401, 24000 Guelma, Algeria

A. Belaadi (✉)
Department of Mechanical Engineering, Faculty of Technology, University 20 Août 1955, El-Hadaiek Skikda, Skikda, Algeria
e-mail: a.belaadi@univ-skikda.dz; ahmedbelaadi1@yahoo.fr

H. Alshahrani
Department of Mechanical Engineering, College of Engineering, Najran University, Najran, Saudi Arabia

M. Bourchak
Aerospace Engineering Department, King Abdulaziz University, Jeddah, Saudi Arabia

M. Jawaid
Laboratory of Biocomposite Technology, Institute of Tropical Forestry and Forest Products (INTROP), Universiti 11 Putra Malaysia, 43400 Serdang, Selangor, Malaysia

 
S. N. Sarmin et al. (eds.), *Wood Waste Management and Products*, Sustainable Materials and Technology, https://doi.org/10.1007/978-981-99-1905-5_9

production as well as its practical applications as a construction material have been examined to evaluate the effects of the emission of greenhouse gases (GHG) and carbon footprints. The circularity of WWBC has been discussed to achieve reduction in material waste and carbon footprint as well as encourage further research to improve the sustainability of construction materials in general.

**Keywords** Bio-concrete · Circular economy · Wood waste · Construction industry · Building waste · Environmental sustainability

## Abbreviations

| | |
|---|---|
| OPC | Ordinary Portland Cement |
| $CO_2$ | Carbon dioxide |
| CE | Circular economy |
| WWA | Wood waste ash |
| WW | Wood waste |
| WA | Wood aggregate |
| WWBC | Wood waste bio-concrete |
| WS | Wood shavings |
| GHG | Greenhouse gases |
| CDW | Construction and demolition waste |
| NWC | Normal weight concrete |
| LWC | Lightweight concrete |
| SF | Silica fume |
| MOC | Magnesium oxychloride cement |
| PFA | Pulverized fly ash |
| ISSA | Incinerated sewage sludge ash |

## 1 Introduction

The environmentally degrading effects of climate change are associated with increased $CO_2$ emissions, greater energy consumption, and the corresponding environmental risks. OPC production represents energy consumption and $CO_2$-emissions-intensive operation [1, 2]. From an ecological aspect, $CO_2$ is identified to be among the most relevant greenhouse gases, mainly responsible for the modification of the planet climate.

In order to reduce GHG emissions and attenuate a changing climate, partially replacing cement by waste products has become a common practice for the development of a durable environment [3, 4]. Using supplements for and alternatives to OPC has been proven to be the best solution for reducing $CO_2$ emissions from cement production, which account for 5–7% of total carbon emissions worldwide [5–7]. To minimize this negative effect of cement production on environmental sustainability, WW has been substituted in cement mortars as an alternative greener disposal method [8, 9].

Reuse wood ashes, however, needs to be adequately controlled considering the fine size of the particles and the risk of the production facility causing atmospheric pollution, resulting in respiratory problems among people living in the vicinity. Many experiments have been carried out on environmentally friendly concrete production by incorporating WWA as an alternative to cement as well as a sustainable material [10–13]. Several researchers have conducted investigations using WW as a cement replacement in concrete production [7, 8, 14]. The findings showed that partially using WWA as a cement substitute is suitable in terms of retaining construction concrete properties. These studies have also offered a solution to eliminate wood ash while replacing the traditional waste management practices with a more environmentally friendly method of producing green concrete [15].

Ramos et al. [14] reported the use of WWA as a replacement material partially to cement a range of 0–20% to verify its ability in concrete use and to evaluate the resistance and durability parameters. The improved strength and durability results affirmed the potential of WWA to enhance durability in building. It was observed that the best formulation was obtained for a 10% substitution of cement. Corinaldesi et al. [16] reported that sawdust addition to mortars resulted in better performance compared with wood chippings, but at substitution levels over 5%, it affected mortars' mechanical properties negatively.

Ince et al. [17] incorporated 1, 2, 3, 4, and 5% wood waste into cement mortars as powders and wood fibers to partially replace cement and sand. They reported that the strength bio-concrete decreased and associated physical properties were found to be suitable as structural materials for applications in construction industry practices. Bourzik et al. [18] investigated the concrete behavior using wood waste powder (WWP) with 0, 1, 1.5, 2, and 2.5% sand substitution rates by wood waste powder to examine compressive strength, density, workability, and ultrasonic pulse rate. The workability, density, and concrete compressive strength were found to decrease as the WWP rates increased. Concrete compressive strength decreased by 49.6%, 66.3%, 71.8%, 72.6%, and 75% for WWP substitution rates of 0%, 1%, 1.5%, 2%, and 2.5% respectively, when compared with a 28-day-old control sample.

The construction industry causes severe environmental damage through its resource and energy consumption as well as waste disposal. The introduction of CE principles changes a company's operations. The acquisition and management of optimized and automated information facilitate the ability to decide on innovative solutions to be used in the construction industry. The best management practices vis-à-vis building materials provide tools to integrate CE into building construction. Galdas et al. [19] used recycled WS in the CE strategy to evaluate GHG emissions over the life cycle of wood bio-concrete (WBC) manufacturing. A variety of WS transportation and recycling scenarios using the LCA methodology were considered. It was checked whether increasing the content of WS waste mitigated climate change. It was concluded that WS, when recycled for usage in bio-concrete, can act as a successful carbon sink as well as provide an entry point for circular and low-carbon manufacturing.

The purpose of this chapter is the identification of the main issues related to CE concept application and its potential opportunities for the building industry. First, it seeks to examine CE's application in the public building sector. It also provides a description of the properties, structural performance, and opportunities for incorporating WW into concrete, using WWA to partially replace OPC, WW to partially replace fine aggregates, or WA as partial replacement for coarse aggregates. This investigation examines in detail the use of WW in concrete mortars with the aim of regenerating WW into construction materials, thus providing true added-value compared with available waste management alternatives while also reducing the use of natural resources.

## 2 Development of CE and LCA in the Built Environment

CE has gained interest from the construction industry as a feasible mean to achieve sustainability [20, 21]. It offers a potential opportunity to resolve the conflict between environmental sustainability and economic development by transforming the existing linear economical model (take, produce, utilize, and trash) into circular economical model (reduce, reuse, recycle, and recuperate) [22–24]. Therefore, CE has the ability to generate a business motivation to operate in a sustainable way [25, 26]. Furthermore, CE has the potential to contribute to greater economic stability by securing resources and avoiding resource depletion by ensuring that materials are maintained at optimal efficiency and utility over their life cycle (Fig. 1).

It was difficult for the building industry to adopt effectively CE principles that are successfully applied in other industries [27]. With the growth anticipated in future years, EC development in the construction industry can be considered a reuse or recycling opportunity for bio-sourced waste, representing a promised scenario for biomaterial circulating strategy [28]. Moreover, the use of bio-sourced waste construction materials contributes highly to mitigating climate variability caused by buildings, considered a major carbon reservoir in urban areas [29, 30]. In this CE approach, the waste recovery to be used as feedstock may be restricted according to its accessibility and the required type of treatment. While some research highlights that the principal constraints to the deployment of CE are political and institutional barriers [31]. Technically, waste reuse and recycling are primarily restricted by recycle processing, local accessibility, and expensive costs of transportation [32]. These factors must therefore be evaluated and measured in order to determine the ecological feasibility of reusing the waste and to assess the benefits of using raw resources.

LCA is potentially a valuable tool for measuring ecological effects, particularly for mitigating climate variability and greenhouse gas emissions. It is widely regarded as a powerful means of quantifying and evaluating the environmental impacts of services and products along their life cycle and has gained widespread use in the building sector [24, 33]. During the last few years, LCA applications have attracted interest

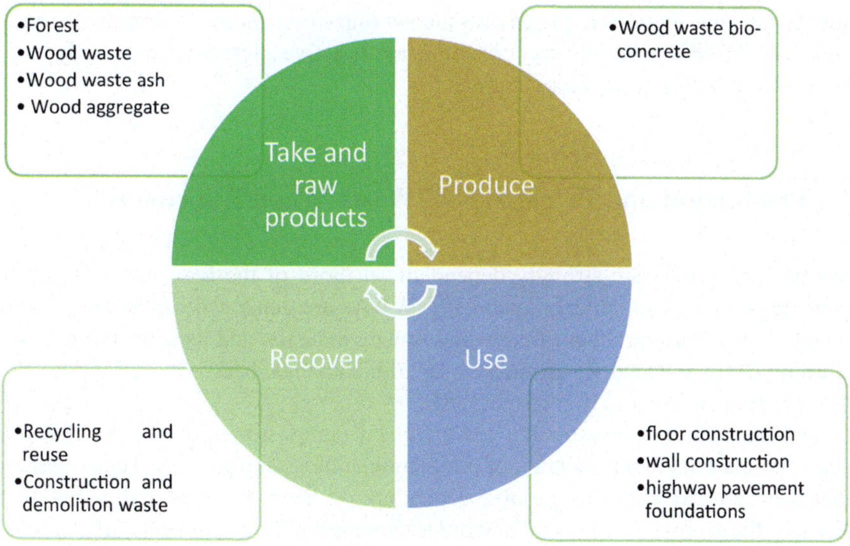


**Fig. 1** Flow chart of adopted CE in building industry using WW

in evaluating CE strategies in the building sector in order to assess their advantages regarding strategies for climate variability mitigation. Of particular interest in the use of recycled construction and demolition waste (CDW). Some researchers [31, 34, 35] reviewed the literature on construction sector CE regarding the mitigation of a changing climate. They concluded that while closure looping may be appropriate GHG emission reduction strategy, this is heavily influenced by its type and recycling process efficiency along with distances of transportation. However, in the literature, there are few examples illustrating LCA introduction to bio-based wastes used in construction materials.

The production of concrete has gained particular consideration in LCA and CE literature, which focuses primarily on using waste materials to replace aggregate or cement. Most studies point out that replacing certain concrete components, particularly Portland cement; can reduce global environmental effects, such as climate change [36]. Caldas et al. [19] investigated the greenhouse gas emissions over the life cycle of two mixtures of bio-concrete based on different contents of wood shavings by implementing LCA. It was concluded that the used WW can be evaluated to be a carbon sink if it is used to generate bio-concrete and that it could therefore be a promising alternative approach for low-carbon and circular material production within the concrete industry. With the aim of minimizing the environmental implications of WW management systems, Hossain et Poon [37] also focused on the recycling and reuse of WW for the production of organic polymer-based particleboard, particleboard bonded to cement, and energy from WW-derived biofuel compared with using raw wood. However, the LCA approach was used to environmentally assess alternative management strategies. The LCA findings revealed substantial contributions to the environmental impacts of the three strategies considered. The WWBC materials may constitute an attractive alternative, especially in the context of building in development nations.

## 3 Production and Properties of Waste Wood Bio-concrete

Material properties are strongly dependent on those of the base material and its production methods and techniques [17]. WWAs are generated by the combustion of sawdust, a waste product of the wood sawing industry, and its conversion to ash. Burning wood or wood waste such as sawmill shavings, bark, sawmill waste, etc., also produces ashes as a by-product [38].

Different authors investigated the chemical compositions of WWAs different forms in relation to their location of production and burning process. They observed that there are two different groups, depending on the main chemical component. Namely, the group with high calcium oxide content and the group with higher silicate oxides content and in every group a more spectrum chemical compounds existed. A major oxides occurring in WWA are the combined oxides ($SiO_2 + Al_2O_3 + Fe_2O_3$; 49.90%) which are therefore in the same category as those with high pozzolanic characteristics CaO (21.60%) and $SiO_2$ (46.90%), with LOI of 24.50%. The rest of the oxides present in WWAs at lower quantities are $Fe_2O_3$, $Na_2O$, $K_2O$, MgO, and $Al_2O_3$ [39].

Pingping et al. [40] investigated the thermal conductivity and durability of magnesium oxychloride cement (MOC) biocomposite panels made from pulverized fly ash (PFA) produced from construction wood waste and incinerated sewage sludge ash (ISSA). Wood-cement panels were found to decrease in thermal conductivity as the fiber percentage increased, mainly for two main considerations. One, this is associated with the smaller wood fiber thermal conductivity (0.2 W/(m k)) against MOC wood paste (2.0 W/(m k)). However, the second reason was increasing the wood fiber content caused the reduction of density, therefore reducing the thermal conductivity. The thermal conductivity was reduced by 13.5%, 26.5%, and 41.5%, respectively, compared to pure MOC pulp when 0, 15, 20, and 25% wood fibers were added (Fig. 2).

## 4 The Microstructure of Waste Wood Bio-concrete

Sigvardsen et al. [41] examined the use of two alternative wood ashes (WAs) as low cement substitutes, in both a raw and a cleaned variety. The morphology of the grains is displayed using SEM images (Fig. 3). This indicates angular, wide particles for the two WAs. A coating of soluble components overlaps the non-treated WAs, Fig. 3a, c, which is viewed as smaller particles overlapping bigger particles. After the WAs were cleaned, Fig. 3b, d, this coating was stripped off. The washed residues

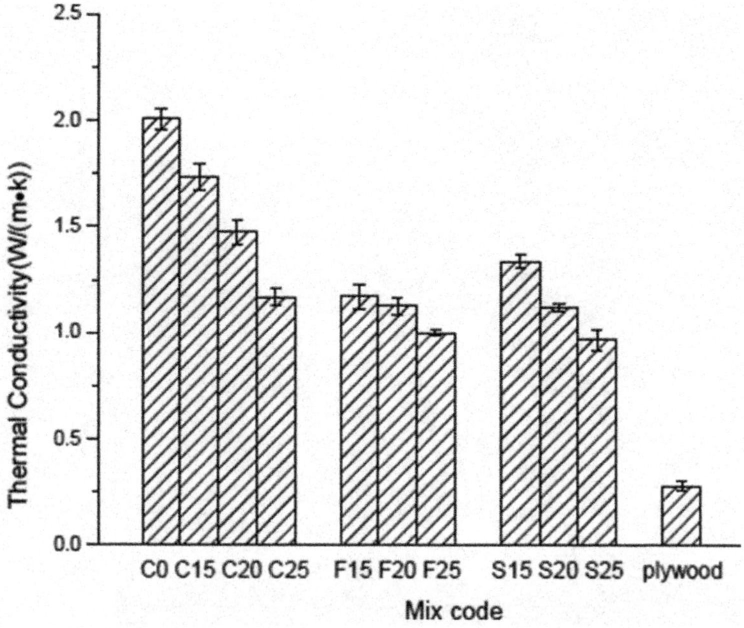


**Fig. 2** Thermal conductivity of plywood, MOC paste, and wood-cement board [40]

obtained by grid combustion (600–1000 °C) (WA1-W) were observed to have larger spherical size particles than the ones obtained by circulated fluidized belt burning (760–930 °C) (WA2-W). The washing operation provided no improvement in the WAs loss on ignition, which was elevated (15%) for all WAs.

Further microscopic analysis of wood waste powder and wood waste fibers, presented in Fig. 4a, b, respectively, showed that a single wood waste material can range in width from a few nanometers to many hundreds of nanometers in length that cluster into larger structures with a complicated interlacing [17]. This wood waste microstructure correlates strongly with research results [42].

Wood waste combustion temperatures influence the yield and chemical composition of WWA produced during sawdust combustion. SEM analysis revealed a variation in physical characteristics such as particle size and shape in ash particles collected across various sources. The specific properties of WWA which were reported from a structural aspect involved lower specific gravity and density than ordinary Portland cement [10, 12]. Similarly SEM micrographs [43] proved that wood waste ash is a heterogeneous particle mixture with various particle sizes, usually having an angular shape. At least partially, however, the wood ash particles appeared porous aggregated particles, which are mainly uncombusted or partially combusted wood chips or bark particles. WW material is produced by mechanical treatment of untreated wood, such as cutting, chipping, sanding, drilling, or powdering using a cutter or similar instrument. They are then screened to the fineness of the sand used. Under

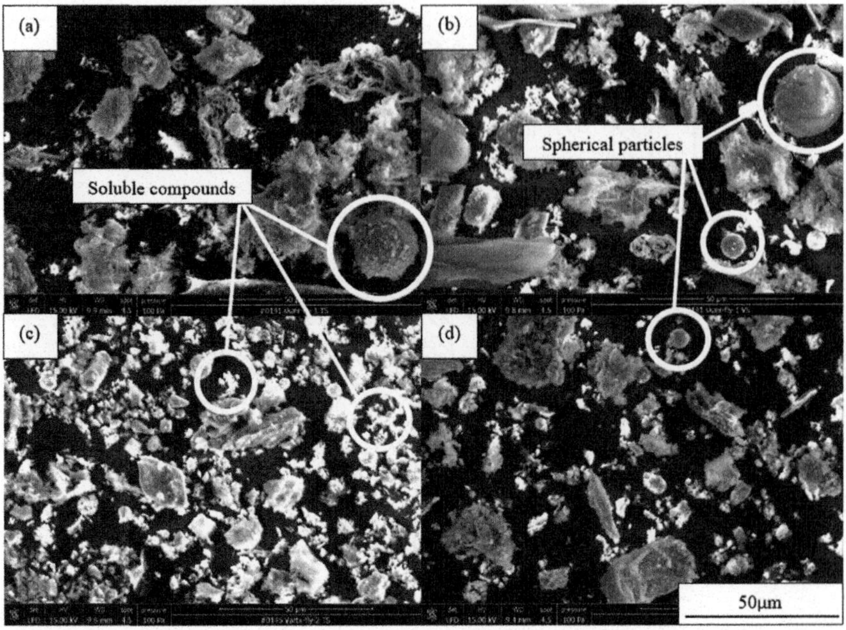


**Fig. 3** Particle morphology of **a** WA1, **b** WA1-W, **c** WA2, and **d** WA2-W [41]

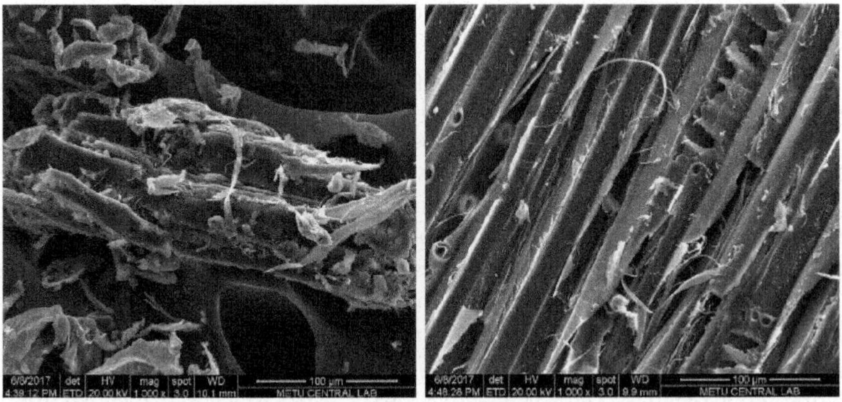

**Fig. 4** Microscope image of **a** the wood powder, **b** the wood fiber [17]

this condition, untreated, WW results revealed that sawdust was found to be more lighter as compared to fine aggregates [44].

## 5 Mechanical Properties of Waste Wood Bio-concrete

Ince et al. [17] examined the bending and compressive strength of cement mortars including powders and wood fibers with different substitution levels (1–5%). As replacement for cement, using WA led to a decrease in cement mortar compressive strength as the substitution rate increased, due to WA's inability to contribute to binder hydration reaction. Consequently, this did have no major part to play in the compressive strength of the mortars (Fig. 5). However, an increase in WA as a substitute for sand was observed to lead to an increase in mortar compressive strength in this investigation due to the wood particle filling action. The increase in the replacement level using wood powders as substitutes for sand and cement was also shown to lead to an increase in compressive strength; however, a minor reduction in the mortar flexural strength was observed as a result of the wood powders' filling action. Reduction in bending strength in mortars wood powder used as a substitute for cement, was found to be induced by the combination of lower hydraulicity within the mixture and the existence of weaker zones of interaction within the matrix. However, replacing the sand with wood fibers provided noticeable increase in cement mortar flexural strength. These fibers within the matrix allowed the bridging of microcracks and thus helped prevent major crack formation and progression, thus delaying failure. The phenomenon is responsible essentially for the improvement observed in the bending strength of cement mortars containing fibers of wood in replacement for sand.

Udoeyo et al. [12] examined the addition of 5–30% WWA by weight of cement on the bending and compressive strength of concrete. Increasing the WWA content decreased the compressive and flexural strength values of the concrete by 9–38%

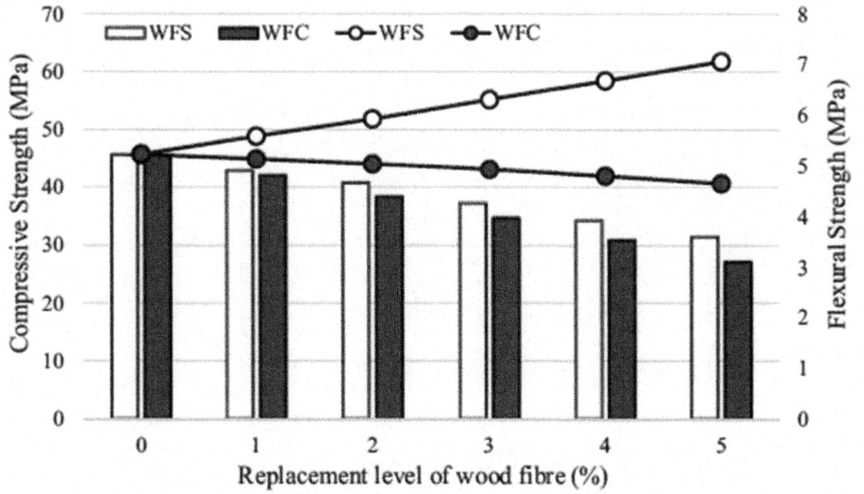


**Fig. 5** Compressive and flexural strength of cement mortars containing wood fiber as substitute cement (WFC) and sand (WFS) [17]

and 2–35%, respectively, compared with the base material properties. An optimum substitution of WWA was 5–10% by weight of cement.

Further, Bourzik et al. [18] used low percentages of WWP to replace sand over concrete for regeneration of the WW into value-added materials for building, with replacement levels of 0.5, 1, 1.5, 2, and 2.5% for this waste. Compressive strength decrease of 75% at 28 days was observed for the concrete incorporating 2.5% WWP in comparison with the concrete control, mainly due to the decrease in concrete density. Furthermore, the degradation of the bond strength of the WWP surfaces to the cement paste can be attributed to this effect. The finding agrees well to those reported in the literature [45, 46].

Furthermore, Ahmed et al. [46] developed an eco-friendly and a thermally effective normal weight concrete (NWC) and lightweight concrete (LWC), using various proportions of sawdust to replace sand. The normal conventional concrete comprises 0, 5, 10, and 15% of sawdust, and the lightweight concrete has 0 and 10% of sawdust per dry total volume of sand. The compressive strength in the two types of NWC and LWC samples at 7 and 28 days was increased with increasing age samples and diminished with an increasing amount of sawdust (Fig. 6). The percentage decrease in 28 days of age compressive strength for NWC with 5%, 10%, and 15% sawdust content was 2.95%, 7.70%, and 12.32% respectively, compared with that of the NWC base material. The LWC sawdust modified specimens showed a decrease in compressive strength of 7.36% at age 28 days. The decrease in compressive strength was associated with individual concrete component strength. At 28 days, the finer aggregates were found to have compressive strengths higher than the requirements for concrete for structural applications as specified in ACI 318, i.e., 17 MPa, and are therefore considered suitable for structural applications.

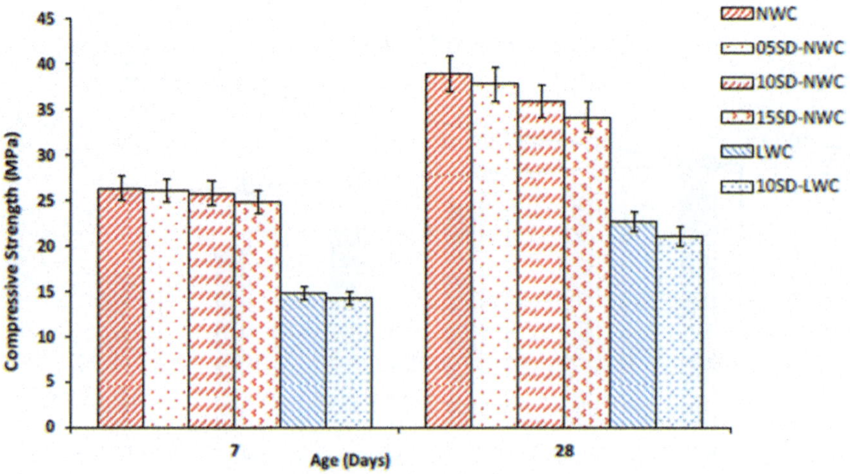


**Fig. 6** Compressive strength of sawdust modified NWC and LWC [46]

Kunchariyakun et al. [47], studied a curing effect on mortar properties containing wood fiber waste (WFW) from 0 to 5 wt% to ordinary Portland cement. Figure 7 illustrates the compressive and flexural strength results under different curing environments and reveals that the compressive strength is reduced by adding WFW proportions at all curing environments. It was found that the mortar incorporating 5% WFW by weight under normal cure at 28 days had the highest ratio of compressive to bending strength. An increasing proportion of WFW decreased compressive strength about 10–70% respectively, across the cure conditions. The autoclave curing application results in small compression strength ratio values comparing to ordinary and low-temperature curing. This is due to the porous wood fiber waste structure, which results in increased water consumption and lost strength. Cement alkalinity and hydration temperature helped to degrade the natural fiber components, leading to a decrease in the strength and the cementitious materials sustainability.

**Fig. 7** Compressive and bending strength of mortar incorporation WFW at different hardening conditions [47]

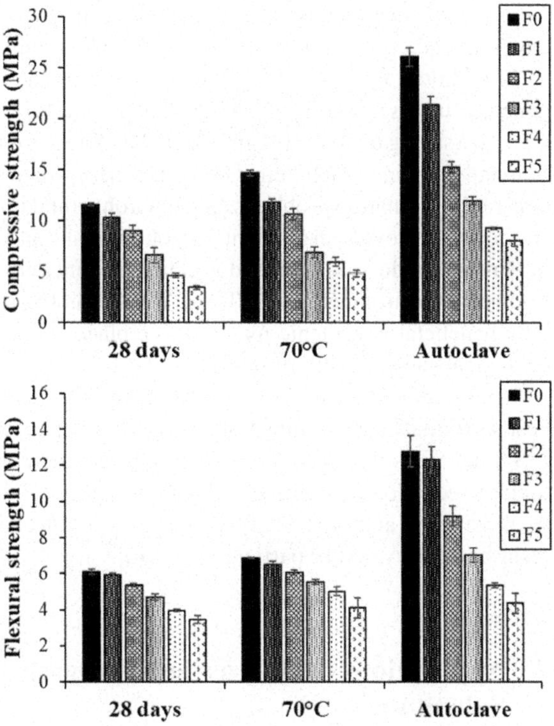

## 6 Ecological Consequences and Perspectives of the Production of *WW* Bio-concrete

Considering the environmental perspective, WWAs can have an advantageous impact on cement manufacturing [48]. However, WWA utilization as a final operation requires appropriate treatment considering the small particle size and the air pollution potentially generated by the combustion process [49]. Additionally, the WWA's acidic environment exposition can result in the liberation of heavy metals, and it can influence the environmentally friendly system. Consequently, more research is needed on the ecological impacts of WWA concrete. In order to investigate the environmental impact of WWA as an additive material, Udoeyo et al. [12] performed a nitric acid leaching test on WWA with a demineralized water solution. The tests revealed WWA as having Zn, Fe, Cr, and Cu and discharged these heavy metals to acid environment. Due to its discharge in nature, WWA may liberate to the environment these heavy metals. Therefore, WWA implementation will mitigate reduction environmental pollution via a need for waste dumping zones.

In addition, Ghorpade [50] studied the influence of adding WWA (0–30%) to concrete. In this research, the effective deployment of WWA in high-quality concrete showed a saving of up to 10% in cement, leading to a cost reduction in cement production and generated GHG emissions. By incorporating WWA as a cement adder in concrete, elimination problems, and environmental impact will be minimized. WWA was found to provide an efficient pozzolanic material to partially replace cement, with no strength reduction, better durability of concrete, and practicality to contribute to durable environment [14]. Thus, WWA has the potential to be an environmentally beneficial pozzolanic material to replace cement and also to contribute to the ecological sustainability of buildings.

In terms of GHG emissions, the MOC panels manufactured with ISSA offered the best performance environmentally. As well, an estimated 36% of the waste wood and ISSA materials were consumed. Considering the considerable GHG emission and toxicity reduction for humans and conservation of raw resources, the MOC panel can be classified as an environmentally friendly product and used to replace conventional resin-based plywood or particleboard [40].

## 7 Applications of *WW* as an Aggregate or Cement Partial Substitute

Pingping et al. [40] explored wood waste construction as wood fiber to produce MOC composite panels. MOC wood composites made with PFA and incinerated ISSA were found to have higher flexural strength, better performance at elevated temperatures, and higher water absorption compared to other wood composites. In addition, the manufacture of MOC wood panels results in reduced GHG emissions compared to plywood, and reduced and less toxic effects compared to traditional resin-based panels by using LCA technique. Application of PFA and ISSA reduced the wood-MOC panels thermal conductivity.

WWA is a more refined material compared with cement and is capable of closing the matrix voids in cured concrete, thus making the penetration of sodium chloride into concrete surfaces difficult. Hence, it is suggested that concrete buildings in marine environments should be constructed with WWA to ensure protection against damage due to severe salinity. Using sawdust provides the sustainable building industry with a raw material source and an environmentally friendly solution to removing sawdust waste, which contributed to the preservation of the existing natural aggregate reserves [46]. Although a decrease in the mechanical properties of concrete occurs with the increase in WW's substitution ratio, WW concrete has many possible applications including floor and wall construction, highway pavement foundations, and miscellaneous construction where resistance is not essential to the construction process [18].

## 8 Conclusion

WWP can be utilized to produce concrete at specific substitution percentages. Its use decreases concrete's proper weight in buildings and preserves the availability of raw materials, like sand, for the construction process. WW was considered a supplementary raw material or an added product to improve a binder's reaction capability and thus had the ability to serve as an alternative construction material partially replacing cement and aggregates for ecological concrete solutions. Remarkable strengths were obtained in the results of many studies, and certain essential properties were investigated. Considerable progress was made in reaching zero-waste technology by adopting CE and LCA while maintaining adequate performance capabilities as raw cementitious material was achieved. Until the writing of these lines, WW has been deployed in concrete manufacturing in limited quantities. WWA concrete incorporating WW blended with a wide range of ashes yielded better results compared to non-WWA concrete. these results are mainly explained by the effects of the synergistic properties within its pozzolans to provide a higher microstructure density. It is concluded that WW has the potential to be a carbon sink when deployed in the manufacture of WWBC and may be an attractive approach to achieve the development of circular and low-carbon materials in the construction industry.

## References


1. Paris JM, Roessler JG, Ferraro CC, DeFord HD, Townsend TG (2016) A review of waste products utilized as supplements to Portland cement in concrete. J Clean Prod 121:1–18. https://doi.org/10.1016/j.jclepro.2016.02.013
2. Chowdhury S, Mishra M, Suganya O (2015) The incorporation of wood waste ash as a partial cement replacement material for making structural grade concrete: an overview. Ain Shams Eng J 6:429–437. https://doi.org/10.1016/j.asej.2014.11.005
3. Udoeyo FF, Dashibil PU (2002) Sawdust ash as concrete material. J Mater Civil Eng 14:173–176. https://doi.org/10.1061/(ASCE)0899-1561(2002)14:2(173)
4. Ince C, Hamza A, Derogar S, Ball RJ (2020) Utilisation of waste marble dust for improved durability and cost efficiency of pozzolanic concrete. J Clean Prod 270:122213. https://doi.org/10.1016/j.jclepro.2020.122213
5. Vu V-A, Cloutier A, Bissonnette B, Blanchet P, Duchesne J (2019) The effect of wood ash as a partial cement replacement material for making wood-cement panels. Materials 12. https://doi.org/10.3390/ma12172766
6. Barcelo L, Kline J, Walenta G, Gartner E (2014) Cement and carbon emissions. Mater Struct 47:1055–1065. https://doi.org/10.1617/s11527-013-0114-5
7. Tamanna K, Raman SN, Jamil M, Hamid R (2020) Utilization of wood waste ash in construction technology: a review. Constr Build Mater 237:117654. https://doi.org/10.1016/j.conbuildmat.2019.117654
8. Cheah CB, Ramli M (2011) The implementation of wood waste ash as a partial cement replacement material in the production of structural grade concrete and mortar: an overview. Resour Conserv Recycl 55:669–685. https://doi.org/10.1016/j.resconrec.2011.02.002
9. Usman M, Khan AY, Farooq SH, Hanif A, Tang S, Khushnood RA, Rizwan SA (2018) Eco-friendly self-compacting cement pastes incorporating wood waste as cement replacement: a feasibility study. J Clean Prod 190:679–688. https://doi.org/10.1016/j.jclepro.2018.04.186
10. Raza MS, Rai K, Kumar D, Ali M (2020) Experimental study of physical, fresh-state and strength parameters of concrete incorporating wood waste ash as a cementitious material 7:267–276
11. Kim MH, Song HB (2014) Analysis of the global warming potential for wood waste recycling systems. J Clean Prod 69:199–207. https://doi.org/10.1016/j.jclepro.2014.01.039
12. Udoeyo FF, Inyang H, Young DT, Oparadu EE (2006) Potential of wood waste ash as an additive in concrete. J Mater Civil Eng 18:605–611. https://doi.org/10.1061/(ASCE)0899-1561(2006)18:4(605)
13. Pavlíková M, Zemanová L, Pokorný J, Záleská M, Jankovský O, Lojka M, Sedmidubský D, Pavlík Z (2018) Valorization of wood chips ash as an eco-friendly mineral admixture in mortar mix design. Waste Manag 80:89–100. https://doi.org/10.1016/j.wasman.2018.09.004
14. Ramos T, Matos AM, Sousa-Coutinho J (2013) Mortar with wood waste ash: Mechanical strength carbonation resistance and ASR expansion. Constr Build Mater 49:343–351. https://doi.org/10.1016/j.conbuildmat.2013.08.026
15. Gerges N, Issa CA, Antoun M, Sleiman E, Hallal F, Shamoun P, Hayek J (2021) Eco-friendly mortar: optimum combination of wood ash, crumb rubber, and fine crushed glass. Case Stud Constr Mater 15:e00588. https://doi.org/10.1016/j.cscm.2021.e00588
16. Corinaldesi V, Mazzoli A, Siddique R (2016) Characterization of lightweight mortars containing wood processing by-products waste. Constr Build Mater 123:281–289. https://doi.org/10.1016/j.conbuildmat.2016.07.011
17. Ince C, Tayançlı S, Derogar S (2021) Recycling waste wood in cement mortars towards the regeneration of sustainable environment. Constr Build Mater 299:123891. https://doi.org/10.1016/j.conbuildmat.2021.123891
18. Bourzik O, Akkouri N, Baba K, Nounah A (2022) Study of the effect of wood waste powder on the properties of concrete. Mater Today: Proc 58:1459–1463. https://doi.org/10.1016/j.matpr.2022.02.518

19. Caldas LR, Saraiva AB, Lucena AFP, Da Gloria MY, Santos AS, Filho RDT (2021) Building materials in a circular economy: the case of wood waste as $CO_2$-sink in bio concrete. Resour Conserv Recycl 166:105346. https://doi.org/10.1016/j.resconrec.2020.105346
20. Hossain MU, Ng ST, Antwi-Afari P, Amor B (2020) Circular economy and the construction industry: existing trends, challenges and prospective framework for sustainable construction. Renew Sustain Energy Rev 130:109948. https://doi.org/10.1016/j.rser.2020.109948
21. Eberhardt LCM, Birkved M, Birgisdottir H (2022) Building design and construction strategies for a circular economy. Archit Eng Des Manag 18:93–113. https://doi.org/10.1080/17452007.2020.1781588
22. Kylili A, Fokaides PA (2017) Policy trends for the sustainability assessment of construction materials: a review. Sustain Cities Soc 35:280–288. https://doi.org/10.1016/j.scs.2017.08.013
23. Akanbi LA, Oyedele LO, Akinade OO, Ajayi AO, Davila Delgado M, Bilal M, Bello SA (2018) Salvaging building materials in a circular economy: a BIM-based whole-life performance estimator. Resour Conserv Recycl 129:175–186. https://doi.org/10.1016/j.resconrec.2017.10.026
24. Benachio GLF, do Carmo Duarte Freitas M, Tavares SF (2020) Circular economy in the construction industry: a systematic literature review. J Clean Prod 260:121046. https://doi.org/10.1016/j.jclepro.2020.121046
25. Maier D (2021) Building materials made of wood waste a solution to achieve the sustainable development goals. Materials 14. https://doi.org/10.3390/ma14247638
26. Adams KT, Osmani M, Thorpe T, Thornback J (2017) Circular economy in construction: current awareness, challenges and enablers. In: Proceedings of the Institution of Civil Engineers—Waste and Resource Management, vol 170, pp 15–24. https://doi.org/10.1680/jwarm.16.00011
27. Pomponi F, Moncaster A (2017) Circular economy for the built environment: a research framework. J Clean Prod 143:710–718. https://doi.org/10.1016/j.jclepro.2016.12.055
28. Jarre M, Petit-Boix A, Priefer C, Meyer R, Leipold S (2020) Transforming the bio-based sector towards a circular economy—What can we learn from wood cascading? Forest Policy Econ 110:101872. https://doi.org/10.1016/j.forpol.2019.01.017
29. Chen S, Chen B, Feng K, Liu Z, Fromer N, Tan X, Alsaedi A, Hayat T, Weisz H, Schellnhuber HJ, Hubacek K (2020) Physical and virtual carbon metabolism of global cities. Nat Commun 11:182. https://doi.org/10.1038/s41467-019-13757-3
30. Deetman S, Marinova S, van der Voet E, van Vuuren DP, Edelenbosch O, Heijungs R (2020) Modelling global material stocks and flows for residential and service sector buildings towards 2050. J Clean Prod 245:118658. https://doi.org/10.1016/j.jclepro.2019.118658
31. Ghisellini P, Ripa M, Ulgiati S (2018) Exploring environmental and economic costs and benefits of a circular economy approach to the construction and demolition sector. A literature review. J Clean Prod 178:618–643. https://doi.org/10.1016/j.jclepro.2017.11.207
32. Norouzi M, Chàfer M, Cabeza LF, Jiménez L, Boer D (2021) Circular economy in the building and construction sector: a scientific evolution analysis. J Build Eng 44:102704. https://doi.org/10.1016/j.jobe.2021.102704
33. Ruiz LA, Ramón XR, Domingo SG (2020) The circular economy in the construction and demolition waste sector—A review and an integrative model approach. J Clean Prod 248:119238. https://doi.org/10.1016/j.jclepro.2019.119238
34. Ginga CP, Ongpeng JMC, Daly MKM (2020) Circular economy on construction and demolition waste: a literature review on material recovery and production. Materials 13. https://doi.org/10.3390/ma13132970
35. Santos S, da Silva PR, de Brito J (2019) Self-compacting concrete with recycled aggregates—A literature review. J Build Eng 22:349–371. https://doi.org/10.1016/j.jobe.2019.01.001
36. Habert G, Miller SA, John VM, Provis JL, Favier A, Horvath A, Scrivener KL (2020) Environmental impacts and decarbonization strategies in the cement and concrete industries. Nat Rev Earth Environ 1:559–573. https://doi.org/10.1038/s43017-020-0093-3
37. Hossain MU, Poon CS (2018) Comparative LCA of wood waste management strategies generated from building construction activities. J Clean Prod 177:387–397. https://doi.org/10.1016/j.jclepro.2017.12.233

38. Berra M, Mangialardi T, Paolini AE (2015) Reuse of woody biomass fly ash in cement-based materials. Constr Build Mater 76:286–296. https://doi.org/10.1016/j.conbuildmat.2014.11.052
39. Adamu M, Tifase AS, Uche OAU (2017) Engineering properties of industrial wood waste ash-concrete. Int J Adv Constr Eng 1:1–11
40. He P, Hossain MU, Poon CS, Tsang DCW (2019) Mechanical, durability and environmental aspects of magnesium oxychloride cement boards incorporating waste wood. J Clean Prod 207:391–399. https://doi.org/10.1016/j.jclepro.2018.10.015
41. Sigvardsen NM, Geiker MR, Ottosen LM (2021) Phase development and mechanical response of low-level cement replacements with wood ash and washed wood ash. Constr Build Mater 269:121234. https://doi.org/10.1016/j.conbuildmat.2020.121234
42. Berger F, Gauvin F, Brouwers HJH (2020) The recycling potential of wood waste into wood-wool/cement composite. Constr Build Mater 260:119786. https://doi.org/10.1016/j.conbuildmat.2020.119786
43. Naik TR, Kraus RN (n.d.) A new source of pozzolanic material. Concrete Int 25
44. Kumar SSKN, Gupta A, Dilip (2014) Low cost construction material for concrete as sawdust. Glob J Res Eng 14(4-E). https://engineeringresearch.org/index.php/GJRE/article/view/1189
45. Raheem AA, Ige AI (2019) Chemical composition and physicomechanical characteristics of sawdust ash blended cement. J Build Eng 21:404–408. https://doi.org/10.1016/j.jobe.2018.10.014
46. Ahmed W, Khushnood RA, Memon SA, Ahmad S, Baloch WL, Usman M (2018) Effective use of sawdust for the production of eco-friendly and thermal-energy efficient normal weight and lightweight concretes with tailored fracture properties. J Clean Prod 184:1016–1027. https://doi.org/10.1016/j.jclepro.2018.03.009
47. Kunchariyakun K, Sinyoung S, Kajitvichyanukul P (2022) Comparative microstructures and mechanical properties of mortar incorporating wood fiber waste from various curing conditions. Case Stud Constr Mater 16:e00855. https://doi.org/10.1016/j.cscm.2021.e00855
48. Yin K, Ahamed A, Lisak G (2018) Environmental perspectives of recycling various combustion ashes in cement production—A review. Waste Manag 78:401–416. https://doi.org/10.1016/j.wasman.2018.06.012
49. Aprianti E, Shafigh P, Bahri S, Farahani JN (2015) Supplementary cementitious materials origin from agricultural wastes—A review. Constr Build Mater 74:176–187. https://doi.org/10.1016/j.conbuildmat.2014.10.010
50. Ghorpade VG (2012) Effect of wood waste ash on the strength characteristics of concrete. Nat Environ Pollut Technol 11

# Application of Wood Waste in Agriculture

**Noorshilawati Abdul Aziz, Nurulatika Minhad, Nur Suraya Abdullah, Fazidah Rosli, Nazatul Asikin Muda, Muhammad Esyam Adip, Noor Azimah Darus, and Mohd Khairi Che Lah**

**Abstract** This article reviews the application of wood waste materials in agriculture sectors. Wood waste has numerous potentials to become a value-added product that can be applied in diverse activities including agriculture, energy production, the furniture industry, and construction material. The products that could be obtained from wood wastes comprise organic amendments such as compost and biochar, poultry bedding, organic mulching, source of renewable energy, and construction and furniture materials. These various sources could enhance crop production, and revenue and resolve food security challenges if they are fully utilised. This systematic review elucidates the diverse areas in which wood waste could be utilised in agriculture and addresses the issues contributing to the underutilisation of wood waste in the agricultural sector. The gathered information will be relevant and beneficial to players and stakeholders in the agricultural sector. In conclusion, wood waste remains a promising and beneficial source that could be recycled as other products and concurrently boost productivity in the agriculture industry. The utilisation of wood waste materials could also generate income, thereby reducing the risk of environmental deterioration and adverse effects on human health.

**Keywords** Wood waste · Agriculture · Mulching · Fertilisers · Poultry

## 1 Introduction

All types of wood or wood-based goods that have reached the end of their useful lives and thus meet the criteria of waste are referred to as "wood wastes" or "waste wood." The term "wood wastes" refers to a group of wastes comprising wasted

N. Abdul Aziz · N. Minhad · F. Rosli · N. A. Muda · M. E. Adip · N. A. Darus · M. K. Che Lah (✉)
Faculty of Plantation and Agrotechnology, Universiti Teknologi MARA Pahang, 26400 Bandar Tun Abdul Razak Jengka, Pahang, Malaysia
e-mail: khairi@uitm.edu.my

N. S. Abdullah
Faculty of Plantation and Agrotechnology, Universiti Teknologi MARA Melaka, 77300 Merlimau, Melaka, Malaysia



S. N. Sarmin et al. (eds.), *Wood Waste Management and Products*, Sustainable Materials and Technology, https://doi.org/10.1007/978-981-99-1905-5_10

wood products from a variety of industries, including the wood processing industry, demolition and building, and private houses and railroad construction [1]. Depending on the quality grade, wood waste is a valuable resource that is widely accessible and can be used for energy production or material recycling.

There are three types of wood wastes namely biomass or raw wood (untreated), slightly treated wood wastes, which result from coating or glue treatments, and hazardous waste which are highly treated wood wastes and subjected to a particular chemical treatment [2]. Malaysia is known for extensive deposits of biomass and wood waste resources that are yet to be fully utilised. Examples of the primary biomass resources that are readily available for utilisation include large-scale agro-industrial products and wastes, agro-cultural wastes, and woody biomass [3]. Specifically, wood mills, sawdust, and fast-growing trees are good sources of wood waste, whereas agro-industrial wastes include remnants from palm oil mills, rice husks from rice mills, and molasses from sugar manufacturing industries [3].

The majority of wood waste can be recycled into economically viable products such as pulp for animal bedding or paper production or mulch that is employed for landscaping. Additionally, wood waste can be developed and transformed into engineered products and used for energy generation [4]. Recycling wood waste entails the conversion of unwanted waste materials into relevant materials. According to [5], the volume of virgin wood required for harvesting from the forest was significantly reduced by using a single piece of wood subjected to a series of various products. Recycling wood waste into products will not only profit the world from an economic standpoint but also from an environmental perspective by eliminating the need to purchase raw materials and an increase in environmental awareness [6].

Recycling scrap wood is becoming more and more important due to increased disposal costs and environmental consciousness. Effective procedures and strategies for managing wood waste are created to benefit the environment, cut expenses, and provide new opportunities. This article thus critically reviews the treatment of wood waste prior to its further application and utilisation in the agricultural sector. The challenges and the perspective of the industry are also discussed.

## 2 Treatment of Wood Waste

Some wood wastes require a pretreatment process to degrade their biomass. Preliminary treatment processes are necessary to ensure the easy conversion of biomass from its initial resistance. This process is performed to ensure biomass conversion efficiency can be improved and make it easily degradable due to its complex physicochemical structure [7]. The complex biomass molecules are fragmented into cellulose, hemicellulose, and lignin during the preliminary treatment [7, 8]. Cellulose degradation assists to increase the surface area and facilitates the degradation process [7].

Pretreatment methods include physical, biological, chemical, physicochemical, and green solvent-based procedures. The waste is subjected to physical pretreatment processes to reduce its particle size and increase its surface area. An increase in biomass surface area can result in greater product output, decreased cellulose crystallinity, increased cellulose conversion efficiency, and increased accessibility of enzymes and microbes during processing [7].

The mechanical method is a type of physical pretreatment, which involves chopping, milling, chipping, shearing, and stirring the biomass. Although the technique is straightforward and inexpensive, it is expensive to implement [7].

The production of pellets from wood waste treatment and post-biomass collection is followed by storing the products at 22 °C and relative humidity of 65% in an air-conditioned chamber until the biomass is stabilised [9]. Prior to pyrolysis, wood residues and wood scraps produced by ripping and pruning are mechanically chopped into nearly similar lengths. It takes about two days of exposure to sunlight and air drying to eliminate the moisture from wood waste [8].

Biochar is another important byproduct synthesised from wood biomass. Biochar is a carbon-rich waste that is generated upon pyrolyzing several forms of biomasses [8, 10]. It has the capacity to amend the soil and improve the soil structure when applied as an aspect of agronomic management. Pyrolysis plays an important role in transforming organic materials into highly stabilised forms of carbon that could be employed as soil additives to enhance soil nutrient retention and carbon storage [11].

The processing of rubber wood sawdust entails an initial sieving, which is then dried for 24 h at 105 °C and incubated in a well-sealed container. The raw material was then dried using a lab-scale milling machine, ground into smaller particle sizes and sieved into 20 $\mu$m for sample preparation. Next, TGA (Mettler Toledo Star SW901) was applied to assess the thermal stability of raw RWSD in a nitrogen environment and 10 ml per min gas flow rate. The sample was subjected to heat at 5 °C/min ambient temperature to 700 °C [10].

## 3 Application of Wood Waste in Agriculture

### 3.1 Mulching

Mulch is a layer of plant residues or other debris that is spread over the soil's surface, either naturally or artificially [12]. By reducing water loss through evaporation, improving nutrient availability, minimising soil erosion, and increasing water and fertiliser efficiency, mulch assists in retaining soil moisture during drought periods. Mulching may also improve soil temperature, provide nutrients to the plant, and prevent the growth of weeds [13]. Organic and inorganic mulch are the two different types of mulch. Mulch that is organic or biodegradable is made of organic materials, whereas inorganic mulch is typically made of plastic-based products [14]. As the organic mulch decomposes over time, it enriches the soil with nutrients. Furthermore, organic mulch disintegrates and permeates the soil to improve organic components, thereby increasing the water-holding capacity of the soil.

Humaiza and Seran [12] studied how mulching materials affected crop growth and productivity. The experiment included the use of paddy straw, wood chips, and coir fibre. The highest performance among all indicators was reflected by coir mulch, followed by shredded wood mulch depicting higher yield and growth indicators for the crop. All of the mulching materials used in the study were leftovers that were obtained for free or at a very minimal cost. Therefore, the farmer should implement the mulching practice since it improves crop growth and productivity by retaining soil moisture and controlling weeds in the field. Furthermore, the use of organic mulch is significant given that it maintains soil moisture, moderates plant moisture status, increases soil temperature, and increases phytonutrient availability [15].

The recent research by Wei et al. [16] on the effects of mulching on the stability of soil aggregates and aggregate binding agents made from wood chips found that using wood chips and wood compost improved soil aggregate stability by increasing the amount of soil protein related to glomalin. Studies have shown that the utilisation of wood compost and wood chips is less expensive and more effective, whereas the sole use of wood chips had a less significant impact [17]. Feldmane [18] investigated the effect of wood chip mulch on sour cherry growth and first yield, and certain cultivars of sour cherries treated with wood chip mulch produced significantly higher yields. In addition, the outcome depicts that the crown volume increased significantly in the third year of growth. These findings are supported by Ni et al. [19] who investigated the impact of mulching on the growth of tea olives and soil properties. The authors documented that mulching was associated with better soil physiology and plant growth, thereby serving as a better treatment option compared to other treatments.

### 3.2 Soil Conditioner and Composting

The application of wood waste as biochar is acceptable in the agriculture sector. Wood biochar is used as one of the fertiliser components and can act as soil amendment [20]. Wood biochar is considered a good amendment for soil as it increases the availability of essential nutrients and persists longer in the soil. However, the effects depend on the soil conditions. Wood biochar is dark in colour and contains high carbon and organic matter, and it is synthesised in an environment with limited oxygen [21]. The process results in the conversion of biomass to solid—highly alkaline biochar. The chemical and physical properties of the material are altered into a stable, porous, and highly rich carbon material. For instance, the carbon content in some of the biochar is about 80% [22].

The carbon content in biochar is resistant to decomposition and thus contributes to the survival time in the soil. Wood biochar also has a porous structure which helps in ventilation and microorganisms' space. Wood biochar is also important as a hormone for plant growth due to its low chemical content and water-holding capacity [20]. Besides, wood biochar also helps in plant growth by supporting the beneficial rhizobacteria and fungi. Several studies have reported the ability of biochar in reducing greenhouse gas emissions, thus mitigating the effect on climate globally [23]. Wood biochar has also been described to be good in reducing crop disease by increasing resistance.

Çığ et al. [20] evaluated the impacts of applying a combination of biochar and plant growth-promoting rhizobacteria (PGPR). In this study, the individual application of biochar increased the fresh weight, dry weight, and the content of potassium (K), calcium (Ca), iron (Fe), sodium (Na), copper (Cu), zinc (Zn), manganese (Mn), and nickel (Ni) elements. Meanwhile, the combined application of PGCR and biochar had a significant effect on soil properties and plant growth.

Shetty and Prakash [21] recorded that the supplementation of biochar obtained from Eucalyptus wood at 20 tan ha was consistent in reducing soil acidity, as well as exchangeable and soluble aluminium in the soil. In addition, wood biochar demonstrated a significant effect in enhancing rice growth. Rice husk and bamboo biochar performed less compared to higher doses of Eucalyptus wood biochar. The researchers provided evidence to support the suitability of using wood biochar in reducing soil acidity and improving soil fertility. Reduction in soil acidity is highly promising if the alkalinity of the biochar is high.

Biochar made from Acacia bark waste and the whole tree was found able to increase the peanut and maize yields and the girth of apple tree [24, 25]. While papermill biochar was found able to increase radish and wheat biomass [26]. The application of pine sawdust biochar also led to a significant increase in the yield, growth, and quality of tomatoes [27].

Most studies on the impact of applying biochar to enhance soil fertility sparingly entailed a single application [28]. It is common to test the application by adding biochar to organic waste, which may be attributed to the capacity of such a mixture in reducing the loss of nutrients and increasing biological activities in the soil. The combination also can prevent organic waste's putrefaction thus reducing bad odours [29, 30]. Lehmann et al. [31] also mentioned that the combination of biochar and organic waste led to the build-up of the soil's organic carbon.

Applying biochar and compost, either alone or in combination, gives different responses to plants [32], resulting from the various amounts of macro and micronutrients in biochar. Nevertheless, the low amounts of available nutrients reflect the need to combine them with fertilisers for a better result. The two main approaches involved loading the fertilisers in a liquid form into the biochar's pore structure or mixed in a solid state. Furthermore, by storing nutrients, biochar can play the role of a slow-release fertiliser [33]. The rate of nutrient release increases as the size of the biochar pores increases and vice versa [34].

The source of biochar also give different effects on plant growth. A study by Dai et al. [35] found that crop residues, straw, and manure biochar had better effects compared to wood biochar. This however contrasts with Shetty and Prakash [21] where wood biochar yielded the best performance in evaluating the various types of biochar application effects. Biochar from hardwood contains potassium (K), forestry residues contain phosphorus (P) and potassium (K) and timber residues contain phosphorus (P) and magnesium (Mg). Besides the source, the mode of mixing biochar with fertiliser also suggested a contrasting result. A separate application and mixing of the biochar with liquid and solid fertiliser are among the mode [28]. Different results are also obtained when using different application methods. All the disparity in results depends on the biochar's physical and chemical properties.

According to Lin et al. [36], humics were present in biochars made from Acacia saligna sawdust. Additionally, Cantrell et al. [37] found that the ash content in biochar formed from the chicken litter was substantially higher (30.7%) than that in biochar made from pine wood chips (1.5%) [38].

The alkalinity in biochar synthesised at higher temperatures is higher with greater water-holding capacity and electric conductivity. These features lead to an increase in root growth but the latter may not necessarily translate into high plant yield [28]. Apart from enhancing physical and chemical properties, biochar had a significant impact by improving microbial properties. The performance of microbial and enzyme activity and nitrogen fixation is better with the application of wood biochar [28]. For instance, the soil microorganism population significantly increased following the application of wood biochar, thereby promoting nitrogen fixation. *Eucalyptus*, pine bark, and sawdust biochar have an alkaline pH and high carbon content due to the presence of lignin [39].

The water availability for plants is increased based on the direct role of wood biochar's porosity in enhancing the water-holding capacity of the soil. Studies have shown an increment of water-holding capacity in coarse-textured soils and medium-textured soils. However, biochar had no significant effect in fine-textured soils. Improvement in water-holding capacity will also lead to the reduction in water and nutrient leaching, thus assisting plant growth in limited water supply areas. Minimising nutrient leaching will also help will reducing water pollution.

Wood biochar also has been used in water and wastewater treatment. It also has been applied in contaminated areas for the remediation process [40, 41]. Vikrant et al. [42] mentioned that biochar was once used to treat phosphorus toxicity in the aquatic environment. Meanwhile, Kończak and Oleszczuk [43] assessed the impact of modified and unmodified soil and sewage on toxicity rates. Resultantly, the supplementation of biochar in soil and sewage sludge reduced the toxicity rate significantly. Dad et al. [44] also reported that treatment with biochar with $FeCl_3$ and $FeSO_4$ led to a significant reduction in cadmium (Cd) toxicity in contaminated soils and radish plants.

Besides biochar, wood waste is also commonly used as compost in the agriculture sector to supply nutrients for plant growth and crop yield. Only wood waste from forests and technological processes and free from any chemicals are used for composting [45]. According to the study, wood waste can be utilised by composting, and the resulting composts are distinguished by having a high level of organic matter and a low amount of heavy metal. McMahon et al. [46] documented that timber waste has been composted using different approaches. While investigating the suitable soil compost, González et al. [47] discovered that chips, bark, and sawdust from three Populus, Eucalyptus, and Salicaceae species, mud from the paper industry, hardboard dust, and raw cork compost are beneficial and compatible with the soil.

Post-consumer wood is also being composted and tested on chrysanthemum growth and flowering [45]. The treatment was a combination of the post-consumer wood compost with microbiological inoculums. The study depicted the effectiveness of the size and number of leaves, which is consistent with the reports by Brito et al. [48] in which positive effects were observed following the addition of acacia compost.

### *3.3 Poultry Bedding*

Many developing nations and rural regions' economies are significantly boosted by poultry, specifically in providing food [49]. In 2013, the production of poultry meat amounted to 104 million tonnes of global meat production annually. The statistics demonstrate the important role of the poultry business in global agriculture [50]. The population of hens increased by 4.5 times, the number of pigs by 2.5 times, and the number of other significant meat-producing animals (such as cattle, sheep, and goats) by 1.5 times in the 50 years prior to the year 2010 [51].

Wood is a renewable and natural resource with unique properties such as effective control of humidity and preventing the growth and proliferation of microorganisms. Wood residues are frequently utilised as poultry litter in several countries. Wood wastes refer "to a group of wastes encompassing wasted wood products from several industries, comprising the wood processing industry, the railroad construction business, construction and demolition, and wood manufacturing" [1]. Wood wastes are sources of diverse products, such as engineered wood products, mulching, energy production (heat and electricity), and animal bedding [4].

A crucial component of poultry production is litter or bedding material that could impact animal welfare, flock well-being, food safety, impact on the environment, and production efficiency [52]. For example, 80% of turkey production in France uses bedding made of wood. Compared to hulls and other materials like sand, this litter produces approximately 50% less NH3, and it has a variety of environmentally friendly disposal options [53]. Wood bedding can reduce the negative environmental impacts on livestock production by absorbing excess moisture and reducing ammonia (NH3) emissions, thus improving animal comfort [54].

Bedding material is a crucial element of floor-based poultry production methods to meet health and welfare standards. The substrate can be either organic or inorganic. Good absorbency, accessibility, comfort, and non-toxicity to birds should be considered as features. For a product to be called litter material, it must fulfil the following requirements: it must be simple to obtain, inexpensive, absorbent, devoid of dust and pollutants, low heat conductivity, and not compact. A good litter material should have the quickest feasible ability to release moisture into the surroundings after absorbing it [55]. It is easy, inexpensive, and absorbent to use wood chips as animal bedding, creating a hygienic, warm, and dust-free habitat [56].

Research for substitute poultry litter materials has increased in the last few decades. Various substitutes for wood by-products have been used that reflect different results for birds' performance and well-being [57]. However, wood waste was still preferred as bedding material for poultry. This statement was supported by the characteristic of wood shaving that had high absorbency, highly adsorbent litter substance, and also enhances welfare through foraging and dust bathing [58, 59].

Šafarič et al. [60] conducted a study to create composite fiberboards by combining wood by-products such as wood shaving with residual feathers. The study disclosed that the type of wood residue and the method of processing had a significant impact on whether the fibreboard samples had a fine or coarse texture. The homogeneity, appearance, and dense structure were higher in the samples synthesised from mixed wood wastes relative to those fabricated using wood shavings, with the latter presenting a rougher makeup. This study found that feathers from poultry and wood scraps were combined to create composite fiberboards, which have excellent thermal insulating properties and are biodegradable. This approach may be appealing to use a significant amount of poultry feather waste for poultry bedding since it combines poultry feathers (approximately 20–70%) with wood wastes in the manufacture of fiberboards.

## 4 Challenges of Wood Waste in Agriculture

Wood is one of the preferred materials for environmentally friendly and environmentally safe building and product manufacture where it gives a great recyclability quality, which may be achieved and applied by cascading usage and introducing sporadic product lifespan [61]. The application of wood as a product produces a high volume of wood waste. If the wood waste is not handled effectively, it can impact negatively the environment. To solve this problem, wood waste can be recycled or reused as other products. Many typical wood wastes may be converted into a variety of commercial and industrial items that can be used in various fields such as energy generation and agriculture. In agriculture, wood waste can be used as mulching, animal bedding, and as organic fertiliser [4]. Such usage impacts positively the performance of plants and animals. Unfortunately, wood waste cannot become the primary source to be used in agriculture despite its exceptional benefits in this sector. There are several challenges in the utilisation of wood waste in agriculture.

The first challenge is in the agriculture sector itself. Wood waste can be used as an organic amendment for crop development in the soil through composting where it can provide nutrients, retain moisture content, increase biological activity in the soil, and repair the physical and chemical characteristics of the soil. These are all important criteria that can help in plant growth and development. Retaining moisture or increasing water-holding capacity is one of the numerous benefits of compost. Without compost, the soil is unable to hold water for a long period and causes water stress, which could affect the physiology and production of the crops [62]. The decomposition of wood waste is feasible but it relies on the type or structure of the wood [63]. While composting of wood waste can be used for crops, several wood wastes from woody tree parts take a long time to decompose compared to other sources. Wood chipping takes around 3–4 years to break down to become a good compost from the decomposition process and is highly time-consuming before it can be used as organic fertiliser. One of the solutions to increase the composting process is by adding or mixing the wood waste with other organic matter and adding beneficial microbes to increase the speed of the composting process. Another method is to increase the composting process by shredding the wood into small pieces. The challenges in composting wood waste could be addressed by applying biochar from wood waste. This problem arises from the loss of important nutrients such as C, H, N, O, and S in the process of pyrolysis. Meanwhile, biochar application preserves K or P and other metals in comparison to other composts containing highly rich nutrient materials and high N content [64].

Another issue related to composting wood waste is the usage of treated wood as a source of compost. Treated woods are forbidden to be used as composting material. Examples of treated wood materials that cannot be composted are wooden furniture, plywood, hardwood, particleboard, and fibrewood. This treated wood was treated by synthetic bonding/binding agents such as resin and or preservative chemicals to maintain their characteristics and as a result, should never be used as a compost heap where it can affect the environment [63, 65]. Treated wood also contains several heavy metals such as Copper (CU) and Zinc (Zn) that can affect the environment. Excess heavy metals can cause pollution to the environment and harm living organisms. In crop production, heavy metal toxicity can disturb the growth and development of crops by causing chlorosis, low biomass heap, inhibition of growth and photosynthesis, altered water balance and nutrient acclimatisation, and senescence, which causes plant death [66]. Research on the application of treated wood compost (i.e., wood chips) revealed the presence of exceeded levels of Cu above the threshold level in soil and plant—a bad indicator for living organisms and the environment [67]. Chemical preservatives including sodium borate and copper chrome arsenate that are used in wood shading and sawdust for bedding materials of the poultry industry may be retained in the tissue of the poultry and pose health risks to human consumers [68].

Other usages of wood waste from other sectors also reduce the opportunity of fully utilising wood waste in the agriculture sector. Nowadays, wood waste is used broadly in energy production. The usage of wood waste as alternative energy production is promising in renewable energy to replace the limited resources of fossil oil. A report in 2021 depicted that approximately 5.2% of wood and wood waste was used in industrial end-use energy utilisation and more than 4% of total industrial energy utilisation [69]. AgMRC from Iowa State University [70] in the United States reported that more than 84% of total energy production was obtained from wood waste in which more than 70% of wood waste power is co-produced with industrial process heat. On the other hand, 88% of the power generated in wood-fired systems was obtained from sawdust accounts, incorporating the utilisation of shop-produced wood wastes or shavings. Production of furniture is also one of the sectors that implements wood waste as a primary source in their production. Wood waste is widely used as plywood and particle board in the furniture industry. Usually, mills sell their wood waste in two forms that are sawdust and mixed residue, and all the residues are used to generate additional income. Approximately 74% of the mills in Kedah traded their wood residue to composite manufacturers, paper companies, and bulk companies, thus generating around 17,000 Ringgit in Malaysia monthly [71].

Despite the current challenges of using wood waste in the agriculture sector as the main product, the agricultural industry could benefit tremendously from the application of wood waste without interrupting its usage by other sectors. This is achievable with good management and strategy for handling wood waste in the future.

## 5 Conclusion

The study has summarised the benefits and application of wood waste in the agriculture sector. Wood waste has several potential applications ranging from renewable technology to construction materials, the furniture industry, and agriculture. In agriculture, wood waste is widely used as an organic amendment in soil, poultry bedding, and mulching. The management of the wood waste application is important to ensure that it is not considered waste without any benefits to humans and the environment. However, there are several challenges in the application of wood waste that need to be addressed. In the future, robust and thorough quality research should be conducted to ensure the full utilisation of waste application in the agriculture sector. Data obtained from such studies will facilitate a better understanding of how to effectively manage wood waste in the agricultural sector.

## References

1. Van Benthem M, Leek N, Mantau U, Weimar H (2007) Markets for recovered wood in Europe; case studies for the Netherlands and Germany based on the BioXchange project. In: Proceedings of 3rd European COST E31 conference, pp 1–12
2. Rizzo C (2010) Wood waste derived fuel: state of the art and development prospects in France focus on construction and demolition wood wastes. [Master of Science Thesis, Royal Institute of Technology, Stockholm]
3. Zafar S (2022) A glance at woody biomass resources. Bioenergy Consult. https://www.bioenergyconsult.com/tag/wood-wastes/
4. Pandey S (2022) Wood waste utilization and associated product development from under-utilized low-quality wood and its prospects in Nepal. SN Appl Sci 4(6):1–8. https://doi.org/10.1007/s42452-022-05061-5
5. World Wildlife Fund WWF (2016) Reuse and recycling. http://forestsolutions.panda.org/solutions/reuse-and-recycling
6. Hartini S, Wicaksono PA, Rizal AMD, Hamdi M (2021) Integration lean manufacturing and 6R to reduce wood waste in furniture company toward circular economy. In: IOP conference series: materials science and engineering, vol 1072, no 1. IOP Publishing, p 012067
7. Awogbemi O, Von Kallon DV (2022) Pretreatment techniques for agricultural waste. Case Stud Chem Environ Eng 100229. https://doi.org/10.1016/j.cscee.2022.100229

8. Wijitkosum S (2022) Biochar derived from agricultural wastes and wood residues for sustainable agricultural and environmental applications. Int Soil Water Conserv Res 10(2):335–341. https://doi.org/10.1016/j.iswcr.2021.09.006
9. Brand MA, Rodrigues TM, da Silva JP, de Oliveira J (2021) Recovery of agricultural and wood wastes: the effect of biomass blends on the quality of pellets. Fuel 284:118881. https://doi.org/10.1016/j.fuel.2020.118881
10. Shaaban A, Sea SM, Mitan NMM, Dimin MF (2013) Characterization of biochar derived from rubber wood sawdust through slow pyrolysis on surface porosities and functional groups. Procedia Eng 68:365–371. https://doi.org/10.1016/j.proeng.2013.12.193
11. Krull ES, Baldock JA, Skjemstad JO, Smemik RJ (2009) Characteristic of biochar: organo-chemical properties. In: Biochar for environment science and technology. Eruihsan Publication Ltd, UK, p 53
12. Humaiza AJM, Seran TH (2022) Effect of irrigation frequency and mulching material on growth and yield of radish (Raphaus sativus L.). J Food Agric 16–36. https://doi.org/10.4038/jfa.v15i1.5265
13. Subrata M, Subham G, Staya NG (2020) Effects of mulching and supplemented irrigation on growth, yield and fruit quality of wood apple (Feronia Iimonia Swingle). Int J Curr Microbiol Appl Sci 9(10):2479–2484. https://doi.org/10.20546/ijcmas.2020.910.297
14. Kader MA, Senge M, Mojid MA, Ito K (2017) Recent advances in mulching materials and method for modifying soil environment. Soil Tillage Res 168:155–166. https://doi.org/10.1016/j.still.2017.01.001
15. Huang GB, Zhang RZ, Li GD, Li LL, Chan KY, Heenan DP, Chen W, Unkovich MJ, Robertson MJ, Cullis BR, Bellotti WD (2008) Productivity and sustainability of a spring wheat-field pea rotation in a semi-arid environment under conventional and conservation tillage systems. Field Crop Res 107(1):43–55. https://doi.org/10.1016/j.fcr.2007.12.011
16. Wei Z, Xiangyang S, Suyan L, Tiantian D, Zhihui F (2022) Effects of organic mulching on soil aggregate stability and aggregate binding agents in an urban forest in Beijing China. J For Res 33:1083–1094. https://doi.org/10.1007/s11676-021-01402-z
17. Scharenbroch BC, Watson GW (2014) Wood chips and compost improve soil quality and increase growth of Acer rubrum and Betula nigra in compacted urban soil. Arboric Urban For 40(6):319–331
18. Feldmane D (2010) The influence of drip irrigation and woodchip mulch on growth and first yield of sour cherries. Agron Res 8(Special Issue II):453–458
19. Ni X, Song W, Zhang H, Yang X, Wang L (2016) Effects of mulching on soil properties and growth of tea olive (Osmanthus fragrans). Plos One 11(8):e0158228. https://doi.org/10.1371/journal.pone.0158228
20. Çığ F, Sönmez F, Nadeem MA, Sabagh AE (2021) Effect of biochar and PGPR on the growth and nutrients content of einkorn wheat (Triticum monococcum L.) and post-harvest soil properties. Agronomy 11(12):2418. https://doi.org/10.3390/agronomy11122418
21. Shetty R, Prakash NB (2020) Effect of different biochars on acid soil and growth parameters of rice plants under aluminium toxicity. Sci Rep 10910:1–10. https://doi.org/10.1038/s41598-020-69262-x
22. Haque MI, Khan RI, Ashraf W, Pendse H (2021) Production of sustainable, low-permeable and self-sensing cementitious composites using biochar. Sustain Mater Technol 28:e00279. https://doi.org/10.1016/j.susmat.2021.e00279
23. Ndoung OCN, de Figueiredo CC, Ramos MLG (2021) A scoping review on biochar-based fertilizers: enrichment techniques and agro-environmental application. Heliyon 7(12):e08473. https://doi.org/10.1016/j.heliyon.2021.e08473
24. Yamato M, Okimori Y, Wibowo IF, Anshori S, Ogawa M (2006) Effects of the application of charred bark of Acacia mangium on the yield of maize, cowpea and peanut and soil chemical properties in South Sumatra, Indonesia. Soil Sci Plant Nutr 52(4):489–495. https://doi.org/10.1111/j.1747-0765.2006.00065.x
25. Eyles A, Bound SA, Oliver G, Corkrey R, Hardie M, Green S, Close DC (2015) Impact of biochar amendment on the growth, physiology and fruit of a young commercial apple orchard. Trees 29(6):1817–1826. https://doi.org/10.1007/s00468-015-1263-7

26. Van Zwieten L, Kimber S, Morris S, Chan K, Downie A, Rust J, Joseph S, Cowie A (2010) Effects of biochar from slow pyrolysis of papermill waste on agronomic performance and soil fertility. Plant Soil 327(1):235–246. https://doi.org/10.1007/s11104-009-0050-x
27. Dunlop SJ, Arbestain MC, Bishop PA, Wargent JJ (2015) Closing the loop: use of biochar produced from tomato crop green waste as a substrate for soilless, hydroponic tomato production. Hort Sci 50(10):1572–1581. https://doi.org/10.21273/HORTSCI.50.10.1572
28. Schmidt HP, Kammann C, Hagemann N, Leifeld J, Bucheli TD, Sánchez Monedero MA, Cayuela ML (2021) Biochar in agriculture—A systematic review of 26 global meta-analyses. GCB Bioenergy 13(11):1708–1730. https://doi.org/10.1111/gcbb.12889
29. Bezerra J, Turnhout E, Vasquez IM, Rittl TF, Arts B, Kuyper TW (2019) The promises of the Amazonian soil: shifts in discourses of Terra Preta and biochar. J Environ Planning Policy Manag 21(5):623–635. https://doi.org/10.1080/1523908X.2016.1269644
30. Frausin V, Fraser JA, Narmah W, Lahai MK, Winnebah TRA, Fairhead J, Leach M (2014) "God made the soil, but we made it fertile": gender, knowledge, and practice in the formation and use of African dark earths in Liberia and Sierra Leone. Hum Ecol 42(5):695–710. https://doi.org/10.1007/s10745-014-9686-0
31. Lehmann J, Hansel CM, Kaiser C, Kleber M, Maher K, Manzoni S, Nunan N, Reichstein M, Schimel JP, Torn MS, Wieder WR, Kogel-Knabner I (2020) Persistence of soil organic carbon caused by functional complexity. Nat Geosci 13(8):529–534. https://doi.org/10.1038/s41561-020-0612-3
32. Agegnehu G, Srivastavab AK, Michael IB (2017) The role of biochar and biochar-compost in improving soil quality and crop performance: a review. Appl Soil Ecol 119:156–170. https://doi.org/10.1016/j.apsoil.2017.06.008
33. Ding Y, Liu Y, Liu S, Li Z, Tan X, Huang X, Zeng G, Zhou L, Zheng B (2016) Biochar to improve soil fertility A review. Agron Sustain Dev 36(2):1–18. https://doi.org/10.1007/s13593-016-0372-z
34. An X, Wu Z, Yu J, Cravotto G, Liu X, Li Q, Yu B (2020) Copyrolysis of biomass, bentonite, and nutrients as a new strategy for the synthesis of improved biochar-based slow-release fertilizers. ACS Sustain Chem Eng 8(8):3181–3190. https://doi.org/10.1021/acssuschemeng.9b06483
35. Dai Y, Zheng H, Jiang Z, Xing B (2020) Combined effects of biochar properties and soil conditions on plant growth: a meta-analysis. Sci Total Environ 713:136635. https://doi.org/10.1016/j.scitotenv.2020.136635
36. Lin Y, Munroe P, Joseph S, Henderson R, Ziolkowski A (2012) Water extractable organic carbon in untreated and chemical treated biochars. Chemosphere 87(2):151–157. https://doi.org/10.1016/j.chemosphere.2011.12.007
37. Cantrell KB, Hunt PG, Uchimiya M, Novak JM, Ro KS (2012) Impact of pyrolysis temperature and manure source on physicochemical characteristics of biochar. Bioresour Technol 107:419–428. https://doi.org/10.1016/j.biortech.2011.11.084
38. Spokas KA, Novak JM, Stewart CE, Cantrell KB, Uchimiya M, DuSaire MG, Ro KS (2011) Qualitative analysis of volatile organic compounds on biochar. Chemosphere 85(5):869–882. https://doi.org/10.1016/j.chemosphere.2011.06.108
39. Domingues RR, Trugilho PF, Silva CA, de Melo ICN, Melo LC, Magriotis ZM, Sanchez-Monedero MA (2017) Properties of biochar derived from wood and high nutrient biomasses with the aim of agronomic and environmental benefits. Plos One 12(5):e0176884
40. Czekała W, Jeżowska A, Chełkowski D (2019) The use of biochar for the production of organic fertilizers. J Ecol Eng 20(1). https://doi.org/10.12911/22998993/93869
41. Gao Y, Fang Z, Van Zwieten L, Bolan N, Dong D, Quin BF, Meng J, Li F, Wu F, Wang H, Chen W (2022) A critical review of biochar-based nitrogen fertilizers and their effects on crop production and the environment. Biochar 4(1):1–19. https://doi.org/10.1007/s42773-022-001 60-3
42. Vikrant K, Kim KH, Ok YS, Tsang DC, Tsang YF, Giri BS, Singh RS (2018) Engineered/designer biochar for the removal of phosphate in water and wastewater. Sci Total Environ 616:1242–1260. https://doi.org/10.1016/j.scitotenv.2017.10.193

43. Kończak M, Oleszczuk P (2018) Application of biochar to sewage sludge reduces toxicity and improve organisms growth in sewage sludge-amended soil in long term field experiment. Sci Total Environ 625:8–15. https://doi.org/10.1016/j.scitotenv.2017.12.118
44. Dad FP, Khan WUD, Tanveer M, Ramzani PMA, Shaukat R, Muktadir A (2020) Influence of iron-enriched biochar on Cd sorption, its ionic concentration and redox regulation of radish under cadmium toxicity. Agriculture 11(1):1. https://doi.org/10.3390/agriculture11010001
45. Schroeter-Zakrzewska A, Komorowicz M (2022) The use of compost from post-consumer wood waste containing microbiological inoculums on growth and flowering of chrysanthemum (Chrysanthemum × grandiflorum Ramat./Kitam.). Agronomy 12(6):1274. https://doi.org/10.3390/agronomy12061274
46. McMahon V, Garg A, Aldred D, Hobbs G, Smith R, Tothill IE (2008) Composting and bioremediation process evaluation of wood waste materials generated from the construction and demolition industry. Chemosphere 71(9):1617–1628. https://doi.org/10.1016/j.chemosphere.2008.01.031
47. González J, Del Pardo K, Martín S (2009) Wood waste characterization for composting. Acta Hortic 843:337–342
48. Brito LM, Reis M, Mourão I, Coutinho J (2015) Use of acacia waste compost as an alternative component for horticultural substrates. Commun Soil Sci Plant Anal 46(14):1814–1826. https://doi.org/10.1080/00103624.2015.1059843
49. Demeke S (2004) Egg production performance of local and White Leghorn hens under intensive and rural household conditions in Ethiopia. Livest Res Rural Dev 16(2):2004
50. Daghir NJ (2009) Nutritional strategies to reduce heat stress in broilers and broiler breeders. Lohmann Inf 44(1):6–15
51. Godfray HCJ, Beddington JR, Crute IR, Haddad L, Lawrence D, Muir JF, Pretty J, Robinson S, Thomas SM, Toulmin C (2010) Food security: the challenge of feeding 9 billion people. Science 327(5967):812–818. https://doi.org/10.1126/science.1185383
52. Dunlop MW, Mcauley J, Blackall PJ, Stuetz RM (2016) Water activity of poultry litter: Relationship to moisture content during a grow-out. J Environ Manag 172:201–206
53. Munir MT, Belloncle C, Irle M, Federighi M (2019) Wood-based litter in poultry production: a review. Worlds Poult Sci J 75(1):5–16
54. Ahn GC, Jang SS, Lee KY, Baek YC, Oh YK, Park KK (2020) Characteristics of sawdust, wood shavings and their mixture from different pine species as bedding materials for Hanwoo cattle. Asian Australas J Anim Sci 33(5):856
55. Garcês APJT, Afonso SMS, Chilundo A, Jairoce CTS (2013) Evaluation of different litter materials for broiler production in a hot and humid environment: 1. Litter characteristics and quality. J Appl Poultry Res 22(2):168–176
56. Panivivat R, Kegley EB, Pennington JA, Kellogg DW, Krumpelman SL (2004) Growth performance and health of dairy calves bedded with different types of materials. J Dairy Sci 87(11):3736–3745
57. Almeida Paz ICL, Garcia RG, Bernardi R, Nääs IA, Caldara FR, Freitas LW (2010) Selecting appropriate bedding to reduce locomotion problems in broilers. Brazil J Poultry Sci 12:189–195. https://doi.org/10.1590/S1516-635X2010000300008
58. Embury IS (2004) Alternative litter materials for poultry. New South Wales Agriculture, Agfact
59. Charles EB (2005) Litter management for confined turkeys. Poultry Sci Tech Guide 41:3–7
60. Šafarič R, Fras Zemljič L, Novak M, Dugonik B, Bratina B, Gubeljak N, Bolka S, Strnad S (2020) Preparation and characterisation of waste poultry feathers composite fibreboards. Materials 13(21):4964
61. Laleicke P (2018) Wood waste, the challenges of communication and innovation. BioRes 13(2):2182–2183
62. Khairi M, Naqib SA, Hasan MM, Jahan MS (2017) Low water input confers sustainable rice production without affecting soil, plant physiological and yields parameters. AJCS 11(08):1068–1077. https://doi.org/10.21475/ajcs.17.11.08.pne653
63. Mokhtar N, Razali SM, Sulaiman MS, Edin T, Wahab R (2022) Converting wood-related waste materials into other value added products: a short review. IOP Conf Ser: Earth Environ Sci 1053:012030. https://doi.org/10.1088/1755-1315/1053/1/012030

64. Siedt M, Schäffer A, Smith KEC, Nabel M, Roß-Nickoll MT, van Dongen J (2021) Comparing straw, compost, and biochar regarding their suitability as agricultural soil amendments to affect soil structure, nutrient leaching, microbial communities, and the fate of pesticides. Sci Total Environ 751:141607. https://doi.org/10.1016/j.scitotenv.2020.141607
65. Dinis MAP (2011) Composting of wood waste materials generated from the Furniture Industry in Portugal. In: Minimizing the environmental impact of the forest products industries, pp 77–81
66. Singh S, Parihar P, Singh R, Singh VP, Prasad SM (2016) Heavy metal tolerance in plants: role of transcriptomics, proteomics, metabolomics, and Ionomics. Front Plant Sci 6:1143. https://doi.org/10.3389/fpls.2015.01143
67. Kamal A, Wichuk K, McCartney D, Londry K, Felske C (2017) Effect of treated wood on biosolids composting. Compost Sci Utiliz. https://doi.org/10.1080/1065657X.2016.1268937
68. Diarra S, Lameta S, Amosa F, Anand S (2021) Alternative bedding materials for poultry: availability, efficacy, and major constraints. Front Vet Sci 8:669504. https://doi.org/10.3389/fvets.2021.66950
69. EIA, Biomass explained: Wood and wood waste (2022). https://www.eia.gov/energyexplained/biomass/wood-and-wood-waste.php. Accessed 01 July 2022
70. AgMRC, Sawdust (2022). https://www.agmrc.org/commodities-products/biomass/sawdust. Accessed 03 July 2022
71. Shafie SM, Othman Z, Hami N (2017) Potential utilisation of wood residue in Kedah: a preliminary. J Technol Oper Manag (Special Issue):60–69

# Potential Use of Residual sawdust—A Versatile, Inexpensive and Readily Available Bio-waste

**Mohd Hazim Mohamad Amini**

**Abstract** Biomass commonly comes from plants or lignocellulosic materials. The main chemical components in plant biomass include cellulose, hemicellulose, lignin and extractives. According to their size and geometry, small-sized woody biomass is called different names. Each form has its suitable application. Larger particles are ideal for standard wood composites such as the oriented strand board and particleboard. Medium-sized particles are ideal for energy pellet, pulp and papermaking applications. Meanwhile, smaller particles are suitable for applications involving chemical reactions such as nanocellulose extraction, liquefaction and bioethanol production. Wood plastic composites usually utilise medium to small-sized particles. This chapter will introduce the sources and various applications of these small woody biomass materials.

## 1 Sources of Material

### *1.1 Industrial Wood Cutting Waste*

There are many sources of biomass particles and sawdust here that could be obtained. It may come from wood processing facilities to make solid wood planks and poles after the cutting process. It was estimated that for every 1000 board feet of lumber produced, 1 tonne of by-products would be created. Woody biomass in the form of sawdust, shavings, slabs and edgings is gathered in sawmills comprising 75% wood and 25% bark portion [1]. Some names are wood fibre, wood flakes, wood chips, wood powder and wood swiths, as shown in Fig. 1.

M. H. Mohamad Amini (✉)
Faculty of Bioengineering and Technology, Universiti Malaysia Kelantan, Jeli Campus, 17600 Jeli, Kelantan, Malaysia
e-mail: hazim.ma@umk.edu.my

 
S. N. Sarmin et al. (eds.), *Wood Waste Management and Products*, Sustainable Materials and Technology, https://doi.org/10.1007/978-981-99-1905-5_11

**Fig. 1** Different forms of woody biomass [1]

### *1.2 Manufactured Particles*

Biomass sawdust can be classified into wood and non-wood biomass material. Biomass particles may be obtained by intentionally reducing the biomass size for specific usage. Wood and non-wood biomass from the plant parts, including logs, branches, tree stumps or stems, were purposely chipped or ground into a smaller size. Certain types of trees such as the Kenaf, Jute, Flax and Hemp are planted for their fibres, called fibre crops. These plants are considered non-woody biomass materials.

Size reduction of biomass could be made by chopping, chipping, grinding, and milling to perform solid disintegration. After chipping, the grinding process will break the wood into smaller particles through mechanical stress: impact, compression, shear and attrition. Intercellular and intracellular failure will loosen the fibre bonding, making smaller wood particles. Different grinding mills can be utilised, including; impact mills, roller mills, ball media mills, air jets mills and shearing attrition mills [2].

## 2 Energy Production

The most basic usage of sawdust is for energy production. Using wood-based materials for energy production has a tremendous environmental advantage. Energy generation using woody materials promotes the carbon cycle. When woody material is burned, it releases energy and carbon dioxide. The carbon dioxide from the environment is deposited back into plants during photosynthesis. The tree will be cut down and again generates carbon dioxide before being absorbed by other trees. As new trees grow, the cycle repeats over and over; that is the carbon cycle [3]. It differs from petroleum fuel burning, where carbon release happens one way without recapturing. Energy generation from wood sawdust can be done through 3 main categories, which are:

i. *Direct burning*
   Energy generation by directly burning forest biomass is a part of the global carbon cycle. It is considered carbon neutral because the same amount of $CO_2$ is released and captured in tree burning and growth [4]. Sawdust may be burned in its original powder form or compressed into pellets and briquettes for easy transportation and handling [5].
ii. *Heat treatment*
   Heat treatment was done to wood sawdust to turn it into charcoal and activated carbon. Pyrolysis involves the thermal degradation of biomass by heat in the absence or limitation of oxygen. Charcoal (solid), bio-oil (liquid) and fuel (gas) products are the output of this process [6]. The solid product, char or charcoal, is commonly compacted into pellets and briquettes for energy storage. Pyrolysis was usually done by heating the sawdust between 300 and 700 °C [7]. Heat treatment at a temperature above 700 °C is called the activation process, producing activated carbon. Considering the high cost of making activated carbon, pyrolysis temperature is enough for energy storage char production.
iii. *Bioethanol production*
   For bioethanol production, three significant steps should be conducted to the sawdust:

   a. Raw material pre-treatment
   b. Hydrolysis of pre-treated raw material
   c. Fermentation of hydrolysed material [8].

Sawdust is a lignocellulosic material which contains lignin, cellulose and hemicellulose. Pre-treatment of sawdust is needed to delignify the material. Delignification is essential to weaken the bond between cellulose and hemicellulose. It is a crucial step to ensure the success of the hydrolysis process. Pre-treatment techniques can be grouped into four categories, including acid pre-treatment (dilute acid, steam explosion, organosolv), neutral pre-treatment (liquid hot water), alkaline pre-treatment (sodium hydroxide, ammonia fibre expansion, soaking in aqueous ammonia) and ozonolysis [9]. Hydrolysis of cellulose and hemicellulose will produce fermentable sugars, including glucose, xylose, arabinose, galactose and mannose. Lastly, fermentation of reducing sugars will produce bioethanol [10].

## 3 Composite

Sawdust is a very suitable material for composite. Because of its tiny particle size, it can be mixed uniformly with polymer and adhesives. Different kinds of wood composite, such as particleboard, fibreboard and wood plastic composite, can be produced using sawdust as the main component.

Particleboard and fibreboard are made from mixtures of wood particles and fibres with adhesive or binder, respectively. Particleboard and fibreboard save a lot of wood waste, turning them into valuable, solid wood alternatives. Particleboard is prepared by mixing sawdust with 6–12% loading of thermosetting adhesives such as urea-formaldehyde (UF), melamine-urea-formaldehyde (MUF), phenolic resins (PF and TF), and isocyanates (pMDI). Larger particles are commonly placed at the core, while smaller particles are placed at the surface. Mixed resin-sawdust are flat-pressed at a high temperature between 100 and 180 °C for the resin to cure, forming a flat panel. The density of commercial particleboard is commonly set at 650–700 $kg/m^3$.

Fibreboard utilises wood fibre which is produced with the pulping process. Thermochemical pulping is usually employed to separate wood particles into individual fibres. Separated fibres were mixed with 2–3% of thermosetting binders and flat-pressed at a high temperature to cure. Fibreboards are categorised according to their density which is Medium Density Fibreboard, MDF (750–800 $kg/m^3$) and Hardboard (900–1100 $kg/m^3$) [11].

Wood plastic composite is the combination of wood and plastic or polymer. The thermosetting or thermoplastic polymer can be used as the matrix, the material that encapsulates the wood particles and holds them together. Thermosetting polymers are commonly available in liquid form. A polymer such as epoxy, unsaturated polyester and polyurethane can be directly mixed with sawdust using a simple mixer. Mixtures were laid into the mould, degassed using a vacuum bag and left to harden. An example of thermosetting wood plastic composite is the sisal fibre, and wood flour composites were made using unsaturated polyester thermosets by Marcovich et al. [12].

To produce wood plastic composite using thermoplastic as the matrix, the processes that might be involved include extrusion, injection moulding, and compression moulding or thermoforming (pressing) [13]. The plastic and wood were heated and blended when the thermoplastic melted. The sawdust-melted polymer mixture is then shaped before cooled down to harden. Wood plastic composite has many advantages, especially in automotive industries. The presence of sawdust in plastic composite increases the mechanical strength and reduces the material weight of automotive parts. Therefore it causes a reduction in vehicle fuel consumption, lowers production cost, increases passenger safety and improves the vehicle's interior parts' biodegradability [14]. Some wood composites made using sawdust are tabulated in Table 1.

## 4 Animal Feed

Sawdust is originally a non-edible substance for animals. Sawdust for animal feed is possible with pre-treatments required to make the produced animal feed pellet digestible. An example of research on sawdust as fish feed has been done by Sharma et al. [25]. A blend of protein-rich yeast *Candida utilis*, enzymatically hydrolysed

**Table 1** Wood composite based on sawdust

| No | Material | Bending strength | Tensile strength | References |
|---|---|---|---|---|
| 1. | Corncob and sawdust particle board using urea-formaldehyde as binder | 82.555 $N/mm^2$ | NA | Akinyemi et al. [15] |
| 2. | Wood sawdust-based particle board using ionic liquid-facilitated fusion process | NA | 10.4 $N/mm^2$ | Orelma Tanaka [16] |
| 3. | Hybrid particleboards using coconut fibre and sawdust | 17.8 $N/mm^2$ | NA | Tawasil et al. [17] |
| 4. | Sugarcane bagasse particleboard | 11.44 $N/mm^2$ | NA | Yano et al. [18] |
| 5. | Particleboard from mixed-wood sawdust and *Cocos nucifera* (Coconut) husks (bonded with UF) | 1.44 $N/mm^2$ | 218.03 $N/mm^2$ | Dadzie et al. [19] |
| 6. | Particleboards produced from wood chip wastes and modified cassava starch | 35.7 $N/mm^2$ | NA | Akinyemi et al. [20] |
| 7. | particle boards from polystyrene-wood wastes | 4.3 $N/mm^2$ | NA | Akinyemi et al. [21] |
| 8. | Wood Plastic Composites From Sawdust And Recycled Polyethylene Terephthalate (PET) | 27.08 $N/mm^2$ | NA | Rahman et al. [22] |
| 9. | Poly(vinyl chloride)/wood sawdust composites | ~39.00 $N/mm^2$ | NA | Sombatsompop and Chaochanchaikul [23] |
| 10. | *Albizia richardiana* King & Prain wood particles and recycled polyethylene terephthalate (PET) wood plastic composite | ~32.00 $N/mm^2$ | NA | Siddikur Rahman et al. [24] |

sulphite-pulped spruce wood, and enzymatically hydrolysed brown seaweed was prepared. The results of the salmon feeding trial showed that the wood could replace parts of a conventional fishmeal diet without harmful effects.

The effect of wood kraft pulp feed on cow's digestibility, ruminal characteristics, and milk production performance has been done by Nishimura et al. [26]. The use of kraft wood pulp feed for lactating dairy cows made the same rumen fermentation characteristics as those in cows given a large amount of roughage. The milk productivity was also unchanged when using the new feed type.

Rabbit feed using sodium hydroxide-treated sawdust has been done by Omole and Onwudike [27]. Food consumption of food containing sawdust treated with 6%

solutions of sodium hydroxide is higher than consumption of sawdust treated with 0–3% sodium hydroxide. Rabbits given sawdust treated with 4 or 5% sodium hydroxide showed the best weight gain and efficiency of feed utilisation than that given sawdust treated with 0 or 1% sodium hydroxide solutions.

A minimal number of research has been done on using sawdust as animal feed. Digestibility of the sawdust becomes the main issue, while pre-treatment of the sawdust makes the process uneconomical and costly to be done commercially.

## 5 Adsorbent

Sawdust can be used as an adsorbent to remove contamination from water. The unmodified wood sawdust is the lowest cost material used to clean up dyes, oils, toxic salts, pharmaceutical substances and many more from the environment. The mechanism of contaminant adsorption onto the sawdust surface involves ion exchange and hydrogen bonding. The cell walls of sawdust mainly consist of cellulose and lignin, where there are abundant hydroxyl groups which can contribute to a high number of binding sites.

There are many factors affecting the adsorption process of contaminants by wood sawdust. Solution temperature, contact time, initial concentration of pollutant, adsorbent dosage and solution pH are among the determinant of adsorption effectiveness. Commonly, too high in temperature will decrease adsorption. A longer contact time allows the contaminant to be adsorbed onto the adsorbent surface. Higher adsorbent dosage made the water cleaning process faster, but for economic reasons, optimum dosage should be investigated to prevent wastage. The pHzpc, or the point where the adsorbent is in neutral charge. The surface charge of the adsorbent is positive when the solution pH is below the pHzpc value, while it is negative at a pH over the pHzpc. The adsorption depends on the charge of the adsorbate. For example, a positively charged metal ion will be more attracted to the adsorbent when the pH of the solution is above pHzpc, due to the negative charge of the adsorbent [28].

Suitable treatments can improve the adsorption performance of wood sawdust. Modification can be done to increase the number of pores or increase chemical binding sites on the surface of the adsorbent. For example, modification using alkaline solution on the sawdust of deciduous softwood–poplar and coniferous softwood-fir were found to result in adsorption increment of 5 times for copper ions and 15 times for zinc ions [29]. In another research on potassium hydroxide, KOH-modified poplar sawdust also showed an increment in metal ion adsorption, up to 94.3% for copper ion and 98.2% for zinc ion, compared to unmodified poplar [30]. A list of sawdust used for the adsorption of contaminants is tabulated in Table 2.

**Table 2** List of sawdust used for adsorption of contaminants

| No | Adsorbent | Adsorbate | Maximum adsorption capacity | References |
|---|---|---|---|---|
| 1. | Pinewood sawdust modified with maleic acid | Cadmium (II) ions | 180.4 mg $g^{-1}$ | Hashem et al. [31] |
| 2. | Ayous (*Triplochiton scleroxylon*) wood sawdust | Paraquat | 41.66 μmol/g | Togue Kamga [32] |
| 3. | Sawdust modified with sulphuric acid and formaldehyde | Chromium (VI) ions | 8.84 mg/g | Chakraborty et al. [33] |
| 4. | Hexadecylpyridinium bromide-treated sawdust | Allura red AC (dye) | 151.88 μmol/g | Saha et al. [34] |
| 5. | *Pinus halepensis* sawdust | Plumbum | 8.64 mg/g | Semerjian [35] |
| 6. | *Eucalyptusglobulus Labill* sawdust | Pb (II) | 4.80 mg/g | Tejada-Tovar et al. [36] |
| 7. | Magnetised *Tectona grandis* sawdust | Methylene blue | 172.41 mg/g | Mashkoor and Nasar [37] |
| 8. | *Picea smithiana* sawdust | Lead (Pb) | 6.35 mg/g | Mahmood-ul-Hassan et al. [38] |
| 9. | *Picea smithiana* sawdust | Chromium (Cr) | 3.37 mg/g | Mahmood-ul-Hassan et al. [38] |
| 10. | *Picea smithiana* sawdust | Cadmium (Cd) | 2.87 mg/g | Mahmood-ul-Hassan et al. [38] |
| 11. | Pine Sawdust | Cadmium Ions | 3.47 mg/g | Liu et al. [39] |
| 12. | Sawdust | Eriochrome Black T (EBT) | 40.96 mg/g | Akhouairi et al. [40] |

## 6 Nanocellulose

Nanocellulose is a material extracted from the plant cell wall. It is cellulose in nanometer size, high strength, excellent stiffness and large surface area [41]. Nanocellulose can be categorised into three main types; nanocrystalline cellulose, nanofibrillated cellulose, and bacterial nanocellulose [42].

i. Nanocrystalline cellulose
   Nanocrystalline cellulose consists of highly crystalline particles. The extraction was commonly done using acid hydrolysis/heat-controlled techniques. The crystals from cellulose fibres were extracted by hydrolysis of amorphous cellulose regions using sulphuric acid, resulting in highly crystalline particles with dimensions depending on the source of raw material. Sulphuric acid hydrolysis gives a negative charge of sulphate half-ester groups onto the surface of the particles, creating electrostatic repulsion between particles that prevents agglomeration [43].

ii. Nanofibrillated cellulose
Nanofibrillated cellulose is bundles of cellulose chain molecules stretched with a long and flexible cellulose nanofibre network. It has a size of 1–100 nm, with alternating crystalline and amorphous domains along the structure. It is produced by delamination of wood pulp through mechanical force after chemical and enzyme treatment. Nanofibrillar cellulose, cellulose nanofibre and cellulose nanofibril are used for microfibrillated cellulose and are commercially available [44]. The surface of nanofibrillated cellulose can be modified through polymer grafting, adding coupling agents and many more.
iii. Bacterial nanocellulose
Bacterial nanocellulose is generated by the bacteria such as *Gluconacetobacter, Sarcina, Aerobacter, Agrobacterium, Alcaligenes, Acanthamoeba, Achromobacter, Azotobacter, Rhizobium, Escherichia, Pseudomonas, Salmonella* and *Zooglea* [45]. It has various biomedical applications due to its biocompatibility and non-toxic properties [46]. Because bacterial nanocellulose is not extracted from wood materials, it will not cover extensively in this chapter.

Nanocellulose has various applications, from the engineering and electronic sector to biomedical purposes. Nanofibrillated cellulose has the potential to be utilised in nanocomposite making. Nanofibrillated cellulose can be solution cast, in situ polymerisation or through melt mixing for blending with thermoplastics. In biomedical applications, nanofibrillated cellulose is used in medicine for drug delivery in human blood. Cellulosic nanofibres were also used as a material for the development of nanocomposite scaffolds which match the strength of the original ligaments or tendons. Due to its large surface area, cellulose nanofibril also has the potential to be used as an adsorbent for contaminations from water. In the papermaking industry, the strong affinity of cellulose nanofibre with cellulosic pulp and their ability to form a 3D network made them an excellent additive to increase paper strength [47].

## 7 Agricultural Cultivation Media

In the agricultural sector, sawdust is commonly used as mushroom cultivation media. Mushrooms can be grown on any sawdust as the media. However, different compositions of sawdust type will affect the mushroom's growth rate and number of fruiting bodies. In South East Asia, rubberwood sawdust is commonly used as it can be found abundantly in rubber plantation agricultural activity [48].

The growth performance of *Pleurotus ostreatus* from various agricultural wastes mixed with rubber tree sawdust in Malaysia has been done by Ahmad Zakil et al. [49]. Different agricultural wastes were mixed with rubberwood sawdust, including the oil palm empty fruit bunch, oil palm press fibre sugarcane bagasse and corn cob. Results showed the highest total average yield of *P. ostreatus* was 207.96 g/bag on when using the combination of sugarcane bagasse and rubberwood sawdust at a ratio of 1:1.

Research has also been done on the cultivation of oyster mushrooms (*Pleurotus ostreatus*) using Moso bamboo sawdust showed a faster growth rate of 3–7 days than conventional cultivation media. When mixed with rice bran, it gives an even better yield of 97.9 ± 3.9 g/bottle and fruiting bodies at 33.6 ± 4.2 no/bottle [50].

## 8 Clay, Cement, Concrete and Building Materials

Sawdust is commonly added to building materials to reduce its original weight. Low thermal conductivity, high sound absorption and good sound insulation are among the wood characteristics favourable for building material [51]. An attempt to produce lightweight concrete using a mixture of sawdust and cement has successfully created a block of concrete that can withstand more than 2.8 N/mm$^2$, despite the increment in water absorption due to the presence of the sawdust [52]. The incorporation of sawdust into unfired clay blocks showed that up to 2.5% of sawdust could be added to make a good block with particle sizes ranging between 600 and 425 μm [53].

Sawdust has been tested for sand replacement in cement blocks for up to 15% content. Volumetric shrinkage and concrete density decrement were observed, with an increment of water absorption when the sawdust percentage was increased. Testing on two types of concrete prepared, the natural weight concrete, NWC and lightweight concrete recorded 34 MPa and 21 MPa compressive strength, respectively. Therefore, these concretes can be used for structural applications. Up to 21.42% reduction in the heating, ventilation and air-conditioning (HVAC) energy was also recorded in sawdust added concretes [54]. High thermal insulation is favourable in tropical countries as it minimises heat transfer from the outer side of the building into the occupied space.

## 9 Liquefaction

Biomass liquefaction is a technique of dissolving woody materials into liquid form for obtaining biofuels, bio-based materials and chemicals. Various wood liquefaction techniques include hydrothermal liquefaction, organic solvent liquefaction, co-solvent liquefaction, microwave-assisted liquefaction and plasma electrolytic liquefaction [55].

### *9.1 Hydrothermal Liquefaction*

Hydrothermal liquefaction of sawdust is conducted near (subcritical) or above (supercritical) the critical point of water, which is 374 °C, under 22.1 MPa pressure. Before liquefaction, pre-treatment is necessary to facilitate access to the internal structure of wood for fractionation. Physical action (pulverisation and irradiation), chemical treatment (alkali, acid, organosolv and ammonia explosion), or combination of them as physicochemical (steam explosion, carbon dioxide explosion and wet oxidation) and biological pre-treatment (enzymolysis). General parameters of hydrothermal liquefaction are conducted at a low-temperature range between 200 and 400 °C and high pressure between 5 and 25 MPa. Sawdust will be converted into a bio-oil fraction, a gas fraction and a solid residue after the liquefaction process [56].

## 9.2 *Co-solvent Liquefaction*

Hydrothermal liquefaction can be done with co-solvent to assist the liquefaction process. The hydrothermal liquefaction of rice straw with methanol co-solvent has been done by Yerrayya et al. [57]. The presence of methanol as a co-solvent to water at 50:50 has improved the yield of bio-crude production up to 36.8 wt.%. Another hydrothermal liquefaction of rice straw with glycerol as co-solvent. Using 5 wt% of $Na_2CO_3$ as catalyst and 260 °C heating temperature for 1 h, the process yielded 50.31 wt% of bio-oil and 26.65 wt% of solid residue [58]. Meanwhile, hydrothermal sugarcane bagasse at 320 °C reaction temperature, 15 min reaction time and 10 wt% potassium hydroxide, KOH as catalyst produce the highest amount of bio-crude of 36.3 wt% [59].

## 9.3 *Organic Solvent Liquefaction*

Organic solvent liquefaction can be done at temperatures 240–270 °C without catalysts. The temperature could be lowered to around 80–150 °C with acidic catalysts. Organic solvent liquefaction produces very high yields of solvent solubles of about 90–95% based on the lignocellulosic weight. The conventional liquefaction of wood takes several hours of treatments with a higher temperature required at 300–400 °C with or without catalysts. Conventional liquefaction also yields less, around 40–60%, due to the conversion of the woody material into gaseous compounds [60].

The possible chemical compounds that can be produced from the organic solvent liquefaction are resins such as phenolic resins (novolac and resol type), polyurethanes and epoxy, which are commonly used in wood composite making [61, 62]. An experiment on the effects of phenol, ethylene glycol (EG) and ethylene carbonate (EC) solvents shows phenol as the optimum solvent for bamboo liquefaction with a yield of up to 99%. The liquefaction utilises hydrochloride acid at a temperature of 180 °C in autoclaves for different reaction periods [63].

### 9.4 *Microwave-Assisted Liquefaction*

As its name mentioned, microwave-assisted liquefaction is the method to increase liquefaction yield by combining chemical processes and microwave irradiation. Using microwave heating brings the energy directly into the target object due to the applied electromagnetic field. Microwave heating ensures faster heat penetration into the bulk material. Therefore the reaction time can be reduced, which may also reduce the cost of operation [64].

Microwave-assisted liquefaction has been researched by Xue et al. [65] using polyethylene glycol (PEG-400) and glycerol mixture as liquefying solvents together with 97 wt% sulphuric acids as a catalyst at a reaction temperature of 140 °C. A high liquefaction yield (97.47%) was achieved in just 5 min to obtain the polyol. In their research, the polyol produced was made into polyurethane resin.

Another research used pinewood particles as the raw material, with polyethylene glycol and glycerin (70/30 w/w) as the liquefaction reagent. The solvent-to-wood ratio was 7:1, and the liquefaction was done using two types of catalysts, which were 3% of sulphuric acid and phosphoric acid. Liquefaction was done at 150 °C with the help of microwave irradiation. It was found that using microwave heating, the wood meal completely dissolved in just 2 min, compared to 30 min using conventional heating [66].

## 10 Conclusions

Sawdust has a wide variety of potential uses, from energy generation to chemical extraction. Sawdust cannot be treated as trash but as a new source of income generation for the company. It is also a sustainable way of using wood resources as efficient as possible to guarantee the future of our forests. Hopefully, this book chapter has given some insights into the usefulness of wood sawdust.

## References

1. Saal U, Weimar H, Mantau U (2019) Wood Processing Residues. In: Wagemann K, Tippkötter N (eds) Biorefineries. Springer International Publishing, Cham, pp 27–41
2. Arif Hakimi Saadon SZ, Osman NB, Yusup S (2022) Chapter 5—Pretreatment of fiber-based biomass material for lignin extraction. In: Yusup S, Rashidi NA (eds) Value-chain of biofuels. Elsevier, pp 105–135
3. Bergman R et al (2014) The carbon impacts of wood products. For Prod J 64(7–8):220–231
4. Routa J et al (2011) Effects of forest management on the carbon dioxide emissions of wood energy in integrated production of timber and energy biomass. GCB Bioenergy 3(6):483–497
5. Obi OF (2015) Evaluation of the effect of palm oil mill sludge on the properties of sawdust briquette. Renew Sustain Energy Rev 52:1749–1758
6. Demirbas A, Arin G (2002) An overview of biomass pyrolysis. Energy Sour 24(5):471–482

7. Shaaban A et al (2014) Influence of heating temperature and holding time on biochars derived from rubber wood sawdust via slow pyrolysis. J Anal Appl Pyrol 107:31–39
8. Balat M, Balat H, Öz C (2008) Progress in bioethanol processing. Prog Energy Combust Sci 34(5):551–573
9. Xu Z, Huang F (2014) Pretreatment methods for bioethanol production. Appl Biochem Biotechnol 174(1):43–62
10. Sarkar N et al (2012) Bioethanol production from agricultural wastes: an overview. Renewable Energy 37(1):19–27
11. Pizzi A, Papadopoulos AN, Policardi F (2020) Wood composites and their polymer binders. Polymers 12(5):1115
12. Marcovich NE et al (2009) Resin-sisal and wood flour composites made from unsaturated polyester thermosets. Compos Interfaces 16(7–9):639–657
13. Gardner DJ, Han Y, Wang L (2015) Wood-plastic composite technology. Curr For Rep 1(3):139–150
14. Ashori A (2008) Wood–plastic composites as promising green-composites for automotive industries! Biores Technol 99(11):4661–4667
15. Akinyemi AB, Afolayan JO, Oluwatobi EO (2016) Some properties of composite corn cob and sawdust particle boards. Constr Build Mater 127:436–441
16. Orelma H et al (2021) Manufacture of all-wood sawdust-based particle board using ionic liquid-facilitated fusion process. Wood Sci Technol 55(2):331–349
17. Tawasil DN et al (2021) Coconut fibre and sawdust as green building materials: a laboratory assessment on physical and mechanical properties of particleboards. Buildings 11. https://doi.org/10.3390/buildings11060256
18. Yano BB et al (2020) Use of sugarcane bagasse and industrial timber residue in particleboard production. BioResources 15(3):4753–4762
19. Dadzie PK, Awotwe-Mensah M, Amoah M (2021) Materials mix ratio and binder type effects on physical and mechanical properties of particleboard from mixed-wood sawdust and Cocos nucifera (Coconut) husks. J Indian Acad Wood Sci 18(2):128–140
20. Akinyemi BA, Olamide O, Oluwasogo D (2019) Formaldehyde free particleboards from wood chip wastes using glutaraldehyde modified cassava starch as binder. Case Stud Constr Mater 11:e00236
21. Akinyemi BA et al (2019) Durability and strength properties of particle boards from polystyrene–wood wastes. J Mater Cycles Waste Manag 21(6):1541–1549
22. Rahman KS et al (2013) Flat-pressed wood plastic composites from sawdust and recycled polyethylene terephthalate (PET): physical and mechanical properties. Springerplus 2(1):629
23. Sombatsompop N, Chaochanchaikul K (2004) Effect of moisture content on mechanical properties, thermal and structural stability and extrudate texture of poly(vinyl chloride)/wood sawdust composites. Polym Int 53(9):1210–1218
24. Siddikur Rahman K et al (2018) Properties of flat-pressed wood plastic composites as a function of particle size and mixing ratio. J Wood Sci 64(3):279–286
25. Sharma S et al (2018) Microbial protein produced from brown seaweed and spruce wood as a feed ingredient. J Agric Food Chem 66(31):8328–8335
26. Nishimura K et al (2019) Effect of wood kraft pulp feed on digestibility, ruminal characteristics, and milk production performance in lactating dairy cows. Anim Sci J 90(2):189–195
27. Omole TA, Onwudike OC (1981) Investigations of the treatment of sawdust for rabbit feeding. 1. Effect of sodium hydroxide treatment. Animal Feed Sci Technol 6(1):43–50
28. Shukla A et al (2002) The role of sawdust in the removal of unwanted materials from water. J Hazard Mater 95(1):137–152
29. Šćiban M, Klašnja M, Škrbić B (2006) Modified softwood sawdust as adsorbent of heavy metal ions from water. J Hazard Mater 136(2):266–271
30. Kovacova Z et al (2020) Influence of wooden sawdust treatments on Cu(II) and Zn(II) removal from water. Materials 13. https://doi.org/10.3390/ma13163575
31. Hashem A et al (2020) Non-linear adsorption characteristics of modified pine wood sawdust optimised for adsorption of Cd(II) from aqueous systems. J Environ Chem Eng 8(4):103966

32. Togue Kamga F (2018) Modeling adsorption mechanism of paraquat onto Ayous (Triplochiton scleroxylon) wood sawdust. Appl Water Sci 9(1):1
33. Chakraborty R et al (2021) Adsorption of hazardous chromium (VI) ions from aqueous solutions using modified sawdust: kinetics, isotherm and thermodynamic modelling. Int J Environ Anal Chem 101(7):911–928
34. Saha TK et al (2020) Adsorption characteristics of Allura red AC onto sawdust and hexadecylpyridinium bromide-treated sawdust in aqueous solution. ACS Omega 5(22):13358–13374
35. Semerjian L (2018) Removal of heavy metals (Cu, Pb) from aqueous solutions using pine (Pinus halepensis) sawdust: equilibrium, kinetic, and thermodynamic studies. Environ Technol Innov 12:91–103
36. Tejada-Tovar C et al (2021) Potential use of residual sawdust of eucalyptus globulus Labill in Pb (II) adsorption: modelling of the kinetics and equilibrium. Appl Sci 11. https://doi.org/10.3390/app11073125
37. Mashkoor F, Nasar A (2020) Magnetized Tectona grandis sawdust as a novel adsorbent: preparation, characterization, and utilization for the removal of methylene blue from aqueous solution. Cellulose 27(5):2613–2635
38. Mahmood-ul-Hassan M et al (2018) Kinetics, isotherms, and thermodynamic studies of lead, chromium, and cadmium bio-adsorption from aqueous solution onto Picea smithiana sawdust. Environ Sci Pollut Res 25(13):12570–12578
39. Liu X et al (2019) Adsorption characteristics of cadmium ions from aqueous solution onto pine sawdust biomass and biochar. BioResour 14(2):4270–4283
40. Akhouairi S et al (2019) Natural sawdust as adsorbent for the eriochrome black T dye removal from aqueous solution. Water Air Soil Pollut 230(8):181
41. Phanthong P et al (2018) Nanocellulose: extraction and application. Carbon Resour Conv 1(1):32–43
42. Lavoine N et al (2012) Microfibrillated cellulose—Its barrier properties and applications in cellulosic materials: a review. Carbohyd Polym 90(2):735–764
43. Abitbol T et al (2016) Nanocellulose, a tiny fiber with huge applications. Curr Opin Biotechnol 39:76–88
44. Abdul Khalil HPS et al (2014) Production and modification of nanofibrillated cellulose using various mechanical processes: a review. Carbohydr Polym 99:649–665
45. Skočaj M (2019) Bacterial nanocellulose in papermaking. Cellulose 26(11):6477–6488
46. Sharma C, Bhardwaj NK (2019) Bacterial nanocellulose: present status, biomedical applications and future perspectives. Mater Sci Eng, C 104:109963
47. Kalia S et al (2014) Nanofibrillated cellulose: surface modification and potential applications. Colloid Polym Sci 292(1):5–31
48. Tan YH, Krishnen G (2008) Comparative growth and fructification of the shiitake mushroom, Lentinula edodes, on rubberwood sawdust, palm fibres and grasses
49. Ahmad Zakil F et al (2022) Growth performance and mineral analysis of Pleurotus ostreatus from various agricultural wastes mixed with rubber tree sawdust in Malaysia. Bioresour Technol Rep 17:100873
50. Yamauchi M et al (2019) Cultivation of oyster mushroom (Pleurotus ostreatus) on fermented moso bamboo sawdust. J King Saud Univ Sci 31(4):490–494
51. Mwango A, Kambole C (2019) Engineering characteristics and potential increased utilisation of sawdust composites in construction—A review. J Build Constr Plan Res 7(3):59–88
52. Omar MF et al (2020) Partially replacement of cement by sawdust and fly ash in lightweight foam concrete. IOP Conf Ser: Mater Sci Eng 743(1):012035
53. Jannat N et al (2021) Influence of sawdust particle sizes on the physico-mechanical properties of unfired clay blocks. Designs 5. https://doi.org/10.3390/designs5030057
54. Ahmed W et al (2018) Effective use of sawdust for the production of eco-friendly and thermal-energy efficient normal weight and lightweight concretes with tailored fracture properties. J Clean Prod 184:1016–1027
55. Kim J-Y et al (2019) Overview of the recent advances in lignocellulose liquefaction for producing biofuels, bio-based materials and chemicals. Biores Technol 279:373–384

56. Jindal MK, Jha MK (2016) Hydrothermal liquefaction of wood: a critical review. Rev Chem Eng 32(4):459–488
57. Yerrayya A et al (2020) Hydrothermal liquefaction of rice straw using methanol as co-solvent. Energies 13. https://doi.org/10.3390/en13102618
58. Cao L et al (2016) Effect of glycerol as co-solvent on yields of bio-oil from rice straw through hydrothermal liquefaction. Biores Technol 220:471–478
59. Yerrayya A et al (2022) Optimization of bio-crude yield and its calorific value from hydrothermal liquefaction of bagasse using methanol as co-solvent. Energy 244:123192
60. Shiraishi N (1992) Liquefaction of lignocellulosics in organic solvents and its application. Emerging technologies for materials and chemicals from biomass. American Chemical Society, pp 136–143
61. Pan H (2011) Synthesis of polymers from organic solvent liquefied biomass: a review. Renew Sustain Energy Rev 15(7):3454–3463
62. Jiang W, Kumar A, Adamopoulos S (2018) Liquefaction of lignocellulosic materials and its applications in wood adhesives—A review. Ind Crops Prod 124:325–342
63. Yip J et al (2009) Comparative study of liquefaction process and liquefied products from bamboo using different organic solvents. Biores Technol 100(24):6674–6678
64. de la Hoz A, Díaz-Ortiz Á, Moreno A (2005) Microwaves in organic synthesis. Thermal and non-thermal microwave effects. Chem Soc Rev 34(2):164–178
65. Xue B-L, Wen J-L, Sun R-C (2015) Producing lignin-based polyols through microwave-assisted liquefaction for rigid polyurethane foam production. Materials 8:586–599. https://doi.org/10.3390/ma8020586
66. Pan H, Zheng Z, Hse CY (2012) Microwave-assisted liquefaction of wood with polyhydric alcohols and its application in preparation of polyurethane (PU) foams. Eur J Wood Wood Prod 70(4):461–470

# The Possibility of Using Wood Peeler Core as The Dye-Sensitized Solar Cells

**Norul Hisham Hamid, Norasikin Ahmad Ludin, and Nur Ezyanie Safie**

**Abstract** The wood peeler core is a waste generated during the peeling process of the veneer. The peeler core size varies by species, spindle type, processing, and lathe machine. For several decades, the wood peeler core waste has been burned in a boiler to generate heat for kiln dryers, and some are selling for agriculture poles. Since wood has become more expensive, converting this waste to a more valuable material is paramount. One of the potentials of the wood peeler core is to use dye-sensitized solar cells (DSSC) to generate renewable energy, particularly electricity. Although the power conversion efficiency of DSSC from wood waste ranges only from 0.29 to 0.58% and is far below the commercial silicon solar cells (about 19%), it is potential as a low-cost material for renewable energy cannot be underestimated.

## 1 Introduction

The wood peeler core is the round portion of the pole that cannot slice into usable veneer and is considered waste from plywood mills. The round portion is generated as a result of the knife that cannot slice the log due to the bolt attached at the end of it. The bolt-in veneer lathe machine is used to hold the ends of the log. The peeling veneer involves rotating a log, or more accurately, a bolt, against a sharp knife to achieve the desired effect. Chucks have been inserted into the bolt's end so it can be rotated. In many different designs, the chuck teeth are responsible for providing the necessary torsion resistance to rotate the bolt with sufficient force to cut the veneer from it.

Generally, the diameter of the core produced after the peeling process depends on the logs' diameter, species, and pre-treatment, such as the steaming process. As a consequence, yield is also impacted by the size of the chuck. On the one hand, a big

N. H. Hamid (✉)
Faculty of Forestry & Environment, Universiti Putra, UPM Serdang, 43400 Selangor, Malaysia
e-mail: h_noroul@upm.edu.my

N. Ahmad Ludin · N. E. Safie
Solar Energy Research Institute, Universiti Kebangsaan Malaysia, 43560 Bangi, Selangor, Malaysia

 
S. N. Sarmin et al. (eds.), *Wood Waste Management and Products*, Sustainable Materials and Technology, https://doi.org/10.1007/978-981-99-1905-5_12

chuck is required to prevent chuck spinout and when a log is heavy when it is first started. On the other hand, a smaller chuck diameter makes it possible to peel the log down to a smaller spindle. Retractable chucks, consisting of a center chuck encircled by an outside ring, are used in the majority of contemporary mills to accomplish both of these goals. At first, the ring and the chuck are inserted, which results in the creation of a chuck with a big diameter.

Because bending resistance is proportional to the diameter cubed, a quick drop in a bolt's resistance to the bending forces applied by the knife and nose bar may be expected as the diameter of the bolt diminishes. Because the bolt will deflect away from the knife if the log is not held in some way, the veneer that is cut from the log ends will be thicker than the veneer that is cut from the center of the log. The spindle that is produced will have the shape of a barrel, which will cause the loss of veneer. This predicament can be improved by utilizing a backing roll to provide support for the rear of the bolt. The log may be rotated more easily with the assistance of motorized backing rolls, which also allows for reduced chuck diameters.

When peeling logs with a tiny diameter, using spindle peelers, which can peel down to 50 mm cores, can drastically cut down on the amount of trash produced. However, the technique is not employed in a significant number of today's mills [1]. The yield of veneers determines the mill's profit, and using retractable chunks significantly improves the veneer recovery; most of the peeler core leaves the lathe machine from the heartwood portion. In order to use the peeler core for any purpose, it is crucial to understand the tree itself from the perspective of hardwood and softwood as well as heartwood and softwood.

## 2 Source of Wood Peeler Core

The wood peeler core is generated at the lathe machine during the slicing process of the tree bolt to the veneer (Fig. 1). The peeler core is the center of the tree bolt, which cannot reach by the lathe knife (Fig. 2).

### *2.1 The Tree*

The two fundamental parts of a plant that is growing are called the shoot and the root, respectively. The primary functions of the root are the absorption of water and nutrients, the mechanical connection of the shoot, and the storage of biochemical compounds. At the same time, the shoot is responsible for the development of the trunk, branches, and leaves. In addition, the trunk is composed of many layers, which include the inner bark, the outer bark, the heartwood, the vascular cambium, the sapwood, and the pith.

The tougher outer bark and skin provide the softer inner bark with mechanical protection and serve to minimize the amount of water that is lost via evaporation.

**Fig. 1** The tree bolt before slicing (left)

**Fig. 2** The peeler core after slicing (right)

The sugars (food) that are produced by photosynthesis and carried from the leaves to the roots or other growing parts of the tree pass through the inner bark of the tree. This is the tissue that makes up the inner bark. The vascular cambium is the layer that sits between the bark and the wood and is responsible for the annual production of both the bark and the wood.

Sapwood is the active, "alive" wood that is responsible for transporting water (also known as sap) from the plant's roots to its leaves. It has not yet gathered the often-color compounds that distinguish the non-conductive heartwood located as a core of darker-colored wood in the center of most trees. Heartwood is distinguished by its location in the middle of most trees. The pith is the only remaining part of the tree's early growth that may be found in the trunk's very core, before the formation of the wood [2].

## 2.2 Hardwood and Softwood

Hardwood is classified as wood from angiosperm trees. The trees have large leaves, and their structures contain vessel components (which are also known as pores) that transmit water all across the wood. The trees also have wide trunks. The development rate of hardwood is much slower than that of softwood, and during the fall and winter months, it continually and gradually loses its leaves. Alder, balsa, beech, mahogany, maple, oak, teak, and walnut are examples of trees that are classified as hardwoods.

The structure of hardwoods is complex because they have a more significant number of primary cell types and are more variable within the cell types. Generally, the hardwood species have a greater density than the softwood. Despite being more expensive than softwood and mostly not easy to process or machine, hardwood will last longer due to its greater density and being selected for high-quality furniture, decks, flooring, and construction requiring heavy-duty applications.

On the contrary, softwood is classified as wood from the gymnosperm tree. The tree usually has needles and cones. Its structure is characterized by rays and tracheid transporting the water and producing a sap. Unlike softwood, hardwood is not characterized by visible pores because of the tracheid. The trees cedar, juniper, Douglas fir, spruce, pine, yew, and redwood are examples of softwoods.

The density of softwood is significantly lower than that of hardwood, despite the fact that softwood grows at a quicker pace. It is less costly than hardwood; around 80% of all timbers come from softwood and are used as construction components (for example, windows and doors), furniture, medium-density fiberboard (MDF), paper, and Christmas trees, among other things. Hardwood has a higher resilience to fire than its softer counterpart, softwood.

## 2.3 Sapwood and Heartwood

The sapwood is the light-colored outer portion of the tree trunk, the water conductivity that occurs from the root to the left and where the excess food is stored. The heartwood is the central core portion of the tree. These two portions can easily recognize by their different color. Generally, the heartwood is darker than the sapwood as the former is older during the growing phases, except for hemlock, aspen, cottonwood, beech, and partially spruce. When the sapwood is continuously formed in the tree, the earliest and inners sapwood will form a new heartwood, and this cycle is repeated every year. As most of the peeler core comes from the heartwood, it is vital to know its properties.

## 2.4 Anatomical Properties

The anatomical properties differ not only by site location but also by the tree portion [3]. The fiber and vessel proportions are higher in heartwood (25.8, 32.1%) than the sapwood (22.7, 25.3%) of *B. Costata* grown in Maoershan, China. However, in the other site location (Muzhaling), only the vessel proportion is higher in heartwood (32.2%) than the sapwood (21.2%). The vessel diameter is wider in the sapwood and narrower in the heartwood of *Caesalpinia echinate.*

The vessel frequency does not differ among sapwood and heartwood. The vessel length is longer in the sapwood and shorter in the heartwood, but the vessel area is larger in the sapwood and lowers in the heartwood. Some heartwood vessels are filled with extractives, which are not present in the sapwood. The vessel-ray pits, fiber wall thickness, and diameter are not significantly different for heartwood and sapwood. The fiber length is longer in the sapwood than in the heartwood [4].

## 2.5 Physical Properties

As reported by Bahmani et al. 2020, the heartwood ($887 \pm 49.2$, $634 \pm 37.2$ kg $cm^{-3}$, $16.60 \pm 2.5$, and $20.4 \pm 3.6\%$) of Sorbus torminalis aged 45 years has higher basic density, oven-dry density, volumetric swelling, and volumetric shrinkage values than the sapwood ($721 \pm 20.4$, $567 \pm 12.7$ kg $cm^{-3}$, $15.1 \pm 1.44$ and $17.7 \pm 1.9\%$). An almost similar trend was found in *Eucalyptus Grandis* aged 10 years grown in Turkey.

The oven-dried density of the sapwood is significantly higher, which is almost 20%, than that of the oven-dried heartwood. Additionally, the shrinkage and swelling values are lower in the heartwood than in the sapwood. The fiber saturation point (FSP) and applicable maximum moisture content (MMCp) are more remarkable in sapwood than in heartwood [5].

In comparison, the density of sapwood and heartwood in Maple (Acer velutinum) trees cultivated in the Caspian Forest in the north of Iran ranges from 0.59 to 0.62 gr/$cm^3$, while the density of sapwood and heartwood in Oak (Quercus castaniefolia) trees ranges from 0.73 to 0.76 gr/$cm^3$. Because of this circumstance, the density of the wood in the heartwood is greater than that of either of the two sapwood types. Both the sapwood and the heartwood of both species have pH values that range from 4.1 to 4.4 on average, therefore there is not a major difference between the two [6].

## 2.6 Chemical Properties

In addition, the structural and chemical components of the Quercus faginea's heartwood and sapwood have been shown to be distinct from one another in terms of the composition of non-polar and polar extracts, as well as antioxidant characteristics. In most cases, the amount of chemicals that can be extracted using a solvent is greater in the heartwood than in the sapwood (on average, 19.0 vs. 9.5%). The lipophilic extractible compounds make up less than 1% of the total, and the vast majority of these are polar compounds, such as those that are soluble in ethanol; they make up 65% of the total extractives in the heartwood and 43% in the sapwood.

There is not a substantial difference in the lignin concentration of sapwood and heartwood (28.1 and 28.6%), which includes their respective sugar compositions. Both sapwood and heartwood include a majority of lipophilic substances that are saturated fatty acids (23.0 and 36.9%). Sapwood also contains an abundance of aromatic compounds (22.9%). The phenolic material concentration of the ethanol-water extractable is rather high (558.4 and 319.4 mg GAE/g extract, for heartwood and sapwood).

Both the heartwood and sapwood have a comparable polyphenolic composition, with larger levels of ellagitannins (168.9 and 153.5 mg tannic acid/g of extract in sapwood and heartwood, respectively) and a shallower level of condensed tannins. The antioxidant activity is quite strong, with an IC50 of 2.6 g/ml for sapwood and 3.3 g/ml for heartwood, respectively. This is in comparison to the IC50 of 3.8 g/ml for Trolox, which is the standard antioxidant.

Additionally, the hygroscopic, viscoelastic, and chemical characteristics of wood cell walls' heartwood in Chinese fir demonstrate no substantial shift in the chemical composition of the wood throughout the sapwood to heartwood transition. This is shown by the absence of a difference in the hygroscopic properties. On the other hand, a greater quantity of extractives, in particular aromatic compounds, are deposited in the heartwood's cell wall matrix [7].

## 2.7 Mechanical Properties

The modulus of elasticity, modulus of rupture, specific gravity, tensile strength, and the strength of dynamic bending for plantation grown of Eucalyptus, poplar, red pine, and Scots pine are significantly different for both sapwood and heartwood. Overall, sapwood has higher mechanical properties than heartwood [8]. The mechanical properties of red pine grown in Taiwan differ with the site location and for sapwood and heartwood [9].

The heartwood has a higher compression strength than the sapwood in the Goseong gun (38.17 MPa and 29.86 MPa) and the Hongcheon gun (38.97 MPa and 34.50 MPa). Identical to the shear strength for both site locations. The heartwood has a higher shear strength than the sapwood in the Goseong gun (10.63 MPa, 8.72 MPa) and the Hongcheon gun (10.75 MPa, 7.14 MPa), respectively.

## 3 Effect of Species-Crude Extractive Ratio on the DSSCs' Performance

Solar energy has a vast potential for the generation of power, particularly in Malaysia, which has a sustained solar radiation supply for the whole year. Dye-sensitized solar cells (DSSCs) are classified as the 3rd-generation solar cells devices, and it has captivated much curiosity from researchers due to their inexpensive cost as well as could generate a soaring photo conversion efficiency of as high as 12% as reported by Gratzel [10–12].

The mechanism of DSSC consists of porous semiconductor loaded with a sensitizer on a glass substrate, a redox couple electrolyte, and a counter electrode [13]. The sensitizer develops into one of the critical components that pique the interest of researchers looking for ways to improve the operation of the gadget. From metal complex sensitizers to metal-free sensitizers and inorganic sensitizers to organic sensitizers, each type of sensitizer possesses both positive and negative qualities.

It has been found that the zinc porphyrin dye and ruthenium, which are the examples of metal complex, achieve greater performance, which is between 12 and 13% efficiency when compared to others [14–18]. However, concerns over the expense of synthesis have led to a heightened interest in the utilization of sensitizers that are not metal complexes, in spite of the performance of these other types of sensitizers is still regarded as being subpar [19].

Sensitizers, which play the role of light harvesters in the mechanism of DSSCs, are critical to the devices' overall function. The most effective sensitizers are those that generate electrons and absorb a broad spectrum of visible light [20]. Natural dyes offer a variety of colors, and as a result, they can provide a substitute sensitizer for organic-based DSSCs which are less costly. The following are the criteria that naturally occurring colors should fulfill:

1. The capacity to captivate a broad range of visible and near-infrared rays.
2. The existence of a functional group that is capable of firmly attaching to the layer of mesoporous oxide.
3. An energy state is in between the highest occupied molecular orbital (HOMO) and lowest unoccupied molecular orbital (LUMO) for effective electron pumping from semiconductor valence to conduction bands [21].

When excited, condescending sensitizers would be capable of injecting electrons with a quantum yield of unity from the dyes into the substances [22]. These sensitizers could also be capable of absorbing photons that penetrates under a threshold wavelength, which is 920 nm, and convert sunlight into electricity at a common air mass (AM) of 1.5. Before selecting pigments as sensitizers, it is vital to check the effectiveness of the procedure as follows:

1. the assimilation of input rays by molecules of the dye;
2. the reformation of photons to electron-hole pairs; and
3. the detachment and assemblage. This is the case when natural dyes are being considered as one of the criteria in the creation of DSSCs [23].

The development of efficient natural sensitizers is of particular importance in this study. In the course of the previous two decades, many researchers have investigated various components of the plant, such as flowers, roots, barks, and, leaves in an effort to identify natural sensitizers that are both efficient and safe to use. On the other hand, the vast majority of the research that are publicly available don't examine the possibility of sensitizers being extracted from the wood of forest trees. The tropical rainforests of Malaysia are renowned for their rich biodiversity and serve as a habitat for various tree species.

It is obligatory to do research on the potential contributions that trees may make to industries besides the furniture sector, including the application of solar energy. The current research investigates the optical characteristics of mengkulang (Heritierae-lata) and rengas (Gluta spp.) wood-derived sensitizers, as well as their performance in double-sided solid-state cells (DSSCs). In addition, the proportions of these sensitizers within the mixtures are changed following analyzing the outcomes of these changes on the device performance.

## *3.1 Properties and Performance of UV-Vis Absorption Spectra*

The rengas sensitizer has a significant capacity for absorption at a short wavelength in the mengkulang and rengas that were derived from waste wood (Fig. 3). The absorption spectra of the mixes of rengas and mengkulang sensitizers with 3 distinct compositions are displayed in Fig. 4. All of the combined sensitizers had absorption spectra that are quite equivalent to that of the rengas sensitizer. This is due to the rengas' chemical property being absorbed onto the $TiO_2$ surface in a more dominating manner than the mengkulang extract.

Figure 4 depicts the absorption spectra of the mixture of extracts that were anchored onto the surface of the $TiO_2$ material. After the grafting procedure, the spectra do not display a noticeable peak; rather, between 400 and 600 nm, there is a broad shoulder that can be observed. The sensitizers' absorption spectra that are anchored onto $TiO_2$ produce a stable and broad shoulder. This enables the harvesting of rays throughout a large spectrum of solar, which eventually results in the creation of significant photocurrent in DSSC devices [24].

Based on the strong peak that can be seen between 400 and 512 nm [25], anthocyanin has been determined to be the direct dye that is produced by the rengas sensitizer. Anthocyanins make up the most numerous and diverse subclass of water-soluble pigments found across the plant world. Anthocyanins are a category of naturally occurring dyes that may be observed in the flowers, fruits, and leaves of plants. They are responsible for numerous hues that fall within the red–blue color spectrum. Carbonyl and hydroxyl groups that are found in anthocyanin have the potential to form bonds with the porous $TiO_2$ film (Fig. 5). Because of this condition, an electron will be transferred from the molecule of anthocyanin to the $TiO_2$'s conduction band.

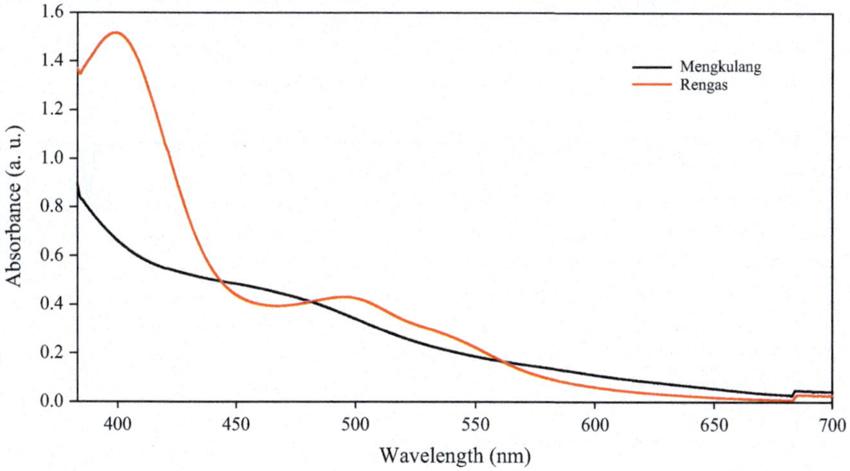


**Fig. 3** The individual extractives' UV-Vis absorption spectra

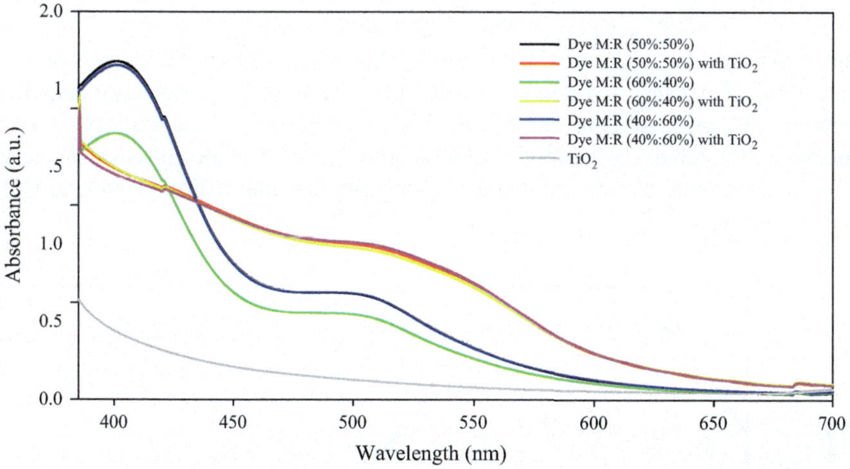


**Fig. 4** (M:R is mengkulang:rengas). UV-Vis absorption of mixed extractives, $TiO_2$ electrode, and extractives anchored onto $TiO_2$

## 3.2 *FTIR Analysis*

The analysis of FTIR spectroscopy method is utilized to examine the functional groups of the sensitizer, with KBr serving as the background reference. The patterns of the IR spectra seen in Fig. 6 are consistent across all of the samples since the functional groups that are present in each sample are the same. The existence of the free hydroxyl group (Ar–O–H)'s vibrations in phenols is responsible for a broader

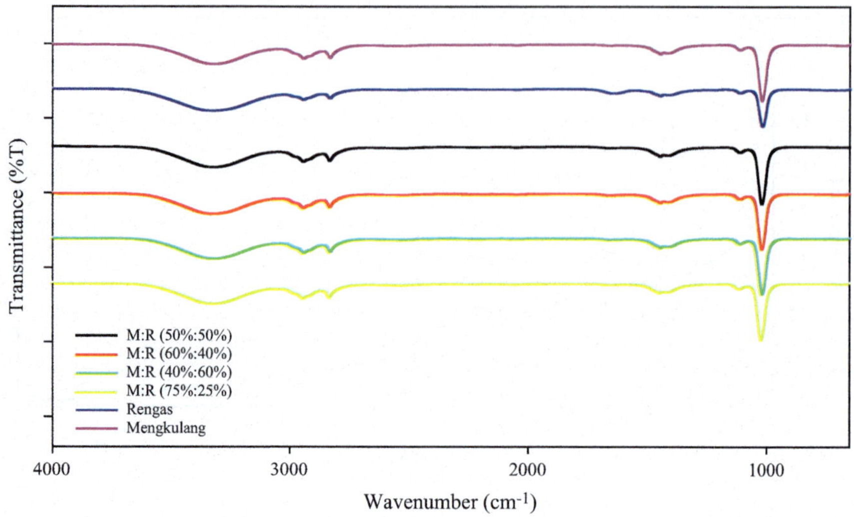


**Fig. 5** FTIR analysis of individual and mixed extractives

band that may be noticed at 3323, 3309, 3324, and 3319 $cm^{-1}$. Meanwhile, the stretching of the $sp^3$ C–H bond is responsible for a modest but powerful peak that appears at 2832 and 2944 $cm^{-1}$. An aromatic stretching of the C=C bond is to blame for the relatively modest shoulder at 1449 $cm^{-1}$. The mixed sensitizers have functional groups that are analogous to one another, like hydroxyl groups, C–H bonds, as well as an aromatic C=C bonds.

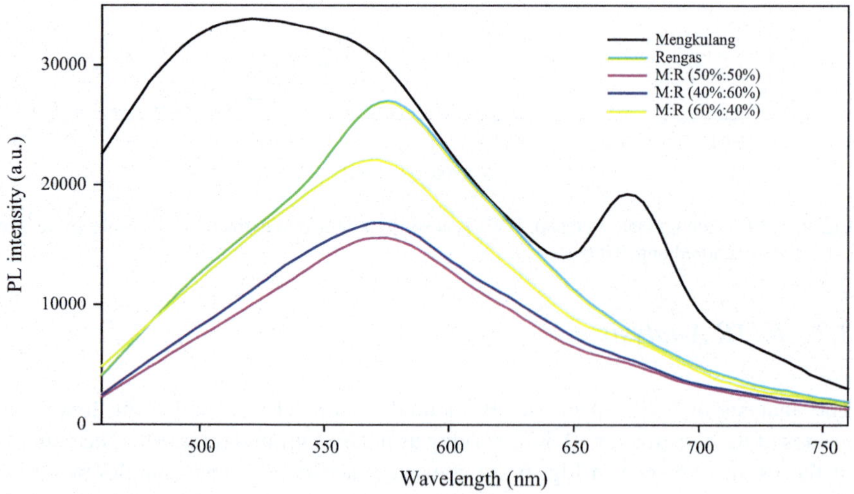


**Fig. 6** Emission spectra of individual and mixed extractives

### 3.3 HOMO–LUMO Calculations and Optical Band Gap

PL is used to quantify the optical band gap, whereas CV analysis is used to compute the HOMO–LUMO values [17]. The relation Eopt = hc/*k* is used to calculate the sensitizers' optical band gap. The h designates Planck's constant, c designates the speed of light, and *k* designates the emission peak calculated from the PL emission spectra. A greater optical band gap value of 2.38 eV is displayed by the mengkulang sensitizer, as measured from the emission peak of 522 nm. Furthermore, the other point to a value of band gap that is comparable, falling somewhere in the range of 2.15 and 2.18 eV.

The peaks of emission found for the M:R (50:50%), M:R (60:40%), and M:R (40:60%) Table 1 provides a measurement of the optical band gap for each of the extractives used in this investigation. When measured in comparison to the rengas sensitizer, the corresponding wavelengths of the mengkulang mixes are 569, 572, and 573 nm. As exhibited in Fig. 7, the mengkulang sensitizer is the only one that displays a wide emission peak; the other extractives all show emission peak trends that are quite similar to one another.

**Table 1** Summary of optical band gap measurement

| Sensitizer | λ max (nm) | $E_g^{opt} = hc/\lambda$ (eV) |
|---|---|---|
| Mengkulang | 522 | 2.38 |
| Rengas | 576 | 2.15 |
| M:R (50:50%) | 569 | 2.18 |
| M:R (40:60%) | 573 | 2.16 |
| M:R (60:40%) | 572 | 2.17 |

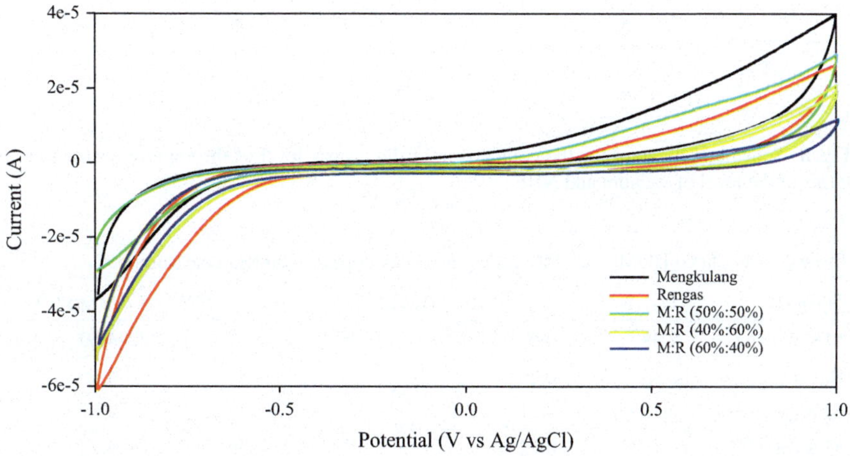


**Fig. 7** Cyclic voltammograms of individual and mixed extractives

In Fig. 8, the cyclic voltammograms for the mengkulang, rengas, and mix extractives are displayed for the reader's perusal. Measurements using cyclic voltammetry are performed at a scan rate of one hundred millivolts per second (mV/s) by using three distinct electrodes, known as, a glassy carbon working electrode, a platinum counter electrode, and an Ag/AgCl reference electrode. Table 2 presents the results of the calculations for the position of HOMO and LUMO levels. The HOMO level, also known as EHOMO, is established by extrapolating the voltammograms of the sensitizers in order to get a conclusion on their respective oxidation potential onset. According to Zhou et al. research's [26], the equation onset EHOMO 14 4:4 Eox) was used (eV).

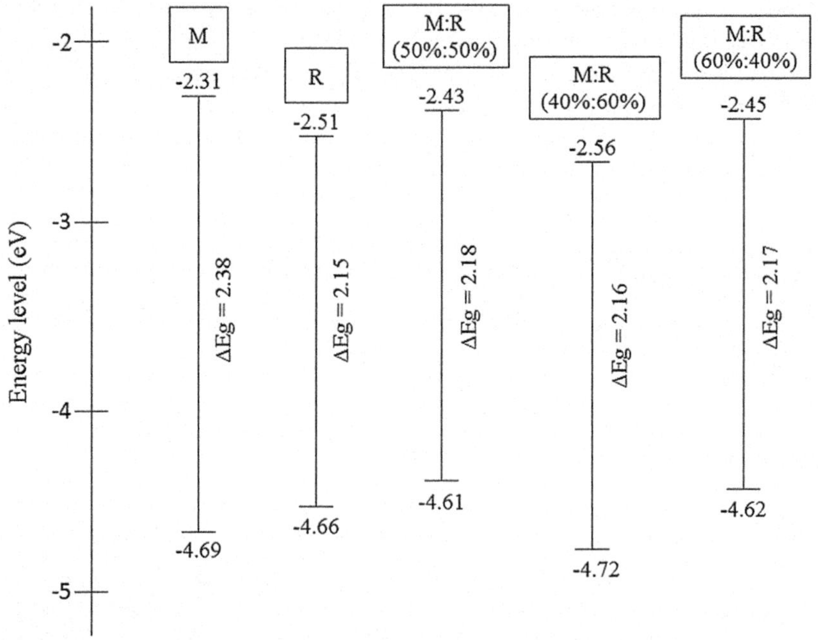


**Fig. 8** Schematic diagram for the comparison of the energy level of the explored extractives in terms of the level of vacuum and NHE

**Table 2** Calculated HOMO–LUMO energy levels and optical band gap measurement

| Sensitizer | $E_g^{opt}$ (eV) | $E_{ox}^{onset}$ versus Ag/AgCl (V) | $E_{LUMO}$ (eV) | $E_{HOMO}$ (eV) |
|---|---|---|---|---|
| Mengkulang | 2.38 | 0.29 | −2.31 | −4.69 |
| Rengas | 2.15 | 0.26 | −2.51 | −4.66 |
| M:R (50:50%) | 2.18 | 0.21 | −2.43 | −4.61 |
| M:R (40:60%) | 2.16 | 0.32 | −2.56 | −4.72 |
| M:R (60:40%) | 2.17 | 0.22 | −2.45 | −4.62 |

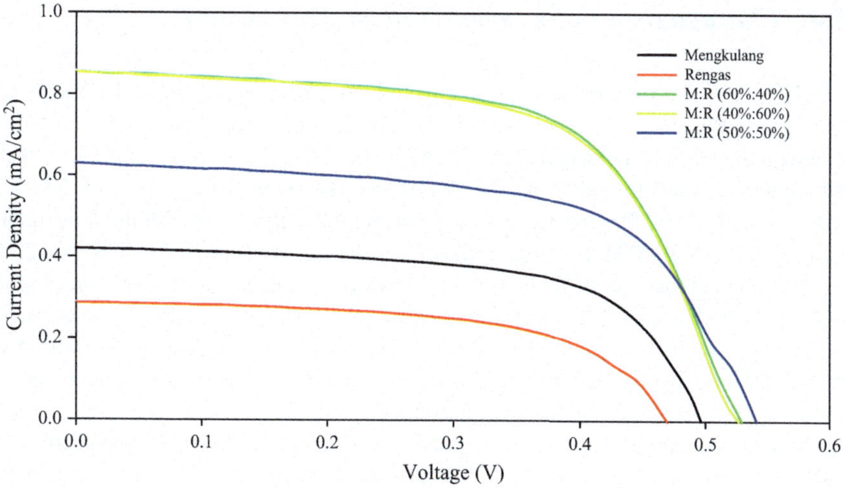


**Fig. 9** Comparison of *I–V* characteristics of individual and mixed extractives

Due to the presence of this condition, it may be deduced that the vacuum level and the derived energy levels are separated by 4.4 eV. The LUMO's position is determined by adding the HOMO level to the optical band gap of sensitizer, denoted via this equation: ELUMO = EHOMO Eopt [27–29]. The supporting electrolyte is 0.1 M lithium perchlorate, which is chemically represented by $LiClO_4$. Figure 9 is a graphic that offers a comparison of the energy levels of the extractives as regards the level of vacuum and the conventional hydrogen electrode. This comparison is depending on the position of HOMO–LUMO that is reported in Table 2 (NHE).

Kumara et al. [30], Ooyama and Harima [31] It has been shown that the $TiO_2$'s conduction band (CB) is situated about −4 to −4.3 eV. When compared to the level of the vacuum, the LUMO position of both individual and combination extractives are found to have a higher positive location from CB of $TiO_2$ (Fig. 9). In order for sensitizers to be able to insert electrons with high quantum yields into the CB of $TiO_2$, it is necessary for the LUMO positions of extractives to be situated in a position that is higher than the energy levels of semiconductors [32]. According to Boschloo and Hagfeldt's research [33], the LUMO level of the extractive and the $TiO_2$'s CB should have a difference of more than 0.2 eV in order to get the required results.

This requirement should be accompanied by the effective regeneration of sensitizer by electron movement from the redox pair in the electrolyte so that can achieve a highly efficient device. This is the case even if the LUMO values for all extractives are more than 0.2 eV. It is necessary for the HOMO position of the extractive to be much lower compared to redox potential in regard to the vacuum level [32]. The acceptable range for this difference is 0.2–0.3 eV. According to the findings of other research, the $I_3^-/I^-$ redox pair's redox potential may be found in the range of −4.6 to −5 eV [12, 13, 34, 35].

### 3.4 Characteristics of Current-Voltage (I–V) Curve

The I-V curves for individual mengkulang and rengas extractives as well as the mixes are depicted in Fig. 10. Table 3 presents the results of calculations and tabulations made on the DSSCs' performances. The M:R (60:40%) gives the good performance, which achieves an efficiency ($\eta$), 0.295% at an open circuit voltage ($V_{oc}$), 0.528 V, a fill factor (FF) of 62.16, and a short circuit current density ($J_{sc}$) of 0.9 mA/cm$^2$ under an irradiance of 1000 W/m$^2$. In addition, these values are achieved at a FF of 62.16.

The mix extractive M:R (40:60%) likewise produces somewhat comparable outcomes, with a $\eta$ of 0.292 percent at an $V_{oc}$ of 0.526 V, a $J_{sc}$ of 0.9 milliamperes per square centimeter, and a FF of 61.78. In addition to this, the mix extractive M:R (50:50%) displays a $\eta$ of 0.205 percent when the Voc is 0.540 V, the $J_{sc}$ is 0.6 milliamperes per square centimeter, and the FF is 63.28.

As a whole, mengkulang extractive achieves a larger $\eta$ of 0.161% when used as the individual sensitizer at an Voc of 0.529 V, a FF of 75.98, and a $J_{sc}$ of 0.4 mA/cm$^2$; these

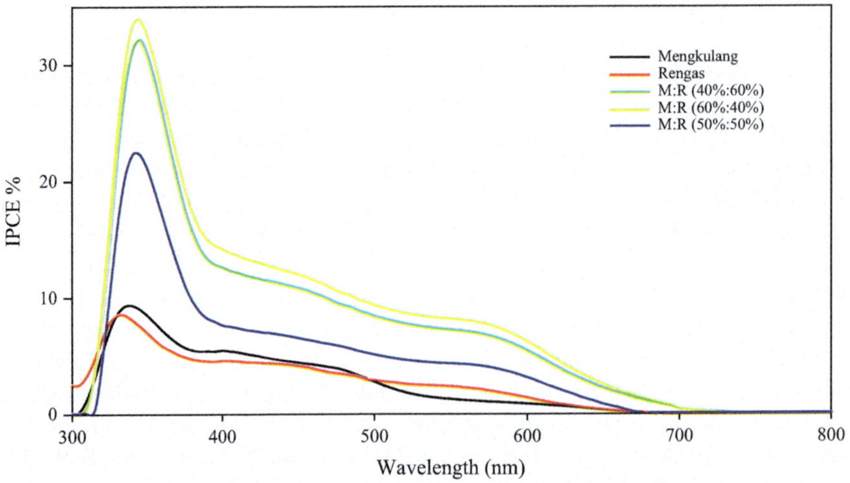


**Fig. 10** IPCE spectra of individual and mixed extractives

**Table 3** Current-voltage ($I$–$V$) characteristics and power conversion efficiencies ($\eta$) of DSSCs sensitized with different types of extractives

| Sensitizer | $V_{oc}$ (V) | $J_{sc}$ (mA cm$^{-2}$) | *ff* | $\eta$/% |
|---|---|---|---|---|
| Mengkulang | 0.53 | 0.40 | 75.98 | 0.16 |
| Rengas | 0.50 | 0.30 | 72.88 | 0.11 |
| M:R (50:50%) | 0.54 | 0.60 | 63.28 | 0.21 |
| M:R (40:60%) | 0.53 | 0.90 | 61.78 | 0.29 |
| M:R (60:40%) | 0.53 | 0.90 | 62.16 | 0.30 |

values are all measured in milliampere-seconds per square centimeter. In contrast, the rengas extractive achieves a $\eta$ of 0.109% at an $V_{oc}$ of 0.5 V, a $J_{sc}$ of 0.3 mA/cm$^2$, and a FF of 72.88. These values were determined by multiplying the $V_{oc}$ by the $J_{sc}$.

The value of $J_{sc}$ is more for the combined sensitizers than it is for the individual sensitizers, which is the reason why the $\eta$ of the mixed extractives is greater. The value of $J_{sc}$ indicates that there is a significant interactivity between $TiO_2$ and the extractive in the photoanode, and their capacity to absorb light, which ultimately effects a high efficiency of electron-injection activity.

## 4 Effect of Ruthenium (N719) Dye as Co-sensitizers on the Performance of DSSC

There has been a significant amount of research employing the possibility of using dye-sensitized solar cells, also known as DSSCs, as an alternative technology for use in solar energy applications. The production of DSSCs is less hazardous to the environment and less expensive than that of traditional silicon-based cells [36]. DSSC devices are made up of a photoelectrode, which consists of a broad bandgap metal oxide anchored by molecules of sensitizer, and a counter electrode, such as platinum. Both of these components are sandwiched between the layers of two conducting glasses.

For a purpose of enabling electronic conduction, a mesoporous metal oxide, including $TiO_2$, ZnO, $SnO_2$, and Nb205 is placed on the conducting glass and then annealed. Among them, the most frequently employed is the anatase $TiO_2$ that possesses a good dielectric constant. This is because it offers effective electrostatic shielding for the inserted electron that inhibits the effect of recombination. It is crucial to escalate the light absorption efficiency [36], and the good refractive index of anatase $TiO_2$ induces effective light scattering inside the porous photoelectrode.

To aid in electron transit and recover depleted sensitizer, the electrolyte is injected into the space between the electrodes. The usual electrolytes used are iodide/triiodide redox couple's dissolves in the organic solvent [37]. The DSSC operating concept is hypothesized to be analogous to photosynthesis; it is essential for the device to be able to collect visible light and transform chemical energy into electrical energy [38]. A component known as the sensitizer is anchored onto the $TiO_2$ surface of the photoelectrode. This component is in charge of the light captivation by a photoelectrode.

After being oxidized in the visible light spectrum (380–800 nm), the sensitizer's job is to distribute electrons throughout the cell in order to fulfill its function. In most cases, the extractive is a metal-organic dye, although it can also be an organic dye or a natural derivative dye. The steady and good $\eta$ are achieved by ruthenium (Ru) polypyridal compounds, such as cis-dithiocyanobis(4,4′-di-carboxy-2,2′-bipyridine) ruthenium(II) (N3), ditetrabutylammonium-cis-bis(isothiocyanates) bis(2,2′-bipyridyl-4,4′-dicarbox).

In spite of the fact that the $\eta$ of Ru complex dyes is more than 10%, their potential is constrained by a number of drawbacks. For example, ruthenium is a very scarce, costly, and ecologically unfriendly metal [40]. Organic dyes are also effective sensitizers due to the fact that their stability can be considered to be good and that they have a substantial molar absorption coefficient [41]. There is no need to worry about a shortage of resources while working with organic dyes because they do not include any metals, and their stereochemical structure may be altered [31].

An improved conversion efficiency is the consequence of adjusting the absorption qualities of the sensitizer and rearrangement of its energy level, both of which are brought about by the alteration of molecular structures. Furthermore, the majority of the molecular modifications of organic dyes need complex synthesis techniques [42]. The DSSC based on porphyrin organic dye obtained a high $\eta$ of 13% [43], making it stand out among the organic dyes.

Natural pigments have a straightforward extraction process, a low cost, and a minimal environmental effect, making them a potentially useful alternative to synthetic sensitizers. These numerous natural pigments appear in several fruits, flowers, and leaves of plants. Including molecules that significantly captivate visible light at distinct wavelengths, which results in diverse hues that can be seen [44]. [Note: A number of naturally occurring pigments, such as chlorophyll [45], anthocyanin [46], xanthophyll [47], and carotene [48], have been the concern of significant research].

In spite of the conversion efficiencies that were attained (3%), which are substantially lower than the criteria set by the industry [49]. A lot of research have discovered the potential of distinct plant species to act as sources of sensitizers. The dye-sensitized solar cells' performance is depending on natural extractives that have been derived from rengas (Gluta spp.) and mengkulang (Heritiera elata) wood is further analyzed in this section.

This discovery sheds information on the electron transport mechanism that exists within the cells, which may then be used to forecast any malfunctions that may occur in the devices. The individual sensitizer, which was extracted from rengas (Gluta spp.) and mengkulang wood (Heritiera elata), the mixture extractive, which was created by the mixture of both individual dyes, and the co-extractive, which was an individual extractive that had been co-sensitized with Ru-N719 dye were three types of extractives that were investigated.

## 4.1 The Dyes and DSSC Device's Optical Properties

The findings of the UV-Vis and IPCE spectra are connected to one another in some way. Both studies show the region of the sensitizer that is capable of converting the photons that it has taken into electric current when exposed to an extremely high intensity of light. The IPCE may be summarized in the following way:

$$\text{IPCE}\ (\%) = 1240 \times J_{\text{ph}}/\lambda \times I \quad (1)$$

while

$$\text{IPCE}(\lambda) = \text{LHE} \times \phi_{\text{inj}} \times \eta_{\text{col}} \quad (2)$$

where $J_{ph}$ (mA $cm^{-2}$), (nm), and I ($cm^{-2}$) represent the short circuit photocurrent density, wavelength, and intensity of the monochromatic light, respectively, the inj refers to the efficiency of electron injection from the excited dye into the $TiO_2$, the col refers to the efficiency of collection of the injected electron at the electrode, and the LHE refers to the light harvesting efficiency of the electrode. IPCE is connected in a roundabout way to the dye absorption that can be seen in the spectra of UV-Vis [31].

Figure 11 depicts the absorption spectra of different extractives, including mengkulang and rengas sensitizer, as they adsorb onto a layer of $TiO_2$ together with the IPCE spectra that were generated from the spectral response. The mengkulang extractive has a more extensive absorption spectrum, and the test wavelengths reveal no prominently strong peaks. On the other hand, the rengas dye exhibits extremely bright peaks at 400 and 576 nm. Because of its limited ability to absorb light across a broad spectrum, rengas dye has a decreased light-gathering capacity [50]. As a consequence of this, the proportion of mengkulang-sensitized DSSC in the IPCE in the lower wavelength range is greater than the percentage of rengas-sensitized DSSC (400–500 nm). On the other hand, rengas-sensitized DSSC displays a respectable IPCE % in the upper wavelength between 523 and 600 nm.

Besides, the M:R dyes with a ratio of 50:50% and M:R dyes with a ratio of 60:40% display the same absorption pattern when used with combination sensitizers, as can be seen in Fig. 12. Between 500 and 550 nm, one can make out a sizable shoulder in

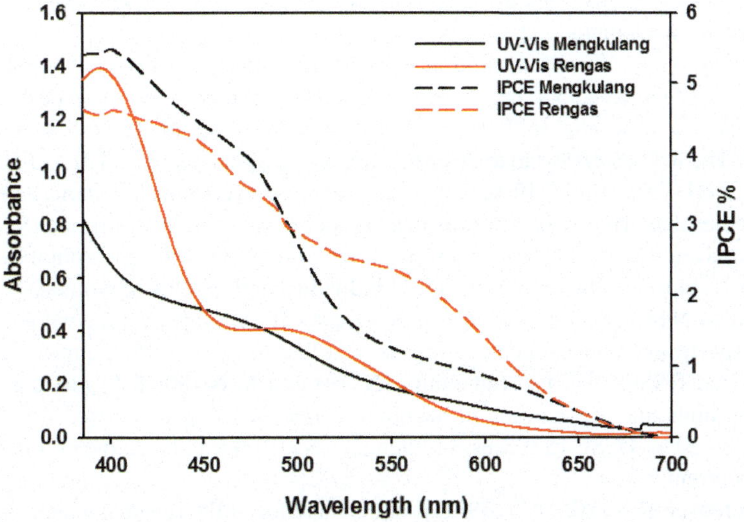


**Fig. 11** UV-Vis absorption and IPCE spectra of individual mengkulang and rengas extractives

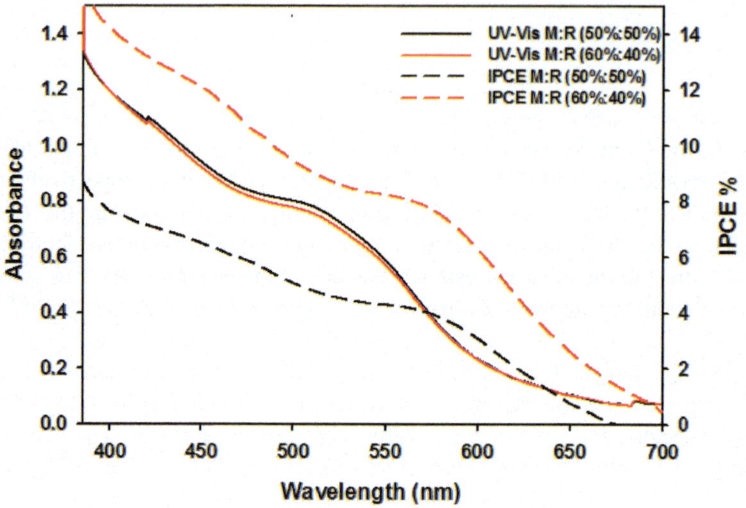


**Fig. 12** UV-Vis absorption and IPCE spectra of the mixture extractives

the spectrum. The M:R (60:40%)-DSSC has the larger IPCE that suggests the cells are capable of converting more photons into current [51]. The region between 550 and 600 nm exhibits the broad shoulder that is characteristic of IPCE spectra for mixed sensitizers.

When compared with the absorption spectrum, the peak is moved to a higher wavelength. Because of this finding, it appears that mixture extractive-DSSCs are able to convert electrons and harvest light more than individual-sensitized DSSCs. This is due to the fact that the absorption and IPCE spectra encompassed the more extensive wavelengths' range.

In addition to this, the absorption spectra for the co-sensitizers have broad shoulders between 400 and 550 nm, and Fig. 13 shows that there is no sharp peak. Sensitizers with a M:N ratio of 80:20% and M:R:N ratios of 40:40:20% follow a similar trend. The R:N (80:20%) extractive features two distinct peaks at 420 and 505 nm, respectively. The Ru-N719 sensitizer acts as a co-dye to anchor more extractive molecules onto $TiO_2$, which results in a broader range of light absorption [52]. This is the reason why the absorption spectrum for co-dyes has wide absorption bands in comparison to the absorption spectra of individual and mixture sensitizers.

The M:N (80:20%)-DSSC possesses a greater IPCE value over the whole visible light spectrum. A strong peak can be seen at 530 nm in the IPCE curves for dye-sensitized DSSCs that use ruthenium (N719) and M:N (80:20%) mixtures. This finding hints that the addition of anchored extractive molecules onto $TiO_2$ for the M:N (80:20%)-DSSC [53] causes ruthenium (N719) dye to boost the production of photocurrent.

Moreover, R:N (80:20%)-DSSC produces a reduced IPCE value, which suggests that the rengas extractive prevents the other extractive from efficiently collecting light

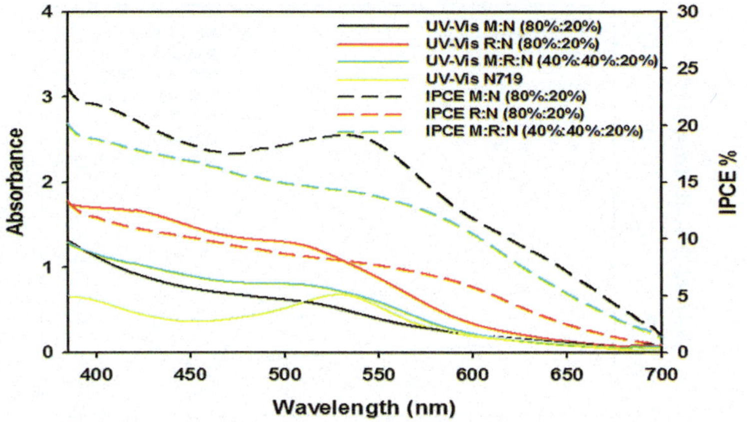


**Fig. 13** UV-Vis absorption and IPCE spectra of co-sensitization dyes and the distinct compositions

and generating current. The IPCE peak for R:N (80:20%) and M:R:N (40:40:20%)-DSSCs that lack a prominent peak at 530 nm, revealed a pattern that is quite similar to one another. Because of their unique capacities for adsorption, various sensitizers produce a spectrum of distinctively diverse absorption bands [54], which in turn results in an IPCE plot that displays a variety of distinctively varied configurations [55].

## *4.2 Photovoltaic Performances*

Figure 14 depicts the solo, mixed, and co-sensitized DSSCs' current-voltage (I–V) curves and Table 4 provides a summary of the photovoltaic performance metrics of each DSSC. It is possible to determine the $\eta$ by utilizing Eq. 3, which states that:

$$\eta = (V_{oc} \times J_{sc} \times \text{FF})/P_{in} \tag{3}$$

The greatest value of voltage recorded in an open circuit is known as the $V_{oc}$, while the maximum current density collected by a cell in a short circuit is known as the $J_{sc}$. The I–V curves that were drawn in Fig. 4 display both of these values. The FF is interpreted as the ratio of the optimal photocurrent and voltage to the product of $V_{oc}$ and $J_{sc}$. The pin is the intensity of the incident light, which is measured in milliwatts per square centimeter.

The value of Voc is nearly constant for individually sensitized DSSCs, ranging from 0.50 to 0.54 V. This indicates the potential difference between the electrochemical potential of electrolyte and the Fermi level of semiconductor does not have

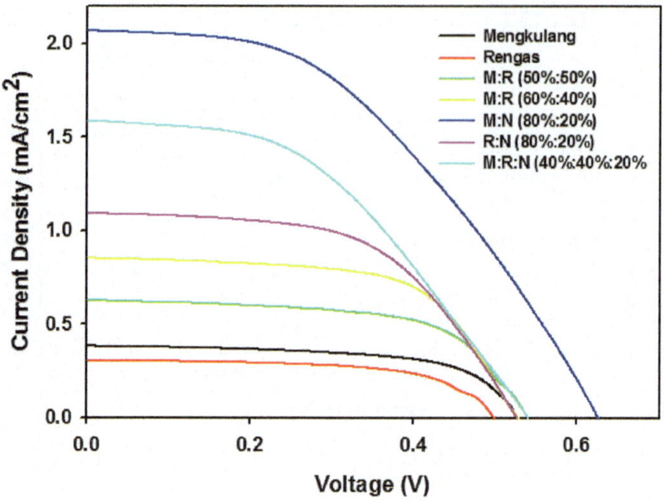


**Fig. 14** I–V curves of DSSC device sensitized with individual, mixture, and co-extractives

**Table 4** Current-voltage (I–V) performance and power conversion efficiency ($\eta$) of DSSCs sensitized with the individual, mixture, and co-extractives

| Sensitizer | $V_{oc}$ (V) | $J_{sc}$ (mA cm$^{-2}$) | FF | $\eta$ (%) |
|---|---|---|---|---|
| Mengkulang (M) | 0.53 | 0.40 | 75.98 | 0.16 |
| Rengas (R) | 0.50 | 0.30 | 72.88 | 0.11 |
| M:R (50:50%) | 0.54 | 0.60 | 63.28 | 0.21 |
| M:R (60:40%) | 0.53 | 0.90 | 62.16 | 0.30 |
| M:N (80:20%) | 0.63 | 2.10 | 44.00 | 0.58 |
| R:N (80:20%) | 0.53 | 1.10 | 57.07 | 0.33 |
| M:R:N (40:40:20%) | 0.55 | 1.60 | 45.23 | 0.39 |
| *N719 | 0.82 | 7.40 | 62.55 | 3.80 |

* The ruthenium (N719)-based DSSC is shown for comparison

a significant amount of variation [39]. When compared to the results for rengas-sensitized DSSC, mengkulang-sensitized DSSC achieves a $\eta$ that is slightly higher, coming in at 0.16% with an $V_{oc}$ of 0.53 V, $J_{sc}$ of 0.40 mA cm$^{-2}$, and FF of 75.98.

In addition, it is anticipated that mixture-sensitized DSSCs would function better than individual-sensitized DSSCs as a result of the absorption increase that was covered in the preceding paragraph. The photovoltaic performance of the M:R (60:40%)-DSSC is improved ($\eta$: 0.30%, $V_{oc}$: 0.53 V, $J_{sc}$: 0.90 mA cm$^{-2}$, and FF: 62.16). Nevertheless, the photovoltaic performance of the M:R (50:50%)-DSSC is somewhat inferior ($\eta$: 0.21%, $V_{oc}$: 0.54 V, $J_{sc}$: 0.60 mA cm$^{-2}$, and FF: 63.28). For the mixture-DSSCs, a good rise in Jsc is achieved, which shows that they are able to create compact electron collection at the surface of the $TiO_2$ layer and that electrons are successfully inserted via the photoanode [31].

The M:N (80:20%)-DSSC demonstrates a greatest performance among the co-sensitized DSSC. This particular DSSC has a $\eta$ of 0.58%, an $V_{oc}$ of 0.63 V, a $J_{sc}$ of 2.10 mA $cm^{-2}$, and a FF of 44.00. The R:N (80:20%) and M:R:N (40:40:20%)-DSSCs exhibit $\eta$ that is comparable to one another. These efficiencies are 0.33% ($V_{oc}$: 0.53 V, $J_{sc}$: 1.10 mA $cm^{-2}$, and FF: 57.07) and 0.39% ($V_{oc}$: 0.55 V, $J_{sc}$: 1.60 mA $cm^{-2}$, and FF: 45.

Their Jsc and Voc values are lesser than the DSSC that has been sensitized by a M:N ratio of 80:20%. This situation indicates that the extractive molecule does not efficiently graft onto the $TiO_2$ layer that has an effect on the movement of electrons from the lowest unoccupied molecular orbital (LUMO) to the Fermi level in $TiO_2$ [44]. The good value of Jsc obtained by the M:N (80:20%)-DSSC indicates that the dense electron collection at the surface of $TiO_2$ causes the shift of Fermi level toward the CB [56].

As a consequence, the value of $V_{oc}$ is higher, which ultimately results in better cell conversion efficiency and process. The addition of ruthenium (N719) dye has an effect on the results, and usage with of the great cell performance demonstrated by co-extractive DSSCs. The photovoltaic performance of the co-sensitized DSSCs is low when compared to the performance attained by the Ru-N719 DSSC as summarized in Table 4. Cell performance can be affected by the kinetics of the electrochemical and photoelectrochemical reactions that occur within the device.

## 4.3 Electrochemical Impedance Study

Electrochemical impedance spectroscopy is utilized further in order to complete the analysis of the DSSC device's electron transport parameter values (EIS). The EIS analysis is performed under one light intensity of AM 1.5, with the circuit state set to open. The theory behind this study, as well as its parameters and computations, have been covered in prior discussions [57, 58]. The diffusion-recombination activity under frontier circumstances is stated as follows [57, 59, 60]:

$$Z = R_{ct}(1/(\omega_{rec}/\omega_d)(1 + i\omega/\omega_{rec}))^{(1/2)} \cot h[(\omega_{rec}/\omega_d)(1 + i\omega/\omega_{rec})]^{(1/2)} \quad (4)$$

The electron diffusion over the semiconductor layer of the photoanode (represented by the symbol $d$) and the electron back-reaction with oxidized redox species (represented by the symbol rec) are the two fundamental processes that are involved in the diffusion-recombination effect. These parameters may be specified as follows:

$$\omega_d = D_{eff}/L^2, \ \omega_{rec} = k_{eff} = 1/\tau_{eff} \quad (5)$$

In the meantime, the diffusion-recombination resistances can be defined as follows:

$$R_{\mathrm{ct}} = \mathrm{con}\ L/D_{\mathrm{eff}} = L^2/(D_{\mathrm{eff}}\ C\mu),\ R_{\mathrm{rec}} = \mathrm{con}\ 1/(L\ k_{\mathrm{eff}}) = \mathrm{Ln}^2/(D_{\mathrm{eff}}\ C\mu) \quad (6)$$

where the constant (con) is defined as follows:

$$\mathrm{con} = (\mathrm{KB}_T)/\left(q^2 A\eta_s\right) \quad (7)$$

where Z (Ω), $R_{ct}$ (Ω), $R_{rec}$ (Ω), $\omega_d$ (Hz), $\omega_{rec}$ (Hz), $\tau_{eff}$ (ms), Cμ, KB (J $K^{-1}$), $T$, $q$, $A$ ($m^2$), and $\eta_s$ designated the impedance, electron transport resistance in $TiO_2$, recombination resistance, characteristic frequency corresponds to the electron diffusion in $TiO_2$. Characteristic frequency corresponds to the electron back-reaction with oxidized redox species in the electrolyte, electron lifetime in $TiO_2$, the chemical capacitance of $TiO_2$, Boltzmann constant, absolute temperature, elementary charge, area of the electrode, and electron density at the conduction band of $TiO_2$, respectively.

Equation 8 [57] may be used to describe the connection that exists between the characteristic frequencies and the diffusion-recombination resistances.

$$\omega_d/\omega_{\mathrm{rec}} = R_{\mathrm{rec}}/R_{\mathrm{ct}} = \mathrm{Ln}^2/L^2 \quad (8)$$

Following is a written representation of the effective electron diffusion coefficient, $D_{eff}$ ($cm^2\ s^{-1}$), and the electron diffusion length, $L_n$ (m), which may be derived from Eqs. 5 and 6.

$$D_{\mathrm{eff}} = (R_{\mathrm{rec}}/R_{\mathrm{ct}})\ L^2 k_{\mathrm{eff}} \quad (9)$$

$$L_n = \sqrt{}(D_{\mathrm{eff}}/k_{\mathrm{eff}}) = \sqrt{}(D_{\mathrm{eff}}\ \tau_{\mathrm{eff}}) \quad (10)$$

The practical electron lifetime, $\tau_{eff}$ (ms), and effective recombination rate constant, $k_{eff}$ ($s^{-1}$) can be expressed in the term of peak frequency, fmax (Hz) obtained from Bode phase plot at the range of 1 Hz–1 kHz [58, 61],

$$\omega_{\mathrm{rec}} = 2\pi\ f_{\mathrm{max}} \quad (11)$$

Hence,

$$\tau_{\mathrm{eff}} = 1/(2\pi\ f_{\mathrm{max}}) = 1/\omega_{\mathrm{rec}} = 1/k_{\mathrm{eff}} \quad (12)$$

The DSSC was sensitized with individual, mixed, and co-extractive sensitizers, and Table 5 provides a summary of the electron transport parameters of the device. The EIS curves were fitted with the NOVA 1.10 software using the analogous circuit displayed in the inset of Fig. 15, different electron transport parameters may be obtained. The sheet resistance of the conducting glass, denoted by Rs, the constant phase element, denoted by CPE, and the interface charge transfer resistance make up the analogous circuit components ($R_{rec}$ and $R_{ct}$).

**Table 5** Electron transport properties of the DSSC device sensitized with individual, mixture, and co-extractive sensitizers

| Sensitizer | $R_{rec}$ (Ω) | $R_{ct}$ (Ω) | $D_{eff}$ ($cm^2$ $s^{-1}$) | $k_{eff}$ ($s^{-1}$) | $\tau_{eff}$ (ms) | $L_n$ (μm) | Con (Ω cm $s^{-1}$) |
|---|---|---|---|---|---|---|---|
| Mengkulang (M) | 150.0 | 46.2 | $4.20 \times 10^{-7}$ | 411.4 | 2.431 | 31.97 | 1.090 |
| Rengas (R) | 331.0 | 19.2 | $1.27 \times 10^{-6}$ | 234.1 | 4.272 | 73.66 | 1.370 |
| M:R (50:50%) | 407.0 | 16.9 | $1.01 \times 10^{-6}$ | 133.2 | 7.508 | 87.06 | 0.962 |
| M:R (60:40%) | 202.0 | 50.0 | $2.14 \times 10^{-6}$ | 1684.7 | 0.594 | 35.66 | 6.040 |
| M:N (80:20%) | 18.3 | 103.0 | $3.20 \times 10^{-9}$ | 57.2 | 17.493 | 7.48 | 0.019 |
| R:N (80:20%) | 26.2 | 139.0 | 5.96 × 10–9 | 100.5 | 9.953 | 7.70 | 0.047 |
| M:R:N (40:40:20%) | 21.0 | 121.0 | $5.49 \times 10^{-9}$ | 100.5 | 9.953 | 7.39 | 0.037 |

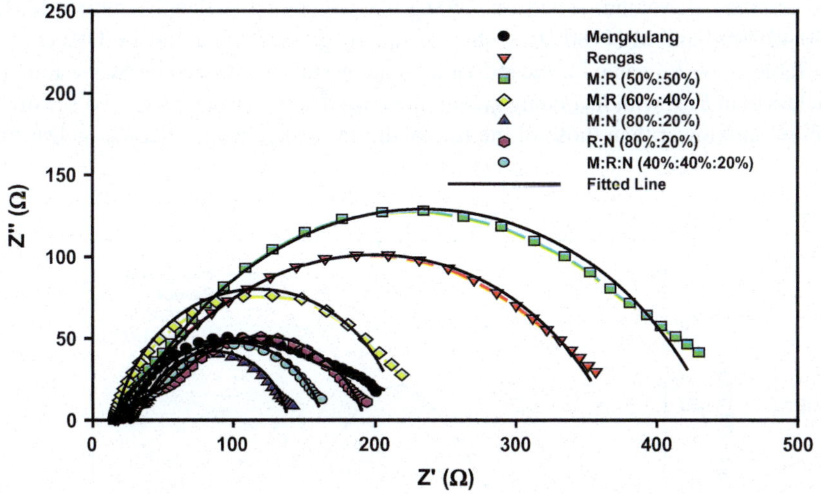


**Fig. 15** Nyquist plot based on the electrochemical impedance of the DSSC devices sensitized with the individual, mixture, and co-extractive sensitizer

Rrec is the transport resistance of electron diffusion at the interface of $TiO_2$/dye, whereas $R_{ct}$ is the electron transfer resistance of the charge recombination process at the interface of $TiO_2$/dye/electrolyte. Double-layer capacitance (Cdl) is being replaced by the capacitance of the double-layer electrode (CPE) in order to provide the divergence of the Cdl from the ideal interfacial capacitance [62, 63]. This deviation is due to the roughness of the electrode that is being employed.

The Nyquist plot is comprised of three semicircles, each of which is present in a different frequency range. The counter electrode is implemented for the transfer of charge, which causes the arc to occur in a high-frequency zone with a range of 1–100 kHz. The middle arc is determined by charge transfer at the interface of $TiO_2$/dye/electrolyte in the intermediate frequency range, which spans from one hertz to one kilohertz. In the meanwhile, an arc with a low frequency is produced as a result of diffusion in the redox electrolyte (0.2–0.75 Hz) [63]. According to the research that has been done [62], DSSCs do not certainly display three distinct curves in a Bode phase plot or three different curves in a Nyquist plot. Rather, this is something that can vary.

The Rrec value at the interface of $TiO_2$ layer/electrolyte was precisely correlative to the second arc's diameter, as shown by the Nyquist plot in Fig. 15. The apparatus that was sensitized with M:R (50:50%) had a diameter that was the longest, followed by rengas, M:R (60:40%), and mengkulang in that order. DSSC that was sensitized with mengkulang had the lesser value of $R_{rec}$ (150). Values of 202, 331, and 407 were achieved by the M:R (60:40%), rengas, and M:R (50:50%)-sensitized DSSCs, respectively.

Figure 16 illustrates the Nyquist plot of the co-extractive DSSC data set. DSSC that was sensitized with R:N (80:20%) had the largest value of Rrec (26.2) and the biggest diameter of the second arc. DSSCs that were sensitized with a ratio of M:R:N (40:40:20%) and M:N (80:20%) showed $R_{rec}$ magnitude of 21 and 18.3. According to Table 2, the $R_{ct}$ value for the individual-and mixture-sensitized DSSCs is found to be less than $R_{rec}$, with $R_{rec}$ being greater than $R_{ct}$. The $R_{rec}$ values of the co-sensitized DSSCs are lower than those of the $R_{ct}$ condition, which means that $R_{rec}$ is less than $R_{ct}$.

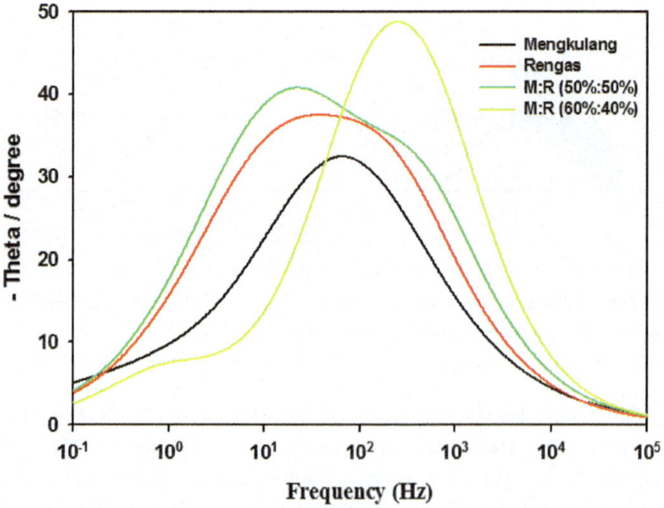


**Fig. 16** Bode phase plot of DSSC devices sensitized with mengkulang, rengas, and mixtures

The $R_{rec}/R_{ct}$ magnitude for co-extractive DSSCs is much less compared to 1, coming in at 0.18, 0.17, and 0.19, respectively, for M:N (80:20%), M:R:N (40:40:20%), and R:N (80:20%). A contour of the center curve veered off course, most noticeably at a higher frequency referred to as the Gerischer impedance [62]. In addition, only the rengas and the M:R (50:50%)-DSSCs were able to achieve a $R_{rec}/R_{ct}$ value that was greater than 10, with respective values of 17.2 and 24.1. As a consequence, the form of the middle arc was that of a perfect circle, as seen in Fig. 15. This outcome is consistent with the findings that Adachi et al. [58] obtained.

The Bode phase plots for the individual, mixed, and co-extractive DSSCs are displayed in Figs. 16 and 17, respectively. According to Eq. 12, the parameter fmax has an effect that is inversely proportional to the eff value across the range of frequency at 1 Hz–1 kHz. fmax represents the charge transfer that occurs at the interface between $TiO_2$ and the dye and electrolyte, and eff is the time constant that describes the amount of time needed for an electron to diffuse through $TiO_2$ before the surplus electrons recombine. M:R (50:50%)-DSSCs gain a lesser frequency magnitude of fmax at an intermediate region of frequency, followed by rengas, mengkulang, and M:R (60:40%)-DSSCs, for individual and mixture-sensitized DSSCs.

Since the fmax of the M:R (60:40%)-DSSC moves to the large-frequency zone, an eff value that is acquired is the shortest possible time at 0.594 ms. Besides, the M:R (50:50%), rengas, and mengkulang-sensitized DSSCs achieve 7.508, 4.272, and 2.431 ms. Because the position of fmax for co-sensitized DSSCs is at the lesser frequency of the intermediate area for M:N (80:20%)-DSSCs, this results in the greatest magnitude of eff, 17.493 ms.

Nevertheless, the R:N (80:20%) and M:R:N (80:20%)-DSSCs achieve a similar exact eff, which is 9.953 ms. This is due to the fact that the location of fmax

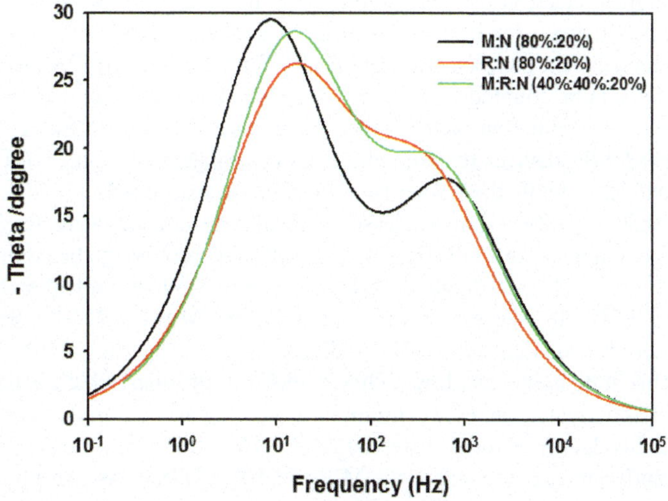


**Fig. 17** Bode phase plot of the co-sensitized DSSCs

that is obtained is the same in both cases. Based on this finding, it appears that the photovoltaic performance of the R:N (80:20%) and M:R:N (80:20%)-DSSCs is comparable. This was covered in the preceding section.

The $TiO_2$ thin film's thickness that was measured (L) for this experiment will serve as the control variable (17.74 m). $L_n$ for M:R (50:50%)-DSSCs is found to be higher than $L$, with values of 31.97, 73.66, 87.06, and 35.66 m. The value of effective electron diffusion length (Ln) is computed using Eq. 10. $L_n$ for mengkulang, rengas, M:R (50:50%)-sensitized DSSCs. This circumstance, in which $L_n$ is greater than $L$, is required for the electrons to be able to diffuse through the photoelectrode.

By returning to Eq. 7, we can see that when $L_n$ is more than $L$ and the value of $R_{rec}$ is finite, $R_{rec}$ must be greater than $R_{ct}$. This indicates that the recombination resistance has to be greater than the charge transport resistance. The greater the value of $R_{rec}$, the less likely it is that the recombination process will take place. The predetermined goal of successfully operating DSSC devices at moderate potentials while maintaining efficiency [58, 63]. The values of Rrec and Rct acquired for M:R (50:50%) and M:R (60:40%)-DSSCs establish a link for both situations. These values were measured for mengkulang, rengas, M:R (50:50%), and M:R (60:40%).

The amount of electron density (s) that builds up on the surface of a photoelectrode may be estimated based on the magnitude of the constant (con) that is determined by applying Eq. 7 to the calculation process. A greater value of estimated con would result in a less distribution of inserted electrons accumulating at the $TiO_2$'s CB, which would be the opposite of what would happen with a lower value. Based on the magnitude of con as summarized in Table 5, M:R (50:50%)-DSSCs are predicted to have the highest electron density, followed by M:R (60:40%)-DSSCs, mengkulang, and rengas.

For the co-extractive DSSCs, it is projected that the M:N (80:20%)-DSSC will have the highest electron density, followed by the M:R:N (40:40:20%)-sensitized DSSCs, and then finally the R:N (80:20%)-DSSCs. This collection of electrons would finally dispersed to the photoelectrode with the effective electron diffusion coefficient ($D_{eff}$), which can be counted by using Eq. 9. Assume that these electrons would be lost with the recombination rate constant ($k_{eff}$) that is the same as the reciprocal of eff [59]. For DSSC devices to attain efficient performance, the values of Deff and s must be increased, while the value of $k_{eff}$ must be decreased [44].

M:R (50:50%)-DSSCs show the lesser value of con (indicates a greater value of s) and the lesser value of keff (133.2 $s^{-1}$), and Deff is determined to be comparatively less, which was $1.01 \times 10^{-6}$ $cm^2$ $s^{-1}$. This is true for both individual and mixture-sensitized DSSCs. In contrast, it is projected that the M:R (60:40%)-DSSC would have a lesser electron density (s). This indicates that the electrons will be lost more quickly. This state is indicated by a shorter electron lifespan, which is denoted by eff, as well as a greater rate of recombination, which was measured at 1684.7 $s^{-1}$.

Despite this, Deff was measured to be $2.14 \times 10^{-6}$ $cm^2$ $s^{-1}$, which is a value that is significantly larger compared to the M:R (50:50%)-DSSC. As a result, a quicker movement of electrons through the $TiO_2$ layer is anticipated, which will ultimately result in an increase in the photocurrent density at the photoelectrode. This condition explains why the Jsc value reached for M:R (60:40%)-DSSC is more than the value of Jsc achieved for M:R (50%:50%)-DSSC, which ultimately leads to a larger $\eta$.

Furthermore, for the co-extractive DSSCs, the M:N (80:20%)-DSSC acquired the lesser value of con (indicates greater s) and the lesser keff (57.2 $s^{-1}$), as opposed to the R:N (80:20%)-and M:R:N (80:20%)-DSSCs, which both obtain the exact magnitude of keff. This is because of the M (100.5 $s^{-1}$). The Deff values that were determined for the M:N (80:20%)-DSSC are very different from one another. The practical electron lifespan (eff) of M:N (80:20%)-DSSC is the highest, which is a primary reason why it has the greatest value for conversion efficiency ($\eta$).

As a result, the electrons that were distributed across the porous semiconductor have sufficient period to move throughout the $TiO_2$ layer before the recombination process happened. Co-sensitized DSSCs have a $\eta$ that is much lesser compared to ruthenium (N719)-DSSCs, despite that this efficiency is greater compared to individual and mixture-DSSCs (Table 4). The $R_{ct}$ value is more compared to $R_{rec}$ ($R_{rec}$ is less than $R_{ct}$) for each and every co-sensitized DSSC.

In this instance, the time required for recombination is substantially less than that required for electron transport. The findings are presented in Table 5, which demonstrates that the value measured for $L_n$ is also lesser compared to $L$ ($L_n$ is less than $L$). In this scenario, the distribution electrons are lost more quickly and have to go through recombination before being able to be inserted into the CB and disperse all the way across the $TiO_2$. In conclusion, the criteria mentioned, are regarded to be unimportant, and it is recommended that they be avoided to attain an effective performance from the cells.

## 5 Conclusion

Natural sensitizers are taken from the rengas (Gluta spp.) and mengkulang (Heritiera elata) wood in order to manufacture the DSSCs. Several different types of sensitizers, including individual, mixture, and co-sensitized sensitizers, are put into play here. If we compare the photovoltaic performance of mixture-DSSCs to that of individually sensitized DSSCs, we see that mixture-sensitized DSSCs have up to a twofold increase in short circuit current density ($J_{sc}$), which indicates an enhanced conversion efficiency. The IPCE curve showed that the M:R (60:40%)-sensitized DSSC had good light harvesting, charge collecting, and dye regeneration efficiency because it obtained a greater photocurrent.

The electrochemical impedance spectroscopy study is utilized to provide estimates on the behavior of electron transport inside the cells. Despite the M:R (50:50%)-DSSC acquired a lesser recombination rate, keff, and a larger estimated electron density, s, the $\eta$ was slightly lesser compared to the M:R (60:40%)-DSSC because the electron diffusion coefficient, Deff, was lesser. This was the cause of the slight decrease in efficiency.

When the prerequisites of $R_{rec} > R_{ct}$ and $L_n > L$ were realized, it was determined that other factors important to the sensitizer's applicability had been reached. These conditions were adhered to by individual and mixture-sensitized DSSCs throughout the course of this research. Co-sensitized DSSCs, on the other hand, were responsible for the phenomena in which $R_{rec} > R_{ct}$ and $L_n > L$. In this particular instance, the boundary condition was regarded as being unnecessary and ought to have been avoided.

In order to be effective, a sensitizer must have a larger electron diffusion coefficient (also known as $D_{eff}$), a less recombination rate (also known as $k_{eff}$), and a larger electron density (also known as s). Due to the fact that the $\eta$ attained by M:N (80:20%), R:N (80: 0%), and M:R:N (80:20%)-DSSCs is larger than that of individual and mixture-DSSCs, these DSSCs are not deemed to be acceptable for use as sensitizers. Based on these data, it appears that mixture sensitizers have a significant amount of untapped potential to enhance the electrical characteristics and overall $\eta$ of sensitizers. Figure 18 provides illustrations of dye-sensitized solar cell prototypes for your perusal.

Lastly, owing to the abundant amount and low cost of wood waste, the DSSC has a strong potential to generate power as a low-cost material for developing nations. This is due to the fact that wood waste is abundant. The technique of extracting the dye is not only simple, but it also calls for less facilities than that of obtaining commercial silicon, which is not a substance that can be replenished and may be harmful to the environment. Even though the $\eta$ of DSSC is lesser compared to commercial silicon-based solar cells, it is possible to utilize a bigger panel of DSSC to achieve the same energy conversion efficiency as a panel made of silicon sun cells. In the future, further research has to be done to isolate just the hydroxyl or anthocyanin groups, which will allow for more efficient electron transmission.

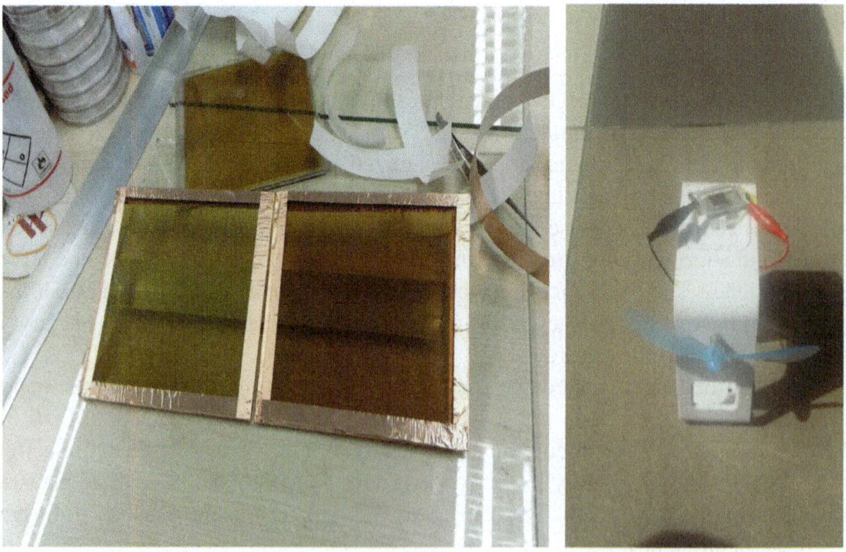

**Fig. 18** The prototype of DSSC made from rengas and mengkulang extractives

## References


1. Theoman H, Erle M, Sernek M (2010) Wood based panels: an introduction for specialist. Brunel University Press, London, UB8 3PH
2. Rowell RM (2005) Handbook of wood chemistry and wood composites. CRC Press, 487pp
3. Zhoa X, Guo P, Zhang Z, Peng H (2019) Anatomical features of branchwood and stemwood of *Betula costata* Frauts from natural secondary forests in China. BioResources 14(1):1980–1991
4. Longoi EL, Romeiro DL, Alves ES (2012) Difference in anatomy and potential hydraulic conductivity between root and stem of *Caesalpina echunata* Lam, (Fabaceae). Hoahnea 39(4):649–655
5. Bal BC, Beltas I (2012) The physical properties of heartwood and sapwood of *Eucalyptus grandis*. Pro Ligno 8(4):35–43
6. Eynollahi Y, Khalthali MB, Tavankar F (2013) Study of physicochemical properties of heartwood and sapwood of oak and maple trees in the Caspian Forest, Iran. Int J Appl Sci Eng Res 2(5):487–492
7. Song K, Yin Y, Salman L, Xiao F, Jiang X (2014) Changes in the properties of wood cell walls during the transformation from sapwood and heartwood. J Mater Sci 49:1734–1742
8. Bektas I, Tutus A, Gultekin G (2020) The effect of sapwood and heartwood difference on mechanical properties of fast-growing tree species. Drvna Industrija 71(3):261–269
9. Kim MJ, Kim JYR, Kim BR (2020) Mechanical characteristics of Korean red Pinus according to province (Goseng, Hongchean, and Bonghwa-gui). J Korean Wood Sci Technol 48(50):666–675
10. Liu B, Li W, Wang B, Li X, Liu Q, Naruta Y, Zhu W (2013) Influence of different anchoring groups in indoline dyes for dye-sensitized solar cells: electron injection, impedance and charge recombination. J Power Sour 234:139–146
11. Liang M, Chen J (2013) Arylamine organic dyes for dye-sensitized solar cells. Chem Soc Rev 42
12. Zang X, Huang Z, Wu H, Iqbal Z, Wang L, Meier H, Cao D (2014) Molecular design of the diketopyrrolopyrrole-based dyes with varied donor units for ef fi cient dye-sensitized solar cells. J Power Sour 271:455–464
13. Dou Y, Wu F, Fang L, Liu G, Mao C, Wan K (2016) Enhanced performance of dye-sensitized solar cell using Bi 2 Te 3 nanotube/ZnO nanoparticle composite photoanode by the synergistic effect of photovoltaic and thermoelectric conversion. J Power Sour 307:181–189
14. Yella A, Lee HW, Tsao HN, Yi C, Chandiran AK, Nazeeruddin MK, Gratzel M (2011) Porphyrin-sensitized solar cells with cobalt (II/III)-based redox electrolyte exceed 12 percent efficiency. Science 334(6056):629–634
15. Hagfeldt A, Boschloo G, Sun L, Kloo L, Pettersson H (2010) Dye-sensitized solar cells. Chem Rev 110:6595–6663
16. Cao Y, Bai Y, Yu Q, Cheng Y, Liu S, Shi D, Wang P (2009) Dye-sensitized solar cells with a high absorptivity ruthenium sensitizer featuring a 2-(Hexylthio) thiophene conjugated bipyridine. Alcohol 6290–6297
17. Leonat L, Sbârcea G, Brañzoi IV (2013) Cyclic voltammetry for energy levels estimation of organic materials. UPB Sci Bull Ser B: Chem Mater Sci 75:111–118
18. Mathew S, Yella A, Gao P, Humphry-Baker R, Curchod BFE, Ashari-Astani N, Tavernelli I, Rothlisberger U, Nazeeruddin MK, Grätzel M (2014) Dye-sensitized solar cells with 13% efficiency achieved through the molecular engineering of porphyrin sensitizers. Nat Chem 6(3):242–247. https://doi.org/10.1038/nchem.1861
19. Smestad GP (1998) Demonstrating electron transfer and nanotechnology : a natural dye—Sensitized nanocrystalline energy converter. J Chem Educ 75(6):752–756
20. Gong J, Liang J, Sumathy K (2012) Review on dye-sensitized solar cells (DSSCs): Fundamental concepts and novel materials. Renew Sustain Energy Rev 16(8):5848–5860
21. Grätzel M (2004) Conversion of sunlight to electric power by nanocrystalline dye-sensitized solar cells. J Photochem Photobiol A 164(1–3):3–14

22. Narayan MR (2012) Review: dye sensitized solar cells based on natural photosensitizers. Renew Sustain Energy Rev 16(1):208–215. https://doi.org/10.1016/j.rser.2011.07.148
23. Lim A, Kumara NTRN, Ling A, Huq A, Chandrakanthi RLN, Iskandar M, Ekanayake P (2015) Spectrochimica acta Part A: molecular and biomolecular spectroscopy potential natural sensitizers extracted from the skin of canarium odontophyllum fruits for dye-sensitized solar cells. Spectrochim Acta Part A Mol Biomol Spectrosc 138:596–602
24. Kumara NTRN, Ekanayake P, Lim A, Iskandar M, Ming LC (2013) Study of the enhancement of cell performance of dye sensitized solar cells sensitized with nephelium lappaceum (F: Sapindaceae). J SolEnergy Eng 135:031014
25. Cardona CM, Li W, Kaifer AE, Stockdale D, Bazan GC (2011) Electrochemical considerations for determining absolute frontier orbital energy levels of conjugated polymers for solar cell applications. Adv Mater 23(20):2367–2371
26. Zhou J, Xie S, Amond EF, Becker ML (2013) Tuning energy levels of low bandgap semi-random two acceptor copolymers. Macromolecules 46(9):3391–3394
27. Ledwon P, Brzeczek A, Pluczyk S, Jarosz T, Kuznik W, Walczak K, Lapkowski M (2014) Synthesis and electrochemical properties of novel, donor-acceptor pyrrole derivatives with 1,8-naphthalimide units and their polymers. Electrochim Acta 128:420–429
28. Nazeeruddin MK, Péchy P, Renouard T, Zakeeruddin SM, Humphry-Baker R, Cointe P, Grätzel M (2001) Engineering of efficient panchromatic sensitizers for nanocrystalline $TiO_2$-based solar cells. J Am Chem Soc 123(c):1613–1624
29. Lim A, Manaf NH, Tennakoon K, Chandrakanthi RLN, Biaw L, Lim L, Ekanayake P (2015) Higher performance of DSSC with dyes from cladophora sp. as mixed cosensitizer through synergistic effect. J Biophys 2015(2015):8
30. Kumara NTRN, Petrovic M, Peiris DSU, Marie YA, Vijila C, Petra MI, Chandrakanthi RLN, Lim CM, Hobley J, Ekanayake P (2015) Efficiency enhancement of Ixora floral dye sensitized solar cell by diminishing the pigments interactions. Sol Energy 117:36–45. https://doi.org/10.1016/j.solener.2015.04.019
31. Ooyama Y, Harima Y (2012) Photophysical and electrochemical properties, and molecular structures of organic dyes for dye-sensitized solar cells. Chem Phys Chem 13(18):4032–4080. https://doi.org/10.1002/cphc.201200218
32. Chandra N, Nath D, Joon H, Choi W, Lee J (2013) Electrochimica Acta Electrochemical approach to enhance the open-circuit voltage ($V_{oc}$) of dye-sensitized solar cells (DSSCs). Electrochim Acta 109:39–45
33. Boschloo G, Hagfeldt A (2009) Characteristics of the iodide/triiodide redox mediator in dye-sensitized. Solar Cells 42(11):1819–1826
34. Gagliardi S, Falconieri M (2015) Experimental determination of the light-trapping-induced absorption enhancement factor in DSSC photoanodes. Beilstein J Nanotechnol 6(1):886–892
35. Tripathi B, Yadav P, Kumar M (2013) Theoretical upper limit of short-circuit current density of $TiO_2$ nanorod based dye-sensitized solar cell. Results Phys 3:182–186
36. Olea A, Ponce G, Sebastian PJ (1999) Electron transfer via organic dyes for solar conversion. Sol Energy Mater Sol Cells 59(1):137–143. https://doi.org/10.1016/S0927-0248(99)00038-0
37. Faccio R, Fernández-Werner L, Pardo H, Mombré AW (2011) Current trends in materials for dye sensitized solar cells. Recent Patents Nanotechnol 5(1):46i61. https://doi.org/10.2174/187221011794474930
38. Koyama Y, Miki T, Wang XF, Nagae H (2009) Dye-sensitized solar cells based on the principles and materials of photosynthesis: mechanisms of suppression and enhancement of photocurrent and conversion efficiency. Int J Mol Sci 10(11):4575–4622. https://doi.org/10.3390/ijms10114575
39. Nazeeruddin MK, Splivallo R, Liska P, Comte P, Grätzel M (2003) A swift dye uptake procedure for dye sensitized solar cells. Chem Commun (Cambridge, England) 1456–1457. https://doi.org/10.1039/b302566g
40. Adeloye AO, Ajibade PA (2014) Towards the development of functionalized polypyridine ligands for Ru(II) complexes as photosensitizers in dye-sensitized solar cells (DSSCs). Molecules 19(8):12421–12460. https://doi.org/10.3390/molecules190812421

41. Zeng Z, Zhang B, Li C, Peng X, Liu X, Meng S, Feng Y (2014) A key point of porphyrin structure affect DSSCs performance based on porphyrin sensitizers. Dyes Pigments
42. Hara K, Sato T, Katoh R, Furube A, Yoshihara T, Murai M, Kurashige M, Ito S, Shinpo A, Suga S (2005) Novel conjugated organic dyes for efficient dye-sensitized solar cells. Adv Func Mater 15(2):246–252. https://doi.org/10.1002/adfm.200400272
43. Lim A, Ekanayake P, Lim LBL, Bandara JMRS (2016) Co-dominant effect of selected natural dye sensitizers in DSSC performance. Spectrochim Acta Part A: Mol Biomol Spectrosc 167:26–31. https://doi.org/10.1016/j.saa.2016.05.024
44. Wang X, Kitao O, Hosono E, Zhou H, Sasaki S, Tamiaki H (2010) $TiO_2$—And ZnO-based solar cells using a chlorophyll a derivative sensitizer for light-harvesting and energy conversion. J Photochem Photobiol A 210(2–3):145–152. https://doi.org/10.1016/j.jphotochem.2010.01.004
45. Ozuomba JO, Okoli LU, Ekpunobi AJ, Science F (2013) The performance and stability of anthocyanin local dye as a photosensitizer for DSSCs. Adv Appl Sci Res 4(2):60–69
46. Kartini I, Dwitasari L, Wahyuningsih TD, Wang L (2015) The sensitization of xanthophylls-chlorophyllin mixtures on titania solar cells. Int J Sci Eng (IJSE) 109–114
47. Shanmugam V, Manoharan S, Anandan S, Murugan R (2013) Performance of dye-sensitized solar cells fabricated with extracts from fruits of ivy gourd and flowers of red frangipani as sensitizer. Spectrochim Acta Part A: Mol Biomol Spectrosc 104:35–40. https://doi.org/10.1016/j.saa.2012.11.098
48. Basheer B, Mathew D, George BK, Nair CPR (2014) An overview on the spectrum of sensitizers: the heart of dye sensitized solar cells. Sol Energy 108:479–507. https://doi.org/10.1016/j.solener.2014.08.00
49. Sayama K, Tsukagoshi S, Mori T, Hara K, Ohga Y, Shinpou A, Abe Y, Suga S, Arakawa H (2003) Efficient sensitization of nanocrystalline $TiO_2$ films with cyanine and merocyanine organic dyes. Sol Energy Mater Sol Cells 80(1):47–71. https://doi.org/10.1016/S0927-0248(03)00113-2
50. Hara K, Horiguchi T, Kinoshita T, Sayama K, Sugihara H, Arakawa H (2000) Highly efficient photon-to-electron conversion of mercurochrome-sensitized nanoporous ZnO solar cells. Chem Lett 29(4):316–317. https://doi.org/10.1246/cl.2000.316
51. Wang ZS, Kawauchi H, Kashima T, Arakawa H (2004) Significant influence of $TiO_2$ photoelectrode morphology on the energy conversion efficiency of N719 dye-sensitized solar cell. Coord Chem Rev 248(13–14):1381–1389. https://doi.org/10.1016/j.ccr.2004.03.006
52. Choi H, Hwang T, Lee S, Nam S, Kang J, Lee B, Park B (2015) The construction of tandem dye-sensitized solar cells from chemically derived nanoporous photoelectrodes. J Power Sour 274:937–942. https://doi.org/10.1016/j.jpowsour.2014.10.125
53. Shahid M, Shahid UI, Mohammad F (2013) Recent advancements in natural dye applications: a review. J Clean Prod 53:310–331. https://doi.org/10.1016/j.jclepro.2013.03.031
54. Kushwaha R, Srivastava P, Bahadur L (2013) Natural pigments from plants used as sensitizers for $TiO_2$ based dye-sensitized solar cells. J Energy, article no. 654953, 8. https://doi.org/10.1155/2013/654953
55. Mora-sero I, Bisquert J (2003) Fermi level of surface states in $TiO_2$ nanoparticles. Nano Lett 3(7):945–949. https://doi.org/10.1021/nl0342390
56. Bisquert J (2002) Theory of the impedance of electron diffusion and recombinantion in a thin layer. J Phys Chem B 106:325–333. https://doi.org/10.1021/jp011941g
57. Adachi M, Sakamoto M, Jiu J, Ogata Y, Isoda S (2006) Determination of parameters of electron transport in dye-sensitized solar cells using electrochemical impedance spectroscopy. J Phys Chem B 110(28):13872–13880. https://doi.org/10.1021/jp061693u
58. Kern R, Sastrawan R, Ferber J, Stangl R, Luther J (2002) Modeling and interpretation of electrical impedance spectra of dye solar cells operated under open-circuit conditions. Electrochim Acta 47(26):4213–4225. https://doi.org/10.1016/S0013-4686(02)00444-9
59. Wang Q, Wang Q, Moser J-E, Moser J-E, Grätzel M (2005) Electrochemical impedance spectroscopic analysis of dye-sensitized solar cells. J Phys Chem B 109:14945–14953. https://doi.org/10.1021/jp052768h

60. Chiu W-H, Lee C-H, Cheng H-M, Lin H-F, Liao S-C, Wu J-M, Hsieh W-F (2009) Efficient electron transport in tetrapod-like ZnO metal-free dye-sensitized solar cells. Energy Environ Sci 2(6):694. https://doi.org/10.1039/b902595m
61. Omar A, Abdullah H, Yarmo MA, Shaari S, Taha MR (2013) Morphological and electron transport studies in ZnO dye-sensitized solar cells incorporating multi- and single-walled carbon nanotubes. J Phys D-Appl Phys 46:8. https://doi.org/10.1088/0022-3727/46/16/165503
62. Kumara NTRN, Ekanayake P, Lim A, Liew LYC, Iskandar M, Ming LC, Senadeera GKR (2013) Layered co-sensitization for enhancement of conversion efficiency of natural dye sensitized solar cells. J Alloy Compd 581:186–191
63. Omar A, Abdullah H (2014) Electron transport analysis in zinc oxide-based dye-sensitized solar cells: a review. Renew Sustain Energy Rev 31(March):149–157. https://doi.org/10.1016/j.rser.2013.11.031

# Effects of Treatments on Eucalyptus Waste to Produce Cement Composites

**Matheus Roberto Cabral, Erika Yukari Nakanishi, Sérgio Francisco Santos, and Juliano Fiorelli**

**Abstract** The reuse of waste has been widely used to optimize the performance of cementitious materials. However, several challenges associated with the composition of wood wastes in cement hydration have been pointed out. Therefore, this research investigated the effects of treatments on strand-type particles made from Eucalyptus waste to produce cement composites. Strands measuring 80 mm × 20 mm × 1 mm were produced from Eucalyptus (*Eucalyptus spp.*) wastes obtained from the pallets industry. Four treatments were studied on the strands, i.e., untreated, cold water, hot water and NaOH. Physical, chemical and microstructural characterizations were performed to assess the effect of these treatments on the strands. The effect of the treatments on the cement hydration was evaluated by assessing the mechanical performance of wood-cement composites. Wettability tests showed that the strands had a surface with a hydrophobic performance except for the strands treated with NaOH. NaOH had the highest values of water absorption after 24 h (186%) while the lowest were those for untreated strands (91%). FTIR showed that the treatments reduced the cellulose and hemicellulose bands. In addition, a reduction in the bands referring to extractives for the treated strands was also identified. For the effects of the treatments on the mechanical properties of the wood-cement composites, it was found that the use of treatments resulted in lower mechanical properties in axial compression. The results for this study showed that the use of treatments does not improve the strand's physical, chemical and morphological performance and also

M. R. Cabral · E. Y. Nakanishi · J. Fiorelli (✉)
Department of Biosystems Engineering, Faculty of Animal Science and Food Engineering, University of São Paulo (USP), Pirassununga, SP, Brazil
e-mail: julianofiorelli@usp.br

M. R. Cabral · E. Y. Nakanishi
Department of Wood and Forest Sciences, Natural Sciences and Engineering Research Council of Canada (NSERC), Industrial Chair On Eco-Responsible Wood Construction, Laval University, Québec, Qc, Canada

S. F. Santos
School of Engineering and Sciences, Department of Materials and Technology, São Paulo State University (UNESP), Guaratinguetá, SP, Brazil



S. N. Sarmin et al. (eds.), *Wood Waste Management and Products*, Sustainable Materials and Technology, https://doi.org/10.1007/978-981-99-1905-5_13

does not improve the mechanical properties of wood-cement composites. Our findings suggest that producing wood-cement with strands type particles does not require the raw materials treatment.

**Keywords** Cold water treatments · NaOH treatments · Hot water treatments · Strands · Untreated · Wettability · Wood

# 1 Introduction

Wood waste is the residual plant material generated from the forest products industry. Manly from forest management operations and wood industry processes (e.g., sawmills, plywood and particleboards industries, cellulose industry, carpentry, furniture factories) [1, 2]. Furthermore, this material can also come from other industrial activities, such as packaging and constructions. The Brazilian wood waste production in 2017 was approximately 19 million m$^3$, representing about 8.5% of world production (i.e., 225 million m$^3$ for the same year). Although Brazil has a low production compared to the world, it is seen growth in the last years as shown in Fig. 1. Compared to other countries (e.g., Canada, the United States and the European Union), Brazil produces and consumes less wood products. Especially for constructions which are built with concrete and bricks. However, this does not diminish the fact that there is a large production and surplus of wood waste worldwide. Making the destination of this by-product to be discussed by several scholars who are proposing its use for energy cogeneration [3, 4].

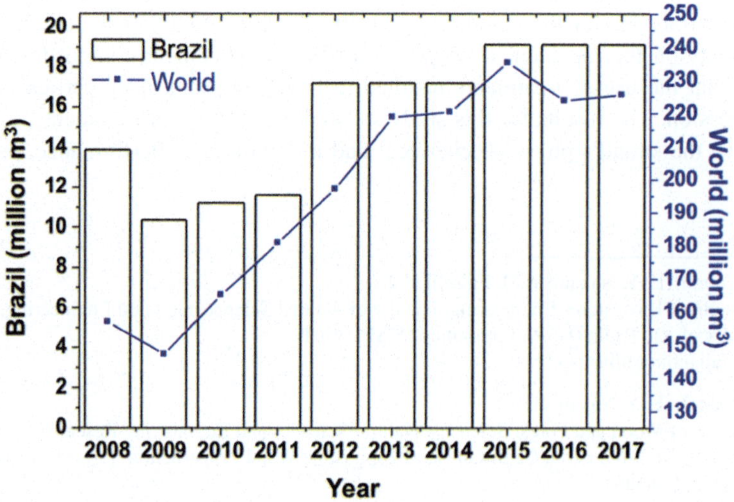


**Fig. 1** Survey of waste wood production for Brazil and for the world

Despite that, applications with higher added value can be explored. Therefore, studies have been conducted for using this by-product to produce both particleboards and wood-cement composites, which have shown suitable performance [5–9].

For cement composites, the introduction gives the possibility of decreasing the cement in the building materials. Moreover, using wood in cement composites also improves the mechanical properties in bending and post-cracking behavior. Making this material an interesting option to those energy-intensive materials conventionally used reinforcements such as fiberglass, polypropylene fiber and steel fiber [10]. Nevertheless, one of the main challenges for the use of wood wastes in cement composites is linked to the wood inhibiting the cement hardening. Previous studies argue that the wood particles used as reinforcement can partially absorb the water used in the mixture to hydrate the cement, as well as the wood chemical compounds degrade and form a protective layer preventing the cement hydration [11–13]. Therefore, a possible route to minimize these problems is treating the wood particles by immersing them in water (cold or hot) or by the use of chemical solutions [14–16]. The main goal of this study was to investigate the effects of treatments on strand-type particles made from Eucalyptus waste to produce cement composites.

## 2 Material and Methods

### *2.1 Materials*

For the experiments, Eucalyptus (*Eucalyptus spp.*) wastes obtained from the pallets industry were used. Then, the waste wood was cut into blocks of approximately 80 mm × 30 mm, which was the dimension necessary to produce the strands particles of 80 mm × 20 mm × 1 mm. The Portland cement (PC) was used as a matrix to produce the composites [17]. This cement is equivalent to PC Type III ASTM containing only 5% of mineral additions.

#### 2.1.1 Treatments

For this study, four treatments were performed (i) untreated, (ii) cold water, (iii) hot water and (iv) NaOH following the methodology proposed by Cabral et al. [12]. Figure 2 shows an overview of the treatment's procedures. The cold and hot water treatments adopted the proportion of 16 L of water to 500 g of the strand. The NaOH treatment was conducted by preparing a solution of 2.5% for 16 L of water for every 500 g of strands. Figure 3 illustrates the final appearance of strands after the treatments.

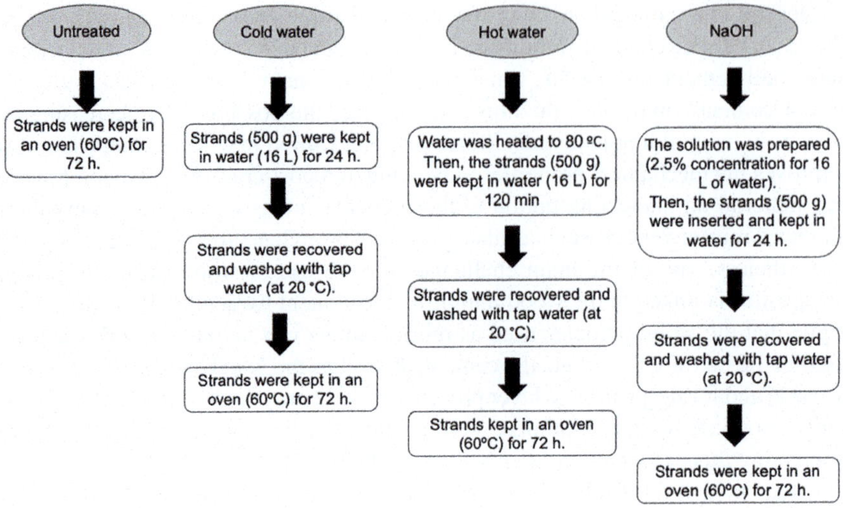


**Fig. 2** Overview of the treatment procedures

**Fig. 3** Final appearance of strands after treatments

## 2.2 *Test Methods*

### 2.2.1 Wettability

Two methods were used (i.e., the contact angle—CA and the water absorption tests) to investigate the treatments' effects on the strand's water absorption capacity. The contact angle is a widely applied technique to measure the wettability capacity of materials [18]. Contact angles lower than 90° represent a surface highly susceptible to wettability (hydrophilic), while angles larger than 90° indicate low wettability (hydrophobic) [19]. For measuring the contact angles, the model CAM 101 dynamic contact angle equipment from KSV Instruments, equipped with a camera and system software (KSV Contact Angle Measurement) for recording the images was used. The strand was fixed in the sample holder and a drop (5 μL) of deionized water was dropped on its surface. During the tests, the images of the drop were captured by a digital camera every 1 s for 30 s and this was conducted in triplicate. For the water absorption tests, measurements were carried out at 2 and 24 h of immersion in water for 20 strands of each treatment. The tests were carried out as follows first, the mass of strands was measured, then, the strands were inserted and kept submerged in a container with water (23 °C). Subsequently, for each evaluation time (2 and 24 h), the strands were removed from the water and using an absorbent paper, the excess water was removed. Finally, the strands were weighed with the aid of an electronic digital scale (model AY220, Shimadzu). The water absorption was performed according to Eq. (1).

$$\text{Water Absorption} = \frac{m_f - m_i}{m_i} * 100 \tag{1}$$

where, Water absorption (%); $m_f$ = strand mass after immersion (g); $m_i$ = strand mass before immersion (g).

### 2.2.2 Chemical

Fourier transform infrared spectroscopy (FTIR) was applied to study the influence of the treatments on the chemical composition of the strands. For conducting the tests, the materials were dried (60 °C) during 72 h. Subsequently, they were chopped using a knife mill with a sieve with an opening of 1 mm in diameter. FTIR is a qualitative technique widely used in the analysis of plant materials, enabling the identification of organic compounds [20]. The FTIR analysis of the powder samples (1 mm size) was performed using the Spectrum One model equipment from the brand Perkin Elmer. This FTIR is equipped with the Spectrum V5.3.1 program. The analyses were run for 128 scans using a spectral resolution of 4 $cm^{-1}$ in the range from 4000 to 650 $cm^{-1}$.

### 2.2.3 Microstructure

Thermogravimetry

To evaluate changes in the thermal stability of the strands due to the treatments, thermogravimetry (TG) and thermogravimetry derivative (DTG) analyses were carried out. The analysis was carried out using the thermal analyzer (TG-DTG) model STA 449 F3 from Jupiter® brand. The strands were dried at 60 °C for 72 h and subsequently, they were cut in a knife mill with a sieve to obtain particles of 1 mm diameter. The analyses were performed with the following parameters: heating speed of 10 °C/min., and a temperature range from 25 to 1000 °C.

X-Ray Diffractometry Technique

The X-ray diffractometry (XRD) was used to determine the crystallinity index (CI) of the strands and thereby evaluate crystallographic changes resulting from the treatments. The samples were prepared as follows, the strands were dried at 60 °C during 72 h and subsequently, they were cut in a knife mill with a sieve to obtain particles of 1 mm diameter. XRD analysis was performed using a Rigaku diffractometer (model MiniFlex 600), using the following parameters: 40 kV and 15 mA, operated with radiation Cu-$K_\alpha$ ($\lambda = 1.54056°$ Å), 2theta from 5° to 50° and with a speed test of 10°/min. The CI values were determined through the empirical procedure proposed by Segal et al. [21] Eq. (2)

$$CI = \frac{I_{002} - I_{am}}{I_{002}} \times 100 \quad (2)$$

where, $l_{002}$ = the intensity value in the plane (002) corresponding to the 22°–23° reflections (2θ); $l_{am}$ = diffraction intensity corresponding to 18°–19°(2θ).

### 2.2.4 Morphological Characterization

The effects of treatments on strand morphology were studied by means of SEM (scanning electron microscopy) and CLSM (confocal laser scanning microscope) methods. For the SEM analysis, a tabletop microscope (model TM3000, from Hitachi manufacturer) with 15 kV of acceleration voltage was used. For the CLSM analysis, it was employed a $L_{ext}$ 3D Measuring Laser Microscope from Olympus brand (model OL-S4100) set with OL-S4100 software.

### 2.2.5 Wood-Cement Composites

To verify the treatments' effects on the composites, the axial compressive strength test was used. The wood-cement materials were fabricated according to the mix design and methodology proposed by Cabral et al. [12]. The water: cement relation of 0.5:1 proposed by Moslemi and Pfister [22] was used. First, the raw materials (strands, cement and water) were separated. Then, the strands (i.e., 15 g) for the specimens were inserted into the steel mold. Subsequently, the Portland cement was blended during 5 min using a mixer (model AG-5), from Metal Cairo. Then, the cement paste was added into the molds. Next, the specimens were placed on a vibrating table for 5 min for compacting. Finally, the composites were kept sealed in plastic bags at 23 ± 2 °C in an air-saturated environment until the 28th day of curing.

Before testing, the finish of the top and bottom of the composites' surfaces was carried out to make them parallel to each other, using a saw with a diamond blade. The tests were conducted with a universal electromechanical machine for mechanical testing (model DL 30,000, Emic) using a speed of 1 mm/min and a load cell of

300 kN. These tests were carried out following the recommendations NBR 7215/1996 standard Portland Cement: Resistance recommendations [23].

## *2.3 Statistical Analysis*

An analysis of variance (ANOVA) was carried out to check whether there were significant differences in the treatments of the strands and the composites. To find a difference, the Tukey test comparison was made between the means at a 5% significance level.

# 3 Results and Discussions

## *3.1 Wettability*

### 3.1.1 Contact Angle

Figure 4 shows the contact angle measurements for each strand treatment. According to Hubbe et al. [24] wettability is described as the capacity of a liquid to spread or be absorbed by a solid material as a consequence of physical and chemical interactions. Using the contact angle technique, it was possible to obtain an indication of the treatment impacts on the wettability of the strands. Once an angle formed by the drop is lower than 90°, the material is hydrophilic and hydrophobic if it is greater than 90° [19].

It was found that for the untreated, cold and hot water strands. The contact angle was greater than 90°, indicating that these surfaces are hydrophobic. On the other hand, for the strands treated with NaOH, a contact angle lower than 90° is seen (Fig. 4). Indicating that NaOH treatment changes the surface of the strands making them hydrophilic. The results found for the studied strands are close to those obtained by César et al. [25], who evaluated the contact angle of strands produced from *Eucalyptus grandis* and *Pinus ocarpa* untreated and treated with distilled water. The researchers reported contact angle values for *Eucalyptus grandis* of 111° (untreated) and 120° (cold water treated), while for *Pinus ocarpa* the angles obtained were 86° (untreated) and 99° (water cold).

### 3.1.2 Water Absorption Tests

Table 1 presents the water absorption values after water immersion (i.e., 2h and 24 h). The water absorption results obtained after 2 h in water ranged from 56.75%

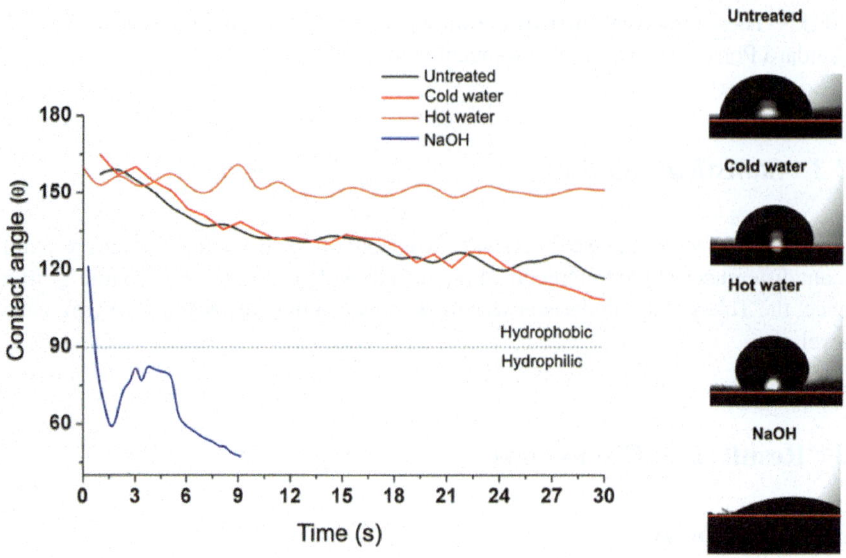


**Fig. 4** Contact angle measurements and images of the drop on the strand surface

(untreated) to 157.78% (NaOH). In addition, it was found that the water absorption values were significantly higher for the strands treated with NaOH ($p < 0.05$). Whereas no significant differences were found for the other treatments ($p > 0.05$). These findings are in accordance with those results obtained by the contact angle, where only the treatment with NaOH reached angles lower than 90° (Fig. 4).

Equal letters in the column imply factor levels with means statistically equivalent to the 0.05 level of significance.

For the water absorption results after 24 h, the values found ranged from 91.89% for the untreated strands to up to 186% for the strands treated with NaOH. Statistical analysis showed no significant differences between the water absorption (24 h) for the strands untreated and those treated with cold water ($p > 0.05$). Furthermore, no significant differences were found between strands treated with cold and hot water ($p > 0.05$). However, for the strands treated with NaOH, the water absorption values were significantly higher than the other treatments ($p > 0.05$).

**Table 1** Mean and standard deviation values for the water absorption of the strands after 2 and 24 h

| Treatment | 2 h (%) | 24 h |
|---|---|---|
| Untreated | 56.75 (9.36) B | 91.89 (12.17) C |
| Cold water | 70.32 (22.31) B | 108.18 (22.22) BC |
| Hot water | 61.89 (13.65) B | 114.23 (18.97) B |
| NaOH | 157.78 (27.32) A | 186.97 (30.77) A |

### 3.1.3 Chemical Characterization

The FTIR spectra for the strands with the absorption bands of functional groups characteristic of wood chemical compounds, such as cellulose, hemicellulose, lignin and extractive are presented in Fig. 5. Among all absorption bands and peaks, in the band of the region 1020–1030 $cm^{-1}$ the C–O bond stretching belonging to cellulose and lignin is seen [26]. In addition, it is also seen the vibrations of the aromatic skeleton in a band between 1500 and 1600 $cm^{-1}$ correspond to lignin [27].

Other bands can be identified: (I) band close to 1740 $cm^{-1}$ related to carboxylic groups (C = O), found in acids and esters of acetic, p-cumeric, ferulic and uronic acids, that are inherent of extractive and mainly hemicellulose [28, 29]; (II) absorption band at about 2900 $cm^{-1}$ region represented by the stretching of cellulose and hemicellulose [28]; (III) band at approximately 3300 $cm^{-1}$ referring to the vibration of hydroxyl groups of cellulose molecules [30, 31]. These FTIR results showed that untreated strands had greater intensity in the bands referring to hemicellulose, lignin and extractive than those treated strands. Therefore, indicating that treating the strands reduced these components in the materials. However, as it removed hemicellulose and extractive that are disturbing compounds for the Portland cement hydration. It was also decreased the cellulose referring band, a component that gives wood mechanical resistance [32, 33].

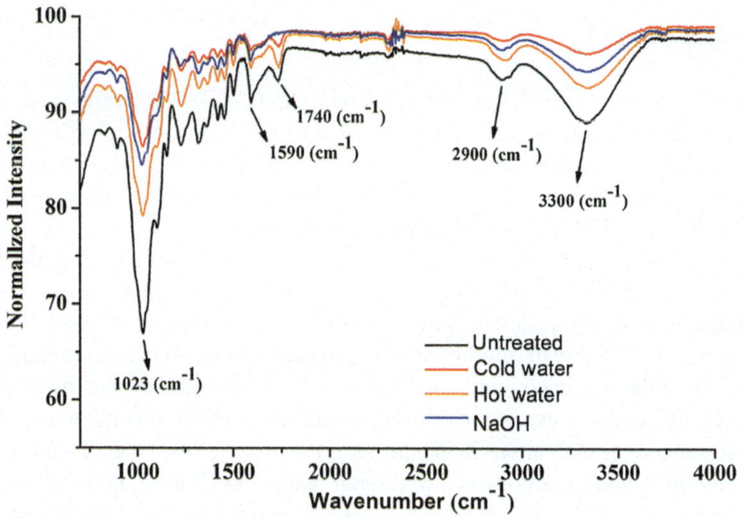


**Fig. 5** FTIR spectra for the strands

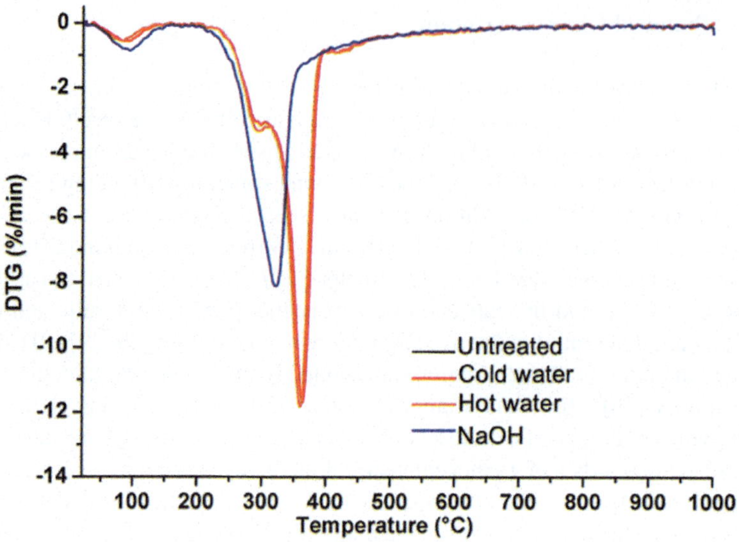


**Fig. 6** Differential thermogravimetric analysis (DTG)

### 3.1.4 Microstructure Assessments

Thermogravimetry

Figure 6 shows the strands' thermogram derivatives (DTG). The initial evident mass loss occurred between 25 and 100 °C, related to the water loss [34]. The thermal degradation of the strands' structural components begins at 200 °C with the decomposition of hemicellulose. Since hemicelluloses have ramified and amorphous structures as well as a little degree of polymerization. Subsequently, the degradation of amorphous and crystalline cellulose occurs [35]. Cellulose has a degradation range between 290 and 390 °C, since it is a semi-crystalline structure, with a high degree of polymerization, composed of intra-and intermolecular hydrogen bonds [36].

However, lignin, due to its complex, interconnected and branched structure, has a wide range of thermal decomposition (200–700 °C) [37]. It is seen that the treatments of the strands, with the exception of those submitted to the NaOH treatment, did not cause significant modifications in the thermogram profile. Since the alkaline solution breaks the inter and intramolecular hydrogen bonds of cellulose, hemicellulose and lignin hydroxyl groups [38]. In addition, according to Chin et al. [39] it is seen that alkaline promotes the elimination of hemicellulose. Through the thermograms shown in Fig. 6, it is seen that the main thermal decomposition of the strands occurred in the range of 200 and 400 °C, potentially showing that the materials under study have a higher fraction of hemicellulose and cellulose.

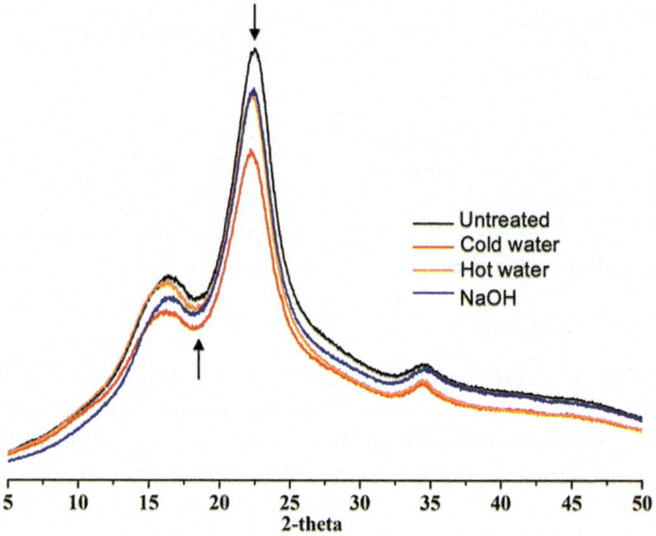


**Fig. 7** Crystallographic analysis (XRD)

X-Ray Diffractometry Results

Figure 7 shows the diffractograms obtained from untreated and treated strands. It is seen that the diffractograms under study present characteristic and wide peaks, as well as a high baseline (background) of the semi-crystalline structure of materials of plant origin [40]. To calculate the crystallinity index (CI), two main peaks were marked, represented in: (I) diffraction angle (2θ) 18° referring to the non-crystalline fraction of wood (hemicellulose and lignin) and (II) diffraction angle (2θ) 22nd referring to the crystalline fraction of wood, referring to the glycoside rings of the cellulosic structure [41]. The black arrows indicate the regions of minimum and maximum diffraction intensity used to calculate the IC Eq. (2). The cellulose IC is a parameter used to estimate the cellulose crystalline compared to the total structure [31, 42, 43].

Therefore, the characterization of the crystallinity index of the strands can help to better understand the internal structure of the strands and their properties. The calculated percentages of the CI are listed in Table 2. The CI for the strand specimens was between 54 and 59%, with the lowest index results for cold water and hot water while the highest index was found for treatment with NaOH. One of the reasons for the slight increase in strands crystallinity for NaOH is related to the partial removal of the amorphous portion of the strands (e.g., hemicellulose and lignin). The removal of the amorphous portion of the plant material promotes the arranging of cellulose chains and consequently, the increase in crystallinity [44, 45]. These results are in accordance with those found in the FTIR analysis (Fig. 5), where reductions were found in the characteristic bands of hemicellulose, cellulose and lignin.

**Table 2** Crystallinity index (CI) according to the respective treatments applied

| Treatment | Crystallinity index (%) |
|---|---|
| Untreated | 58 |
| Cold water | 54 |
| Hot water | 54 |
| NaOH | 59 |

Morphological Characterization

*Scanning Electron Microscopy*

Figure 8 gathers the representative micrographs for the SEM analyses of the strands' surface obtained from a population of 30 micrographs for each treatment. Through SEM micrographs the anatomical structures of the strands were identified as transverse pores and fibers [46, 47]. It is also possible to identify which part of the wall is damaged. This damage was probably caused during the strand production process. It is also noted that treating the strands with cold and hot water caused a "curving" in the damaged surface walls of the fibers, possibly due to the wetting and drying process. However, when evaluating the SEM micrographs of the NaOH strands, it is seen that the surface is quite different in comparison to the other strands. It is possible to see that the treatment with NaOH "cleaned" the surface of the strand. In addition, it is also noted that there was a detachment of the fibers from the strand wall caused by the partial removal of hemicellulose and surface impurities (e.g., extractives) [48, 49]. These results are in accordance with those found in past studies [50–52], indicating that treatments with hot water and NaOH implicated in a greater portion of the cellulose exposed on the wood surface.

*Confocal Laser Scanning Microscope*

Table 3 describes the surface roughness (Sa, arithmetic mean of three-dimensional roughness) mean values for the strands obtained by the surface analysis with the CLSM. It was found that there were no differences in the average surface roughness between the strands ($p > 0.05$). A probable reason for this result is due to the surface complexity of these materials as shown in Fig. 9. By analysis the Fig. 9, it is seen that the strand's surface is complex and irregular, indicating then that hysteresis phenomenon might have contributed to the contact angle, in the evaluation of wettability assessment.

Equal capital letters in the column imply factor levels with means statistically equivalent to the 0.05 level of significance.

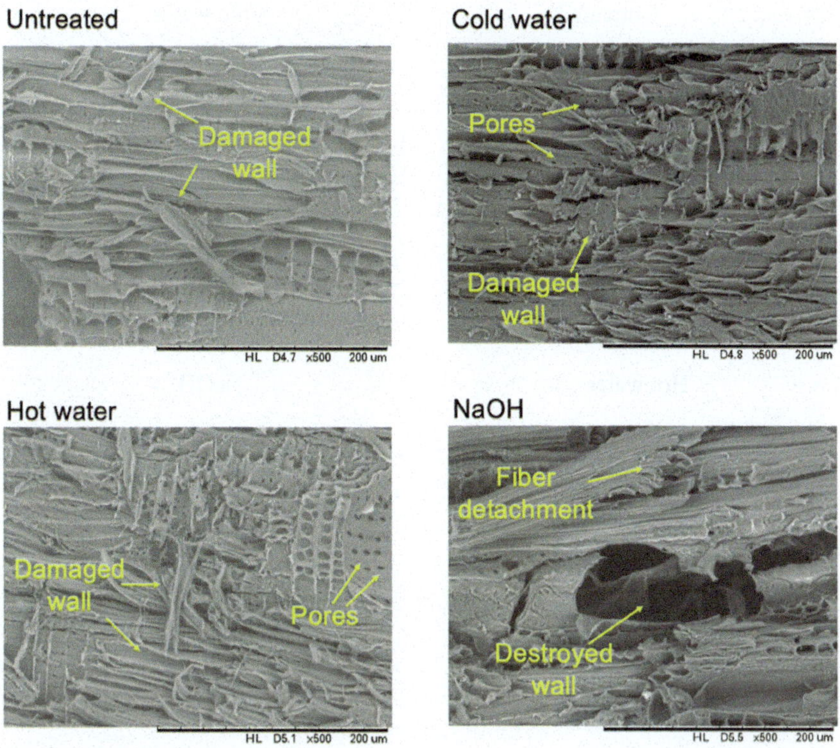


**Fig. 8** Micrographs of strands

**Table 3** Mean values and standard deviations of the surface roughness (Sa) of the strands

| Treatment | Sa (μm) |
|---|---|
| Untreated | 10.59 (2.66) A |
| Cold water | 7.70 (0.77) A |
| Hot water | 16.53 (2.89) A |
| NaOH | 18.28 (5.18) A |

### 3.1.5 Wood-Cement Composites

Table 4 presents the maximum rupture stress for the 28-day composites obtained through the axial compression strength assessment. It can be seen that the maximum rupture stress of the untreated strands composites did not show statistically significant differences compared to the samples produced only with Portland cement (neat cement) ($p > 0.05$). On the other hand, the mean values comparison showed that untreated composites were statistically higher than other studied treatments. It is noted that the comparison between the composites manufactured with strands treated

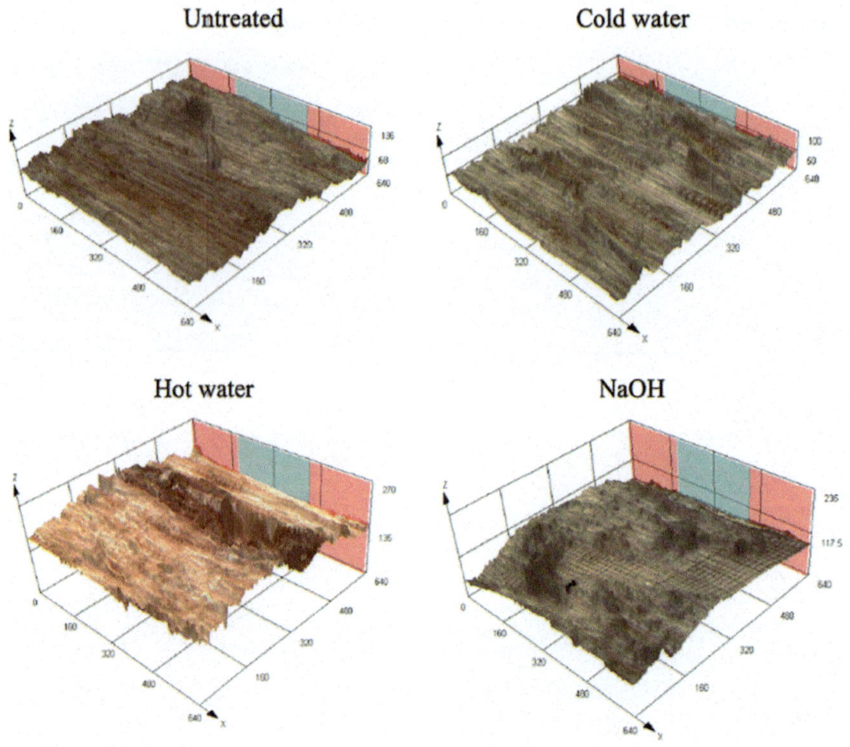


**Fig. 9** Topographic images of strands surface obtained via confocal laser scanning microscope

(i.e., cold and hot water as well as NaOH) did not show a statistical difference compared to Portland cement composites (control) ($p < 0.05$).

Equal capital letters in the column imply factor levels with means statistically equivalent to the 0.05 level of significance.

A possible reason for these results might be due to the strands' geometry, as pointed out by Latorraca [53], the influence of wood particles on the hydration of Portland cement is affected its geometry. The author states that there is an almost linear association between the size of the particles and the effect on cement hydration.

**Table 4** Mean values and standard deviations of the maximum stress obtained from the axial compression test at 28 days

| Treatment | Maximum rupture stress (MPa) |
|---|---|
| Untreated | 22.80 (4.13) A |
| Cold water | 19.34 (3.57) B |
| Hot water | 13.10 (3.90) B |
| NaOH | 13.80 (7.90) B |
| Neat cement | 34.56 (7.34) A |

As the size of the wood particle decreases, its effect on cement hydration becomes greater due to the greater surface area.

## 4 Conclusion

The strands presented a surface with a hydrophobic character, except for the strands treated with NaOH, which presented a hydrophilic character, presenting the highest values of water absorption after 24 h (186%). This may have been influenced by the hysteresis phenomenon, caused by the irregularity on the surface of the strands, as shown by the roughness analysis using the MVCL technique.

By the FTIR spectrums, the results showed that treating the strands with cold and hot water as well as NaOH decreased the cellulose and hemicellulose bands. The hot water treatment also reduced the lignin band. In addition, this decrease in the strands treated with NaOH increased the crystallinity index (CI). Probably, due to the arranging of the cellulose chains after removing the amorphous portion of the strands. In addition, a reduction in the bands referring to extractives for the treated strands was also identified through the FTIR analysis. For the treatments' effects on the mechanical properties of the cement composites, it was found that the use of treatments resulted in lower mechanical properties in axial compression.

The results for this study showed that the use of treatments does not improve the strand's performance, and it also does not improve the mechanical properties of wood-cement composites. Our findings suggest that producing wood-cement with strands type particles does not require the raw materials treatment.

**Acknowledgements** The authors are grateful to the Brazilian financial support from Coordination for the Improvement of Higher Education Personnel (CAPES) Financial code 001, the São Paulo Research Foundation (FAPESP), [Grant# 2016/07372-9 and Grant#2017/18076-4] as well as to the National Council for Scientific and Technological Development (CNPq) [Grant# 312,151/2016-0 and Grant#304,106/2017-8]. The authors also thank the Laboratory of Structural Characterization (Dema/UFSCar) for the general facilities.

## References

1. Garcia CA, Hora G (2017) State-of-the-art of waste wood supply chain in Germany and selected European countries. Waste Manag 70:189–197. https://doi.org/10.1016/J.WASMAN.2017.09.025
2. Corinaldesi V, Mazzoli A, Siddique R (2016) Characterization of lightweight mortars containing wood processing by-products waste. Constr Build Mater 123:281–289. https://doi.org/10.1016/j.conbuildmat.2016.07.011
3. Ghaffar SH, Fan M (2013) Structural analysis for lignin characteristics in biomass straw. Biomass Bioenergy 57:264–279. https://doi.org/10.1016/j.biombioe.2013.07.015

4. Fournel S, Palacios JH, Morissette R et al (2015) Influence of biomass properties on technical and environmental performance of a multi-fuel boiler during on-farm combustion of energy crops. Appl Energy 141:247–259. https://doi.org/10.1016/j.apenergy.2014.12.022
5. Okino EY, Souza MRd, Santana MA et al (2004) Cement-bonded wood particleboard with a mixture of eucalypt and rubberwood. Cem Concr Compos 26:729–734. https://doi.org/10.1016/S0958-9465(03)00061-1
6. Bui THH, Hoang AD, Kim YT (2019) An application of full analytic hierarchy process to identify the best use of Dong Ngo closed-landfill site in Bac Ninh city, Vietnam. Environ Prog Sustain Energy 38. https://doi.org/10.1002/ep.13273
7. Ferraz JM, del Menezzi CHS, Teixeira DE, Martins SA (2011) Effects of treatment of coir fiber and cement/fiber ratio on properties of cement-bonded composites. BioResources 6:3481–3492. https://doi.org/10.15376/biores.6.3.3481-3492
8. Cabral MR, Nakanishi EY, Santos SF et al (2021) Orientation effect on the physical and mechanical properties of strand cement boards. Constr Build Mater 275:122121. https://doi.org/10.1016/J.CONBUILDMAT.2020.122121
9. Barbirato GHA, Junior WEL, Martins RH et al (2022) Experimental evaluation and numerical modeling of the mechanical performance of OSB sandwich panels manufactured with trapezoidal core. Constr Build Mater 326:126721. https://doi.org/10.1016/J.CONBUILDMAT.2022.126721
10. Page J, Khadraoui F, Boutouil M, Gomina M (2017) Multi-physical properties of a structural concrete incorporating short flax fibers. Constr Build Mater 140:344–353. https://doi.org/10.1016/j.conbuildmat.2017.02.124
11. Chakraborty S, Kundu SP, Roy A et al (2013) Effect of jute as fiber reinforcement controlling the hydration characteristics of cement matrix. Ind Eng Chem Res 53. https://doi.org/10.1021/ie300607r
12. Cabral MR, Nakanishi EY, Franco MSR et al (2020) Treatments of residual pine strands: characterization and wood-cement-compatibility. J Wood Chem Technol 40:396–407. https://doi.org/10.1080/02773813.2020.1828474
13. Yasuda S, Hirano J, Tange J et al (1989) Manufacture of wood-cement boards iii cement-hardening inhibitory components of western red cedar heartwood. J Wood Chem Technol 9:123–133. https://doi.org/10.1080/02773818908050289
14. Cabral MRMR, Nakanishi EYEY, dos Santos V et al (2018) Evaluation of pre-treatment efficiency on sugarcane bagasse fibers for the production of cement composites. Arch Civ Mech Eng 18:1092–1102. https://doi.org/10.1016/J.ACME.2018.02.012
15. Peschard A, Govin A, Pourchez J et al (2006) Effect of polysaccharides on the hydration of cement suspension. J Eur Ceram Soc. https://doi.org/10.1016/j.jeurceramsoc.2005.02.005
16. Castro V, Araújo RD, Parchen C, Iwakiri S (2014) Avaliação dos efeitos de pré-tratamentos da madeira de Eucalyptus benthamii Maiden & Cambage no grau de compatibilidade com cimento portland. Rev Árvore 38
17. Associação Brasileira de Normas Técnicas-NBR 5733 (1991) Cimento Portland com Alta Resistencia Inicial. Rio de Janeiro
18. Tran LQN, Fuentes CA, Dupont-Gillain C et al (2011) Wetting analysis and surface characterisation of coir fibres used as reinforcement for composites. Colloids Surf Physicochem Eng Asp 377:251–260. https://doi.org/10.1016/j.colsurfa.2011.01.023
19. Yuan Y, Lee TR (2013) Contact angle and wetting properties. Surf Sci Tech 51:3–34. https://doi.org/10.1007/978-3-642-34243-1
20. Lopes WA, Fascio M (2004) Esquema para interpretação de espectros de substâncias orgânicas na região do infravermelho Wilson Araújo Lopes* e Miguel Fascio. Quim Nov 27:670–673
21. Segal L, Creely JJ, Martin AE, Conrad CM (1959) An empirical method for estimating the degree of crystallinity of native cellulose using the x-ray diffractometer. Text Res J 29:786–794. https://doi.org/10.1177/004051755902901003
22. Moslemi AA, Pfister SC (1987) The influence of cement wood ratio and cement type on bending strength and dimensional stability of wood-cement composite panels. Wood Fiber Sci 19:165–175

23. Associação Brasileira de Normas Técnicas (1996) NBR 7215—Cimento Portland—Determinação da resistência à compressão, pp 1–8
24. Hubbe MA, Gardner DJ, Shen W (2015) Contact angles and wettability of cellulosic surfaces: a review of proposed mechanisms and test strategies. BioResources 10:8657–8749. https://doi.org/10.15376/biores.10.4.Hubbe_Gardner_Shen
25. César AA da S (2011) ESTUDO da interação adesivo-partícula em painéis OSB (Oriented Strand Board)
26. Rodrigues BVM, Ramires EC, Santos RPO, Frollini E (2015) Ultrathin and nanofibers via room temperature electrospinning from trifluoroacetic acid solutions of untreated lignocellulosic sisal fiber or sisal pulp. J Appl Polym Sci 132:n/a-n/a. https://doi.org/10.1002/app.41826
27. Morán JI, Alvarez VA, Cyras VP, Vázquez A (2008) Extraction of cellulose and preparation of nanocellulose from sisal fibers. Cellulose 15:149–159. https://doi.org/10.1007/s10570-007-9145-9
28. Guimarães JL, Frollini E, da Silva CG et al (2009) Characterization of banana, sugarcane bagasse and sponge gourd fibers of Brazil. Ind Crops Prod 30:407–415. https://doi.org/10.1016/j.indcrop.2009.07.013
29. Sain M, Panthapulakkal S (2006) Bioprocess preparation of wheat straw fibers and their characterization. Ind Crops Prod 23:1–8. https://doi.org/10.1016/J.INDCROP.2005.01.006
30. Oh SY, Dong IY, Shin Y et al (2005) Crystalline structure analysis of cellulose treated with sodium hydroxide and carbon dioxide by means of X-ray diffraction and FTIR spectroscopy. Carbohydr Res. https://doi.org/10.1016/j.carres.2005.08.007
31. Poletto M, Ornaghi Júnior HL, Zattera AJ (2014) Native cellulose: structure, characterization and thermal properties. Materials (Basel). https://doi.org/10.3390/ma7096105
32. Xu F (2010) Structure, ultrastructure, and chemical composition, 1st edn. Elsevier
33. Neves PD, Cabral MR, Santos V et al (2021) Technical assessment of leaf fibers from curaua: an amazonian bioresource. https://doi-org.acces.bibl.ulaval.ca/101080/154404782021 1902897. https://doi.org/10.1080/15440478.2021.1902897
34. Martin AR, Martins MA, Da Silva ORRF, Mattoso LHC (2010) Studies on the thermal properties of sisal fiber and its constituents. Thermochim Acta 506:14–19. https://doi.org/10.1016/j.tca.2010.04.008
35. Oliveira F, da Silva CG, Ramos LA, Frollini E (2017) Phenolic and lignosulfonate-based matrices reinforced with untreated and lignosulfonate-treated sisal fibers. Ind Crops Prod 96:30–41. https://doi.org/10.1016/J.INDCROP.2016.11.027
36. Sanchez-Silva L, López-González D, Villaseñor J et al (2012) Thermogravimetric-mass spectrometric analysis of lignocellulosic and marine biomass pyrolysis. Bioresour Technol 109:163–172. https://doi.org/10.1016/j.biortech.2012.01.001
37. Galina NR, Romero Luna CM, Arce GLAF, Ávila I (2018) Comparative study on combustion and oxy-fuel combustion environments using mixtures of coal with sugarcane bagasse and biomass sorghum bagasse by the thermogravimetric analysis. J Energy Inst. https://doi.org/10.1016/J.JOEI.2018.02.008
38. Mohammed M, Rahman R, Mohammed AM et al (2022) Surface treatment to improve water repellence and compatibility of natural fiber with polymer matrix: recent advancement. Polym Test 115:107707. https://doi.org/10.1016/J.POLYMERTESTING.2022.107707
39. Chin SC, Tee KF, Tong FS et al (2020) Thermal and mechanical properties of bamboo fiber reinforced composites. Mater Today Commun 23:100876. https://doi.org/10.1016/J.MTCOMM.2019.100876
40. Labidi K, Korhonen O, Zrida M et al (2019) All-cellulose composites from alfa and wood fibers. Ind Crops Prod 127:135–141. https://doi.org/10.1016/J.INDCROP.2018.10.055
41. Gonçalves APB, de Miranda CS, Guimarães DH et al (2015) Physicochemical, mechanical and morphologic characterization of purple banana fibers. Mater Res 18:205–209. https://doi.org/10.1590/1516-1439.366414
42. Agarwal UP, Ralph SA, Reiner RS, Baez C (2018) New cellulose crystallinity estimation method that differentiates between organized and crystalline phases. Carbohydr Polym 190:262–270. https://doi.org/10.1016/J.CARBPOL.2018.03.003

43. Gümüskaya E, Usta M, Kirci H (2003) The effects of various pulping conditions on crystalline structure of cellulose in cotton linters. Polym Degrad Stab 81:559–564. https://doi.org/10.1016/S0141-3910(03)00157-5
44. Roy A, Chakraborty S, Kundu SP et al (2012) Improvement in mechanical properties of jute fibres through mild alkali treatment as demonstrated by utilisation of the Weibull distribution model. Bioresour Technol 107:222–228. https://doi.org/10.1016/J.BIORTECH.2011.11.073
45. Sinha E, Rout SK (2008) Influence of fibre-surface treatment on structural, thermal and mechanical properties of jute. J Mater Sci 43:2590–2601. https://doi.org/10.1007/s10853-008-2478-4
46. Lube VM (2016) Effects of moisture-induced thickness swelling on the microstructure of oriented strand board. https://doi.org/10.14288/1.0339891
47. Conners T (2015) Distinguishing softwoods from hardwoods
48. Goda K, Sreekala MS, Gomes A et al (2006) Improvement of plant based natural fibers for toughening green composites—Effect of load application during mercerization of ramie fibers. Compos Part A Appl Sci Manuf 37:2213–2220. https://doi.org/10.1016/J.COMPOSITESA.2005.12.014
49. Jähn A, Schröder MW, Füting M et al (2002) Characterization of alkali treated flax fibres by means of FT Raman spectroscopy and environmental scanning electron microscopy. Spectrochim Acta—Part A Mol Biomol Spectrosc. https://doi.org/10.1016/S1386-1425(01)00697-7
50. Rosa MF, Chiou B, Medeiros ES et al (2009) Effect of fiber treatments on tensile and thermal properties of starch/ethylene vinyl alcohol copolymers/coir biocomposites. Bioresour Technol 100:5196–5202. https://doi.org/10.1016/J.BIORTECH.2009.03.085
51. Rahman MM, Khan MA (2007) Surface treatment of coir (Cocos nucifera) fibers and its influence on the fibers' physico-mechanical properties. Compos Sci Technol 67:2369–2376. https://doi.org/10.1016/J.COMPSCITECH.2007.01.009
52. Choudhury A, Kumar S, Adhikari B (2007) Recycled milk pouch and virgin low-density polyethylene/linear low-density polyethylene based coir fiber composites. J Appl Polym Sci 106:775–785. https://doi.org/10.1002/app.26522
53. Latorraca (2000) Eucalyptus spp. na produção de painéis de cimento-madeira

# Microwave Treatment on Wood Waste Product-A Review

**Mohammad Farsi, Mohammad Jawaid, Amir Amini, Masoud Ebadi, and Majid Shahbabaei**

**Abstract** This chapter presents an extensive review of the scientific literature associated with various microwave treatments on wood waste products. First, the basic concepts of microwave radiation and its applications in wood waste product fabrication are reviewed. Then, an extensive literature review of the most significant experimental research papers is provided, divided into two microwave heating treatment uses: wood drying and wood waste products performance improvement. Next, the post-treatment of wood-plastic composites (WPCs) by microwave irradiation as a case study was reviewed and a real example of WPCs samples was discussed. Finally, the chapter concludes with a proposal of doing future research studies concerning the impact of microwave technology on some important properties of wood waste products, i.e., resistance to biological agents, fire, environmental conditions, and so on.

## 1 Introduction

Wood is a complex natural composite material made of cellulose, hemicelluloses, lignin, and extractives [1]. Each of these main ingredients has its unique structure. For example, cellulose is a natural linear homopolymer (polysaccharide), in which D-glucopyranose rings are connected with β-(1–4) glycosidic linkages. Different from

M. Farsi (✉) · A. Amini
Department of Wood and Paper Science and Technology, Sari Branch, Islamic Azad University, 7Th Kilometer Khazar Blvd, Sari, Iran
e-mail: moh_farsi@iausari.ac.ir

M. Jawaid
Laboratory of Biocomposite Technology, Institute of Tropical Forestry and Forest Products (INTROP), Universiti Putra Malaysia, Serdang, Selangor, Malaysia

M. Ebadi
Department of Wood and Paper Science, Tarbiat Modares University, Noor, Iran

M. Shahbabaei
Department of Mechanical Engineering, University of Alberta, Edmonton, AB, Canada

 
S. N. Sarmin et al. (eds.), *Wood Waste Management and Products*, Sustainable Materials and Technology, https://doi.org/10.1007/978-981-99-1905-5_14

**Table 1** Global production of forest products in 2020 [6]

| Product | Unit | 2020 | Production | | |
|---|---|---|---|---|---|
| | | | Change (%) compared to | | |
| | | | 2019 (%) | 2000 | 1980 |
| Roundwood | million $m^3$ | 3 912 | −1 | 12% | 25% |
| Wood fuel | million $m^3$ | 1 928 | −1 | 7% | 15% |
| Industrial roundwood | million $m^3$ | 1 984 | −2 | 17% | 37% |
| Wood pellets and other agglomerates | million tons | 50 | 3 | | |
| Sawnwood | million $m^3$ | 473 | −3 | 23% | 12% |
| Wood-based panels | million $m^3$ | 367 | −1 | 107% | 280% |
| Plywood | million $m^3$ | 118 | 2 | 103% | 200% |
| Particle board, OSB, and fiberboard | million $m^3$ | 250 | −2 | 109% | 335% |

cellulose with several sugar moieties, hemicelluloses are mostly branched polysaccharides with lower molecular mass with a degree of polymerization (DP) of 50–200. Lignin as a randomly branched polyphenol is the most complex polymer among naturally occurring high molecular weight materials made up of phenyl propane (C9) units. All of these materials characterizes as polar polymers witnessed by dipole polarization.

As an important renewable resource, Wood has a broad range of material properties. Relatively high strength and stiffness, natural appearance with interesting texture, good insulation properties, easy machinability, low density, flexibility during the processing with no harm to equipment, good mechanical properties, sustainability, environmentally friendly, and low cost are some of their unique structural and chemical characteristics [2–4].

Owing to unique physical and chemical properties, wood and wood fibers are used in a diverse range of applications such as construction, furniture, packaging, energy production, etc. Hydrophilicity and poor compatibility of those materials with polymer matrices having hydrophobicity characteristics significantly reduce the final performance of the products for use in industries [5].

According to the statistics published by FAO,[1] worldwide wood consumption reached 4802 million cubic meters in 2020 worth around 244 billion dollars (Table 1), which included sawdust, wood panels, sawn wood, and wood pellets [6]. Global wood products trade increased by 143% up to US$244 billion between 1990 and 2019 [6]. The largest contributor to this trade expansion has been related to economic growth and demand from China to import wood products to China which increased by 760% up to US$49 billion [7].

During the secondary production process, wood suffers a lot of waste in the form of sawdust, wood chips, wood bark, sawmill scraps, and hard chips. Today, wood waste is considered an important economic and environmental issue in the world. A large amount of wood waste is produced every year in the world, which requires

[1] Food and Agriculture Organization (FAO).

disposal, reuse, or other end-of-life management. Dovetail Partners [8] estimated that more than 64 million metric tons of wood waste were generated in 2010 with 28 million metric tons recovered, 27 million metric tons still available for recovery, and 10 million metric tons non-recoverable. Bergeron [9] estimated that 50 million cubic meters of wood waste are generated each year in the EU.[2] Currently, wood waste can be used as fuel or building products. A large part of wood waste can be recycled or turned into usable products through chemical, physical, or mechanical treatments. Wood-plastic composites (WPCs) are one of the most important products made from wood waste.

## 2 Wood-Plastic Composites (WPCs)

WPC polymer constitutes wood waste, plastic, and chemical additives such as lubricants, coupling agents, nucleating agents, pigments, and UV stabilizers. Compared with pure plastics, the incorporation of wood fiber into the plastics led to improved flexural and tensile stiffness [10].

In 2019, the global WPC market size was worth 5.3 billion dollars. It is estimated that the market will grow up to 11.4% CAGR[3] from 2020 to 2027 [11]. Advances in the development of sustainable construction materials coupled with rising repair and renovation activities are driving the market growth. 51 and 22% of WPCs market growth are related to Europe and North America, respectively, while the rest is attributed to Asian and African countries [12].

Forecasts for the turnover of WPCs in the world may reach 13.47 billion dollars by 2026. In Asia, China exclusively possesses 61.8% of the market for WPCs [12]. Reportedly, Asia, Europe, and North America, respectively, share 10, 20, and 70% of the WPCs market in the world with a total global sharing of 900,000 tons [13, 14].

Rail and decking products are the main WPCs market in the USA while the automotive industries hold the main market in the EU [15]. As seen in Fig. 1a, the highest production and consumption of WPCs in the world is in the field of construction followed by automobiles, even though its growing consumption is evident by 2024. As Fig. 1b shows, Asia Pacific and specifically China will reach the USA in the global production of WPCs by 2024. Also, the forecast for 2027 indicates that Europe will overtake America in the production of WPCs [16, 17].

Figure 2 compares the application of WPCs including interior panels, headliners, car dashboards, car roofs, seat panels, parcel shelves, acoustic panels, fencing, decking, railing, cladding and siding, park benches, and so on [18, 19]. There are various methods to produce WPCs. The most important approach is the extrusion and injection molding processes [20].

Generally, there are several advantages for WPCs over mixed plastic-filler materials including low cost and processing difficulties with no expense on mechanical

[2] European Union (EU).

[3] Compound Annual Growth Rate (CAGR).

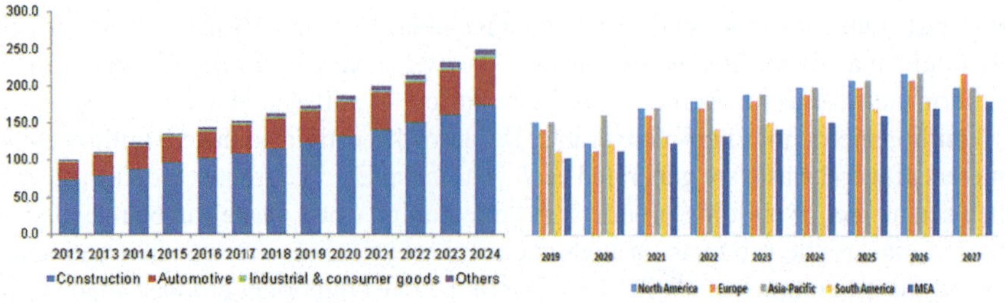


(a) (b)

**Fig. 1** Markets and countries dealing with wood/natural fiber-thermoplastic composites [16, 17]

**Fig. 2** Some applications of wood-plastic composites

properties [21]. In comparison with natural woods, WPCs withstand harsh environments with better structural stability [22] as well as indicate good resistance to termites and fungi [23]. Exploiting wood and plastic wastes to produce high-performance products is the most important hypothesis behind the fabrication of WPCs [24].

Despite having huge advantages, there are several issues with the use of WPCs, as follows:

1. The WPC density is usually twofold the raw wood [25]. Fortunately, some research showed a great reduction in their weight by creating foam structures.
2. The long-term exposure of WPCs to ultraviolet (UV) radiation during outdoor use.
3. Thermoplastic composites usually lack function properly during the long duration of loading since polymers having a linear morphology can strongly respond to time and temperature or creep during loading. Even if, adding fibrous filler into

the polymer matrix may—to some extent—reduce the creep response during loading [26], the problem still remains.

4. Due to the high potential of the destruction of wood fibers at high temperatures, the processing of cellulose fibers in thermoplastics is challenging, thereby limiting their use.
5. Due to the presence of the wood fibers in the polymer field, they act as points of stress concentration and initiation of cracking and failure and thus reduce the impact resistance of WPCs.

In addition to the above-mentioned issues, there are more important problems in the use of WPCs which limit their applications. The high hydrophilicity of natural fibers is against the hydrophobicity characteristics of thermoplastic polymers which makes them incompatible since the mechanical properties of composites strongly depend on the adhesion between fibers and polymers. By improving the bonding and adhesion properties between the fibers and polymers this challenge can be tackled.

## 3 Improvement of the Interface Region of WPCs

Different methods are needed to modify the characteristics of WPCs in the interphase region with improved adhesion. By modifying the polymer matrix using various coupling functional groups, the interfacial strength and mechanical properties of the products were improved. Maleic anhydride grafted styrene-ethylene-butylene-styrene (SEBS-g-MA) [27] and maleic anhydride grafted polyolefins such as HDPE-g-MA [28], PP-g-MA [29], and LDPE-g-MA [30] are the most common examples of reported researches in the literature. Another approach for the enhancement of interfacial adhesion in the natural fiber-reinforced thermoplastic matrix is the fiber treatment before mixing with polymer. Some of these treatments have physical nature and some of them are chemical. Furthermore, plasma, microwave irradiation, and corona treatments of the fibers and wood waste products are some physical treatments reported in the literature with enhanced polymer-matrix adhesion [31–33].

Employing nanoparticles is another strategy to improve the physical and mechanical properties of the composites. Polymeric composites can be turned into nanocomposites by integrating with nano-sized particles which led to the improved mechanical strength of composites [34–37]. Nanocomposites made of carbon nanotubes (CNTs), nanoscale clays, and nanoscale $SiO_2$ received great excitement for industrial applications [38–40].

Nano-based composites present a new venue for the enhancement of wood properties on the horizon of wood modification processes for different applications. Wood-polymer nanocomposites could be a promising approach to obtaining effective products with better physical, thermal, and mechanical properties [41].

For example, utilizing silicate minerals in multilayer played as an in-situ reinforcement to substantially enhance the physical and mechanical properties of various thermosets and thermoplastics at low levels [40, 42–46], thanks to their excellent features such as rich intercalation chemistry, high mechanical stability, high aspect ratio for individual platelets, abundant in nature as a low-cost material, high gas barrier ability, and strong absorbability [42, 43, 47, 48].

## 4 Microwave Treatment

In the usual heating methods, heat flows mainly through conduction, convection, and radiation. In this way, the inherent properties of the material and the heating rate affect the temperature change in the material. Fiber burning, non-uniform temperature distribution, chemical damage to fibers, longer curing time, higher thermal gradient, poor efficiency, high energy consumption, waste generation, and expensive equipment are some of the disadvantages related to the conventional thermal techniques used for manufacturing fiber-polymer composites [49].

Since World War II, major advances have been made in the use of microwaves for heating applications. Afterward, it realized that microwaves have the potential to create rapid and high-efficiency heating in materials. Today, the main applications of microwaves include food processing, wood drying, plastic and rubber processing, ceramic baking, preheating, and post-curing. Microwaves broadly include electromagnetic waves in the frequency range of 300 MHz–300 GHz. Household and industrial microwave ovens usually operate at a frequency of 2.45 GHz, a wavelength of 12.2 cm, and an energy of 5–10 × 2.1 eV [50]. However, all materials can not be heated quickly by microwaves. Materials can be classified into three groups: conductors, insulators, and heat absorbers [51].

Depending on the frequency, dielectric heating can be categorized into two technologies, i.e., radio frequency and microwave technology. Radio frequencies below 100 MHz are generated with open-wire circuits and applied between metallic electrodes. Microwaves are emitted from vacuum tubes direct into the materials through metallic tubes called waveguides and can be generated at 85–94% electrical efficiency [52–54]. Compared to conventional heating in which thermal energy is transferred to the bulk of the material by radiation and/or thermal convection. Microwave heating can transfer thermal energy to the bulk of the material by conduction.

Theoretically, electromagnetic waves consist of two components: electric and magnetic fields. These two fields oscillate vertically relative to each other and are perpendicular to the direction of propagation. The effect of the magnetic field on wood is negligible therefore is not taken into consideration for practical purposes because in dielectric materials magnetic permeability is comparable to the free space. Besides, the effect of the electric field on wood is very high. The high-frequency electromagnetic waves can polarize charges in wood material [55]. A monochromatic electromagnetic wave (Fig. 3) is a sinusoidal wave defined by frequency ($f$) and wavelength (λw). The wavelength is related to frequency through the speed of propagation ($c$) as Eq. 1 [56]:

$$c = \lambda \mathrm{w} \cdot f \; (\mathrm{ms}^{-1}) \tag{1}$$

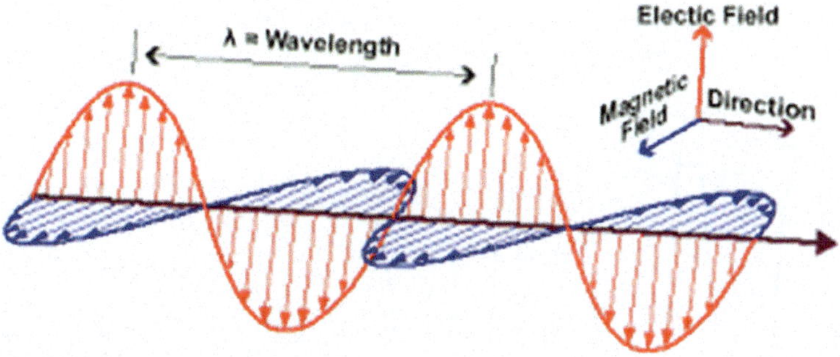


**Fig. 3** Monochromatic electromagnetic wave

The interaction of an electromagnetic field with material may cause several responses, and microwaves can be reflected, absorbed, or transmitted. Reflective materials tend to be bulk metals with many free electrons [53]. Transparent materials tend to have low conductivities associated with members of the glass and ceramics family [57]. Absorbing materials consist of all those which exhibit dielectric character [55].

When an electromagnetic field is penetrating a dielectric material, i.e., wood or any natural fibers, the energy is gradually absorbed by the material due to the polarity. The electromagnetic field strength at the surface is therefore decreasing exponentially during the penetration. The attenuation inside the material can be explained by the exponential Eq. 2 [55]:

$$E(Z) = E_0 \cdot \mathrm{e}^{-\alpha z}\left(\mathrm{V\ m}^{-1}\right) \tag{2}$$

where $\alpha$ is the attenuation factor, and $z$ is the distance of the dielectric from the surface. The effect of attenuation and phase shift on the wave is shown in Fig. 4.

The penetration depth (Dp) depends on the dielectric properties of wood, density, moisture content, temperature, grain orientation, etc. The Dp is defined as the thickness of the material when the transmitted power is reduced to $1/e$ of its original value. It means that the material would absorb approximately 63% of the incident electromagnetic power [58, 59]. The penetration depth is given by Eq. 3:

$$\mathrm{Dp} = 1/2 \cdot \alpha(m) \tag{3}$$

Material with large thickness and high loss factor may cause heating occurs only in surface layers. To prevent this phenomenon, proper electromagnetic heating must be chosen so that enough time is provided for a subsequent heat flow between the surface and core layers.

Processing with microwave radiation has a great potential to improve the current methods of drying wood and processing WPCs. The ability of microwave rays to penetrate and directly interact with materials is abundantly used in the food and

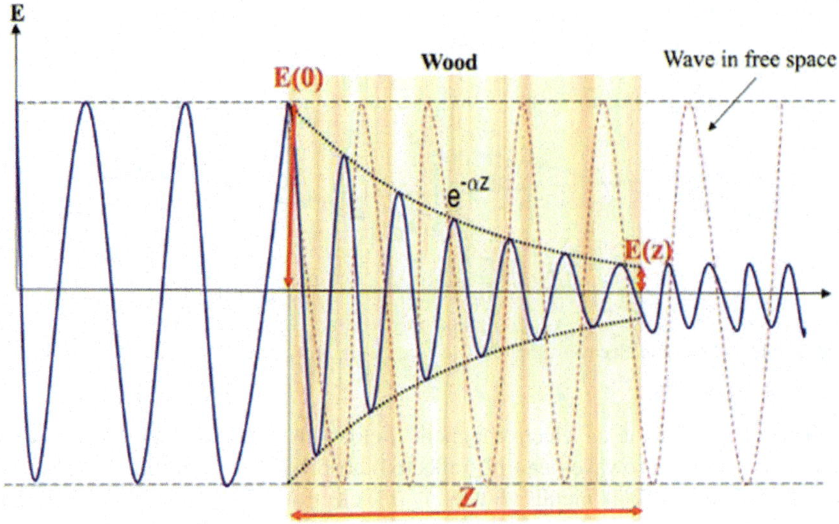


**Fig. 4** Wave attenuation: electromagnetic wave transmitted into wood

pharmaceutical industries, and recently in the production of composites in laboratory dimensions. The heating generates from the microwave beam in the composites while reducing the thermal gradient and rapid heat transfer throughout the thickness of the material, thereby improving the production time and productivity [60].

Heating with microwave radiation is volumetric and is not limited to surfaces; thermoplastic polymer materials can be processed faster with microwave beams [61]. Processing with microwaves in composites, by the convective heat transfer method, has several advantages: faster heating, increased adhesion of fibers to the matrix, and more controllability. Heating with microwaves strongly depends on the dielectric properties of the material [62]. The energy of the microwave is directly transferred into the very thin boundary layer of the polymer matrix [63].

## 5 Wood Drying with Microwave Irradiation

One of the important applications of microwaves is wood drying. Although 6 values of microwave frequencies are used industrially, only two of them (0.922 and 2.54 GHz) are used in heat treatment and wood drying [64]. Microwaves affect wood resistance, moisture content (MC), weight, and thickness of wood species. The temperature of wood during microwave heating depends on several factors such as the power of microwave radiation, time of treatment, dimensions of material, MC, frequency, and wood permeability [65–69].

The main advantage of this method in wooden materials is the rapid heat of the material throughout the whole cross-section due to the dielectric properties of wood and water [55, 70]. Another advantage is the ability of microwave heating to be used in a continual process, which will be beneficial to flow technology production. The principle behind microwave heating is based on the polar characteristic of molecules and their ability to absorb and transform microwave radiation into heat [55, 71]. Permanent dipoles of molecules begin to move with the same frequency as the electromagnetic field. Therefore, rapid changes in the field polarity cause vibration and rotation of molecules, which transforms the microwave energy into frictional heat [71, 72]. Some researchers investigated the effect of microwaves irradiation on raw wood, as given below.

For example, BinHua et al. [73] summarized the possible applications of microwave technology for the wood industry, like wood drying, curing of adhesives, defect detection of wood, MC measuring, and volatile organic compound (VOC) emission reduction. This technology can be used also for the plasticization of wood [69] and the acceleration of chemical reactions, e.g., microwave acetylation [70, 75, 76]. Research by Mori et al. [77] showed that the surface temperature of wood spices and internal layers of wood by fiber optical temperature sensor reached 90–110 °C and 100–130 °C, respectively, when a microwave power in the range of 0.6–2.4 kW at a frequency of 2.45 GHz and time of treatment of 1–3 min is applied. Based on wood drying, Brodie [78] showed that heat treatment of wood with microwaves reduces the density and MC of wood, therefore resulting in decreasing the cost and energy consumption during the wood drying process.

By drying *Picea Abies* wood through traditional and microwave irradiation methods, Hansson and Antti [71] investigated the wood resistance according to the moisture content, density, and fiber direction. Their results showed that the resistance performance can be controlled by the moisture content, the number of annual rings, weight, thickness, and width of wood species. A reduction in drying time was observed when the microwave method is applied. Employing two types of high and low-intensity microwaves, Balboni et al. [79] studied the effect of microwave treatment on drying defects, density, MOE, MOR, compression strength parallel to the grain, and shear strength of *Eucalyptus Macororhyncha* wood samples. They found that low-density microwaves reflected better impacts on moisture content distribution, and wood density, while MOE and MOR values remained statistically unchanged. He et al. [80] investigated the permeability and drying properties of *Eucalyptus Europhyla* wood samples by controlling the radiation power, radiation time, and initial moisture content. Their results showed that by increasing the power and time of radiation, the wood permeability increased in both longitudinal and transverse sections, followed by decreasing the wood moisture content. Researchers believe that microwaves can accelerate the quality of wood drying having reduced drying time by 65%. By applying microwave irradiation, the change in the internal structure of wood was demonstrated through SEM micrographs. As shown in Fig. 5a–c vessels and ray cells of the eucalyptus were filled with many tyloses. After the microwave treatment was implemented, the tyloses in the vessels were ruptured (Fig. 5d). The shape of the ray cells along the tangential direction of wood remained the same, but the tyloses were considerably reduced followed by minor cracks in the middle of the lamella (see red arrows shown in Fig. 5e). Pit membranes in the radial direction of wood were seriously damaged after applying microwave irradiation with 15 kW for the 90 s (see red arrows shown in Fig. 5f); this led to increased pathways for the liquids and gases to get in or out of wood during the impregnation or drying process. It was concluded that the microwave modifications led to increasing the wood permeability by up to a factor of 2.3 and a shortened drying time by a maximum of 65%.

Vongpradubchai and Rattanadecho [81] reported that moisture content has a significant effect on temperature development since microwave energy is mainly absorbed by water in the wood. Work by Studhalter et al. [69] showed that green timber reflects the lowest and highest temperature distribution on the surface and internal parts, respectively, when there is a large temperature gradient. Zielonka and Dolowy [68], Antti and Perré [82], Zielonka and Gierlik [83], and Brodie [84] demonstrated that when thick material is heated by microwaves, the highest temperature is located at a distance away from the surface. From a heat transfer point of view, the temperature increases from the surface to the critical points and vice versa, even though it linearly decreases away from the points [68]. From the above-mentioned research works, it is concluded that microwave modification may be an energy-saving process for drying raw wood with a short drying time.

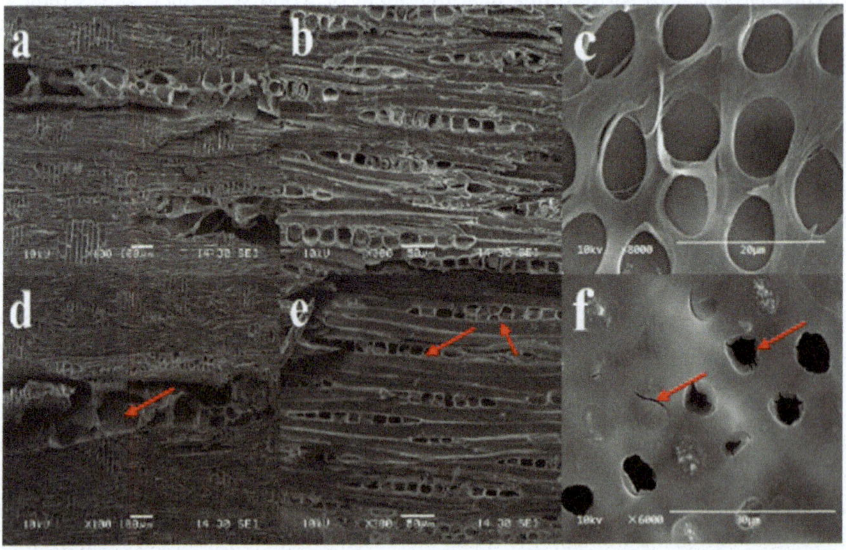


**Fig. 5** SEM micrographs of reference and microwave-treated wood; (**a**)–(**c**) are the radial-section, tangential-section, and pits of control wood, respectively; (**d**)–(**f**) are the radial-section, tangential-section, and pits of microwave-treated wood, respectively

## 6 Microwave Treatment of Wood-Plastic Composites

The motivation behind the use of microwaves in the process of heating biobased composites is the hydrophilicity characteristics of plant-based reinforcements. The interaction between the microwaves and polar water molecules with asymmetric charges present in the plant-based reinforcements results in the partial dissipation of heat energies in the bio-composite materials [85]. The amount of energy dissipated varies depending on the concentration, structure, reinforcement/matrix humidity degrees, temperature, and microwave radiation frequency [86].

Microwaves have been utilized in the development of polymeric compounds for the last three decades [87–90]. Microwave treatment is a potential approach that outperforms conventional techniques in the production and processing of composites [63]. Moreover, microwaves in the form of convection having electromagnetic energy sound more useful compared to thermal energy with heat transfer. Microwaves are not limited to the surface as they are volumetric thus polymeric materials can be processed faster with a microwave [61]. From a heat transfer point of view, energy is transferred because of thermal gradients, while microwave heating is the process of transferring electromagnetic energy to thermal energy and is considered energy conversion rather than heat transfer. Therefore the way that energy is delivered in the microwave heating process could be beneficial to many potential applications related to material processing. Upon the energy concept that explains the generation of heat throughout the volume of the material, microwaves can penetrate materials. As the energy transfer is independent of the heat diffusion from the surfaces, uniform and fast heating of thick material can be achieved via microwaves [91]. Microwaves propagate through space at light speed and, usually, the frequency ranges swap from 300 MHz to 300 GHz [92].

Fast heating, fiber-matrix adhesion, and higher controllability are some of the advantages of microwave processing using convective microwave methods. Microwave heating strongly depends on the dielectric characteristics of the polymers. The transfer of the energy of microwaves into narrow materials takes place at the boundary layers [90, 93, 94]. This offers multiple advantages in the material processing of ceramics, metals, and composites in regard to conventional heating methods, including unique microstructure and properties, improved product yield, energy savings, and reduction in manufacturing cost and synthesis of new materials [95]. The microwaves do not interact with the majority of polymeric materials due to the lack of dipolar moment. Thus, additives acting as heating susceptors can be used to prepare materials capable of absorbing microwaves [92]. Some of the research in this regard is given below.

Erchiqui et al. [86] studied the thermoforming of PP/WF composite by employing microwave and infrared radiation heating sources. The WPC samples treated by microwave radiation reflected a uniform temperature distribution in comparison with infrared radiation treatment. The time duration of microwave heating has a direct relation with the reinforcement concentration and the sample size. Another new method of using microwaves on WPCs was done by Tewari et al. [14]. They used the microwave-assisted compression molding (MACA) method to make WPCs, and at the same time, they used alkaline chemical treatment of pine cones to improve the bonding. The properties of composites were controlled through physical, mechanical, and thermal tests. Their results indicated a significant effect of the MACA method in the strengthening of WPCs. They also showed that the dielectric constant of pine cone filler is 1.6 while the dielectric constant of rHDPE at 2.45 GHz is 2.3. Having the complex permittivity of the material during microwave interaction ($\varepsilon^*$), dielectric constant ($\varepsilon^{'}$) as the energy absorbed by the material, and dielectric loss factor ($\varepsilon^{\varepsilon}$) as the ability of the material in dissipating microwave energy, one can calculate the dielectric response of materials (representing the interaction between microwave and the nonmagnetic material), as shown in Eq. 4:

$$\varepsilon^* = \varepsilon' - j\varepsilon^{\varepsilon} \tag{4}$$

Dielectric loss tangent or energy dissipation factor (δ) is a function of $\varepsilon^{'}$ and $\varepsilon^{\varepsilon}$ via Eq. 5:

$$\tan\delta = \varepsilon^{\varepsilon}.(\varepsilon^{-1}) \tag{5}$$

Dipole polarization is represented as a parameter contributing to the dielectric response which is the dominating energy transmission mechanism in nonmagnetic materials at the molecular scale such as polymeric composites, ceramics, etc. [96]. A pine cone filler having a higher dielectric constant can be heated by interacting with the microwave. This led to heat transmission to the rHDPE and adjacent pine cone via conduction, resulting in melting rHDPE pellets and consequently bond formation between the pine cone and matrix. Next, the mixed melted rHDPE and pine cone filler were exposed to room temperature to get cool down. Erchiqui et al. [86] conducted a numerical study on the potential of microwave radiation with 140 °C heating temperature for thermoforming WPCs. They also compared infrared and microwave heating modes on pine cone/polypropylene composite sheets and control polypropylene. Their results showed that microwave irradiation creates somehow uniform temperature distribution in the sheets compared to the infrared method.

## 7 Post-treatment of WPCs by Microwave Irradiation: Case Study Outline

Another application of microwave irradiation is to modify the characteristics of composites after their production process. Some researchers have addressed this issue in recent years. For example, the work by Yuan et al. [97] showed the potential of microwave irradiation to modify carbon fibers immersed in water. Their results indicated that a large number of oxygen-containing groups which are exposed to exfoliation of the surface layers were placed on the carbon fibers when microwave irradiation is applied. Although the tensile strength of carbon fibers was slightly lower, the shear strength of carbon fibers in epoxy resin composite significantly increased. Work by Zhang et al. [94] reported the fabrication of PES/pine wood powder composites using selective laser sintering (SLS) with a layer thickness of 0.1 mm, preheat temperature of 83 °C, and internal power of 13 W. A CNT was utilized as an additional reinforcement aimed to improve the mechanical and physical properties of the produced WPCs. Next, a microwave oven at 385 W power and 2.45 GHz frequency was used for WPCs post-treatment. The 10 s was an optimal treatment time to produce samples with the highest bending strength as high as 15.7 Mpa, while microwave treatment for less than 60 s improved the bending strength by 64.2%.

Nightingale [98] compared autoclaved composites with microwave post-cured composites and fully microwave-cured composites by performing physical and mechanical tests. The bending test results showed that the post-cured composites by the microwave possessed the lowest bending strength compared to the composites cured by the normal autoclave. Boey and Lee [99] and Yue and Boey [100] reported that the elastic modulus of microwave-cured composites is higher compared to the conventional thermal method, however they have lower tensile strength. Comparing the mechanical behavior of microwave-cured epoxy resin composites over thermal curing, Bai et al. [101] concluded that microwave-cured resin showed slightly higher strength and tensile modulus due to greater homogeneity and proper temperature distribution. The effects of microwave post-curing and traditional autoclave heating methods on the mechanical and microstructure properties of composites were investigated by Adeodu et al. [102]. They found that there is a direct relationship between tensile strength and the weight percentage of fillers according to post-curing methods. Also, the composites produced by the microwave post-curing method showed fewer defects compared to the composites produced by the autoclave method. Moreover, the microwave post-curing heating methods improved the strength of the microstructure of composites compared to autoclave heating. Zlobina et al. [33] used microwave modification to improve the resistance properties of the composite incorporated with carbon and glass fibers for aviation applications. The effect of the microwave beam on the composite at a specific power level increased the bending strength. At the same time, the microwave beam caused significant changes in the microstructure of the composite. They realized that the homogeneity and density of composites increased since microwave treatment was implemented. Chavooshi et al. [103] investigated the effect of microwave radiation heat treatment on the mechanical and morphological properties of polypropylene/MDF[4] powder nanocomposites. They used nano clay and MAPP as coupling agents to improve the bonding and properties of the composite. Their results showed that the mechanical properties of samples treated with microwave radiation were better than other samples. From FE-SEM for the nanocomposites treated with microwave, they observed that MDF powder is more optimally mixed with the polymer matrix, thus the encapsulation of MDF powder by the polymer is better.

---

[4] Medium Density Fiberboard.

Owing to the importance of microwave treatment in the improvement of the mechanical properties of the composite, the following section aims to fully explain our research recently conducted regarding microwave heating on WPC by injection molding method. The purpose of this research is to investigate the improvement of the mechanical, physical, and morphological properties of composites treated with microwave radiation in the presence of nano-silica as an accelerator of microwave absorbance. Recent advances in nanotechnology provide a source of excellent microwave-absorbent materials due to their exceptionally high dielectric constant which allows for absorbed electromagnetic energy to be dissipated into heat [104]. Wood flour, high-density polyethylene (HDPE), and Nanao-$SiO_2$ (i.e., 0, 1, 2, and 3 wt%) were used as nanomaterials to fabricate WPCs. When the samples were fabricated, they were exposed to 900 W microwave radiation for 7 min before performing the physical and mechanical tests.

## 7.1 Materials and Method

HDPE matrix with the specification of 18 g melting flow index in 10 min and 190 °C melting point supplied from Tabriz Petrochemical Co., Iran. The 0.8% Maleic anhydride grafted with polyethylene (MAPE) was employed as a coupling agent with 0.4 g melting flow index in 10 min which was supplied from Merck Co., Germany.

In this study, the wood flour was supplied by a local factory based in the North of Iran and was further pulverized by a lab-scale grinder. The produced powder was used as the filling fiber by passing through 60 and 80-mesh sieves. $SiO_2$ nano-sized particles (with an average diameter of 12 nm, a density of 0.37 g/cm$^3$, an apparent ratio of 208, and a specific surface area of 200 m$^2$/gr) supplied by Evonik Degussa Co., Germany, were utilized as reinforcement. The XRD pattern of Nanao-$SiO_2$ used in this study with the 2θ of 23.2° is shown in Fig. 6. The polymer and $SiO_2$ nanoparticles were used right after receiving them. An air-circulating oven was used to dry the lignocellulose materials at 100 °C for 24 h to reach a moisture content of <1 wt% before their use.

## 7.2 Preparing the Composites

The prepared composites are given in Table 2. The concentration of $SiO_2$ nanoparticles varied from 0, 1, 2, and 3 wt% at a constant 50% WF for all the samples.

The mixing phase was implemented at 150, 155, 155, 155, 150, and 145 °C and 60 rpm with 1500 MPa die pressure in a counter-rotating twin screw extruder (Brabender® Plasti-Corder®). Afterward, the sample was grounded using WIESER WGLS200/200 Model semi-industrial crusher and then transferred to the injection molding machine. Five test samples were prepared for each treatment at 160 °C and 45 rpm loading speed, 80 bar injection pressure, 75 s cooling time, and 200 bar

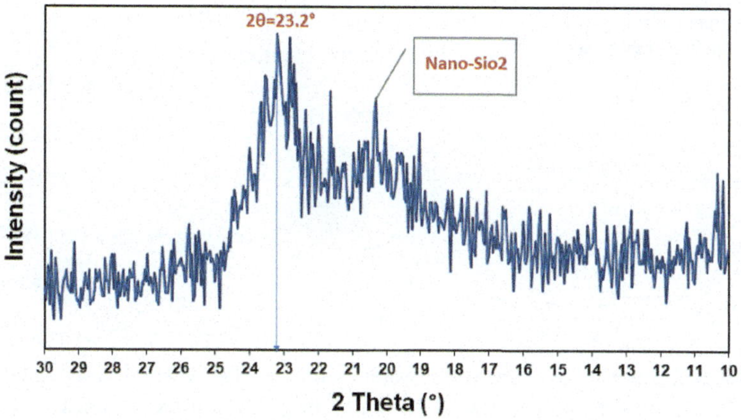


**Fig. 6** X-ray diffraction pattern of the nanao-$SiO_2$

**Table 2** Composition of evaluated formulations

| Number | Composite formula | HDPE (wt%) | Wood flour (wt%) | $SiO_2$ nanoparticles (wt%) | MAPE (wt%) |
|---|---|---|---|---|---|
| 1 | 47HDPE/50WF/0NS/3 M | 47 | 50 | 0 | 3 |
| 2 | 46HDPE/50WF/1NS/3 M | 46 | 50 | 1 | 3 |
| 3 | 45HDPE/50WF/2NS/3 M | 45 | 50 | 2 | 3 |
| 4 | 44HDPE/50WF/3NS/3 M | 44 | 50 | 3 | 3 |
| 5 | 47HDPE/50WF/0NS/3 M* | 47 | 50 | 0 | 3 |
| 6 | 46HDPE/50WF/1NS/3 M* | 46 | 50 | 1 | 3 |
| 7 | 45HDPE/50WF/2NS/3 M* | 45 | 50 | 2 | 3 |
| 8 | 44HDPE/50WF/3NS/3 M* | 44 | 50 | 3 | 3 |

* Treated with microwave irradiation

loading pressure. The samples were restrained minimum of 40 h in an environment with 23 °C temperature and 50% relative humidity followed by the 99-618ASTM standard. A 900 W microwave oven (NN-CD997S model, Panasonic Co.) was utilized for the samples heat treatment (Fig. 7). Five, seven, and ten minutes of irradiation were considered to find an optimal microwave irradiation period. Controlling the color and morphology of the samples after irradiation and following Chavooshi et al. [103], 7 min of microwave irradiation was chosen as an optimal period of irradiation.

Five samples from each combination were selected and dried in an oven for 24 h at 100 ± 3 °C to control the water absorption and thickness swelling of the samples following the ASTM D7031-04 standard. Measuring the weight and thickness of the dried samples with an accuracy of 0.001 g and 0.001 mm, respectively, the samples then were immersed in distilled water at a temperature of 20 ± 2 °C for 24 h. The thickness and weight properties of the samples were measured after rinsing the

**Fig. 7** Image of microwave device and its components

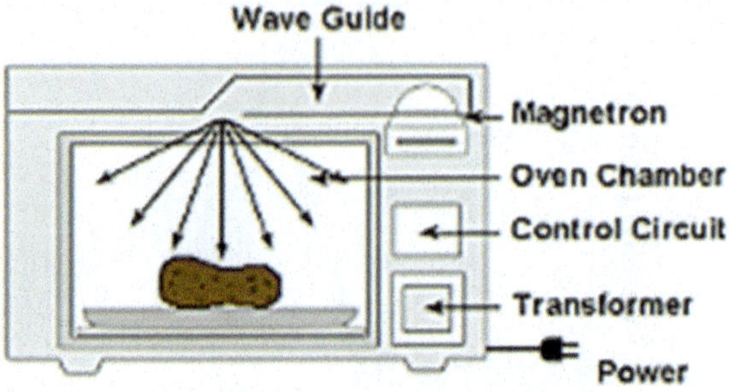


samples with an excessive amount of water on their surfaces. The following equation (Eq. 6) calculates the percentage of the water absorption of the samples:

$$\mathrm{WA}(t) = \frac{W(t) - W_0}{W_0} \times 100 \tag{6}$$

where WA ($t$), $W_0$, and $W(t)$ denote the water absorption at time $t$, oven-dried weight, and weight of the sample at immersion time $t$, respectively.

Also, the percentage of the thickness swelling is calculated via Eq. 7.

$$\mathrm{TS}(t) = \frac{T(t) - T0}{T0} \times 100 \tag{7}$$

where TS($t$), $T_0$, and $T(t)$ are the thickness swelling at time $t$, and the initial thickness and thickness of the samples at time $t$, respectively.

Flexural strength and modulus were taken following standard ASTM D747 using the UK Instron 6025 model. At a crosshead speed of 5 mm/min, the tests were implemented. An Izod impact strength test was performed following ASTM D-256 using Zwick 5102 model machine.

The structure of the composites is controlled by a scanning electron microscope (SEM) (LEO Oxford Co.). The fracture surfaces of the specimens were sputter-coated with gold after conducting the impact test right before the analysis is initiated. All images were taken at an accelerating voltage of 26 kV.

## *7.3 Results and Discussion*

Figure 8a, b show the 48 h water absorption and thickness swelling of composites comprising wood flour, $SiO_2$ nanoparticles, and HDPE in two states of untreated and microwave-treated, respectively. Based on the 50% constant amount of wood fiber used in all compositions, the water absorption and thickness swelling properties of the tests show different behavior due to the presence of $SiO_2$ nanoparticles and the microwave treatment. The composites in the absence of $SiO_2$ nanoparticles reflect

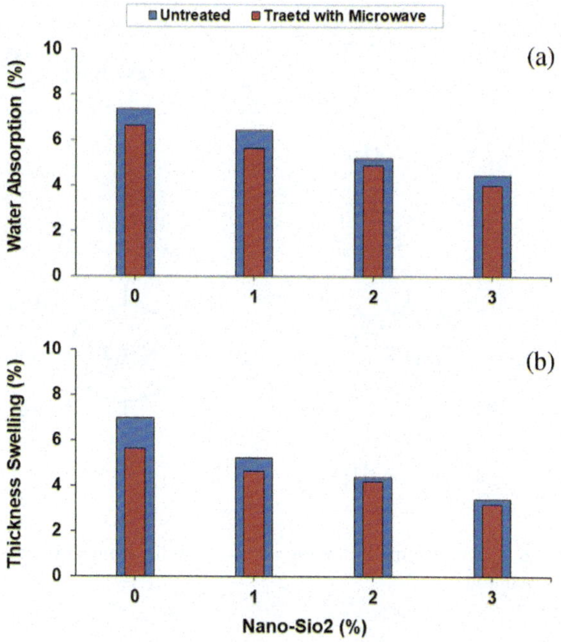


**Fig. 8** Effect of $SiO_2$ nanoparticles and microwave irradiation on water absorption and thickness swelling of WPCs

the highest adsorption and swelling properties compared to the samples filled with $SiO_2$ nanoparticles.

The formation of hydrogen bonding between $SiO_2$ nanoparticles and the hydroxyl groups of wood flour (Fig. 9) reduces active sites for water molecules to form hydrogen bonds with hydroxyl groups. In other words, there is a kind of competition between the hydroxyl groups of wood flour and the hydroxyl groups of $SiO_2$ nanoparticles to form hydrogen bonds. Therefore, the bond between wood flour and $SiO_2$ nanoparticles reduces the chances for water molecules to form hydrogen bonds in this competition.

Chemically, according to Eq. 8, $SiO_2$ nanoparticles have hydroxyl groups, hydrogens attached to hydroxy groups, and siloxane groups. The first two compounds are hydrophilic and the latter is hydrophobic. Therefore, it suggests that the $SiO_2$ nanoparticle is polar and hydrophilic. Although $SiO_2$ nanoparticle has the moisture adsorption property because of its hydroxyl groups, when they are in contact with wood flour to form hydrogen bonds, the hydroxyl groups of both materials are practically unavailable, leading to the reduced water absorption of the composites [105].

$$\begin{aligned}
&Si(OC_2H_5)_4 + H_2O \rightarrow Si(OC_2H_5)_3OH \\
&+ C_2H_5OH \rightarrow \equiv Si - O - Si \equiv + H_2O \equiv Si - O - H \\
&+ H - O - Si \equiv \rightarrow Si - O - Si \equiv + C_2H_5OH \equiv Si - OC_2H_5 \\
&+ H - O - Si \equiv
\end{aligned} \tag{8}$$

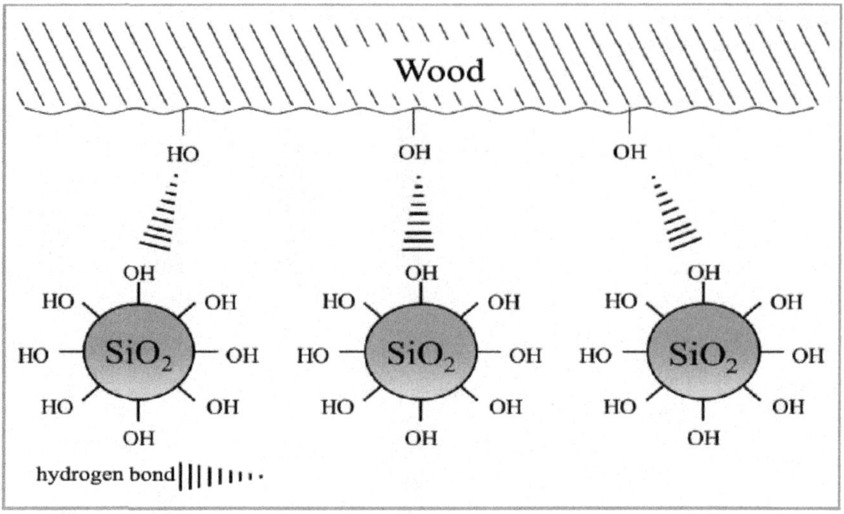


**Fig. 9** The scheme of linking the hydrophilic groups of nano-silica with the hydroxyl groups of wood flour

According to Espert et al. [106], water absorption in natural fiber-plastic composites could be obtained through moisture absorption by either the cell wall of the lignocellulosic material or the capillary process through the existing voids gaps between the plastic and natural fibers. The former is because of the inherent characteristics of cellulose fibers and the latter is due to the weak interaction between wood fibers and plastic and the formation of possible defects during its production process. Among the samples treated with microwave radiation, the lowest amount of thickness swelling is related to the samples containing 3 wt% $SiO_2$ nanoparticle, which is due to the small dimensions of nano-silica particles which can fill the empty spaces in the matrix, thereby blocking a large number of capillary tubes [107]. Therefore, the presence of nano compounds inside the empty spaces and their interactions with water to form hydrogen bonds reduces the water absorption of the composites. This is in agreement with the research by Ismaeilimoghadm et al. [107] and Hosseini et al. [108]. Figure 8a, b also show that the water absorption and thickness swelling decrease further by performing the microwave treatment. The microwave radiation contributes to softening again the thermoplastic polymer material and thus the cellulose fibers are better enclosed and encapsulated by the matrix [103]. This leads to blocking the porosity of the matrix and the pores of the capillary tubes of cellulose fibers which are the potential spots for water penetration. In Fig. 8a, it is found that the microwave-treated composites with 3 wt% $SiO_2$ nanoparticles and 4% water absorptivity show a 66% absorption reduction in comparison to treated samples with no $SiO_2$ involved.

The average values of flexural strength and modulus of WPCs containing different amounts of $SiO_2$ nanoparticles in two states of untreated and treated with microwave radiation are presented in Fig. 10a, b, respectively. Figure 10b shows the increase in the flexural modulus by adding $SiO_2$ particles. Figure 10b shows the highest flexural modulus of 116.86 MPa for the composites containing 3 wt% nano-silica treated by microwave radiation. While the lowest flexural modulus belongs to the microwave-treated samples without $SiO_2$ nanoparticles (control sample) with a value of 769.84 MPa. The modulus of elasticity of composites depends on several parameters such as the amount of fiber used, the orientation of fibers, the connection of fibers and the polymer matrix in the interphase region, the modulus of elasticity of the components, and its density [45]. According to the nano-silica information, the elasticity of $SiO_2$ nanoparticles has been reported between 66.3 and 74.8 GPa. Therefore, the replacement of wood fibers with an elasticity modulus of about 5 to 8 GPa with $SiO_2$ nanoparticles can have a significant effect on the flexural modulus of the composites. Moreover, the higher ratio of length to diameter (L/d) of the $SiO_2$ nanoparticles used in this research plays an important role in the increase of the flexural modulus of WPCs. These results are consistent with the research of Nourbakhsh et al. [45]. Also, by using a microwave treatment, the flexural modulus increased from 685.72 to 1116.86 MPa compared to the control sample. These results are in agreement with Zlobina et al. [33]. Another reason for the connection improvement between the fibers and the polymeric matrix, especially in composites reinforced with natural fibers, is the development of curing the matrix during post-treatment with a microwave beam which is heat transferred from the fiber surface to the internal volume of the composite [109]. In the curing process by the microwave beam, wood fibers and $SiO_2$ nanoparticles, because of their high dielectric constant [104, 109], absorb a large amount of microwave power in the form of heat which is transferred from the outer surface of the fibers to the volume of the polymer matrix. This leads to the formation of a stress-free interface area around the fibers [110, 111]. With microwave beam treatment, the flexural modulus of the composites containing 1, 2, and 3 wt% $SiO_2$ nanoparticles are 29.3, 44, and 62.87%, respectively, which are greater than that of the control sample (Fig. 10b). The elasticity modulus of 17.65, 6.14, and 10.93% are also observed for the samples containing 1, 2, and 3 wt% $SiO_2$ nanoparticles, respectively, which are higher than that of the untreated samples. In connection with the effect of nano compounds on the performance of microwave waves, research by Chavooshi et al. [103] showed that these waves together with nano clay particles improved the mechanical properties of composites. On the other hand, nanomaterials are excellent absorbents for microwaves beam and due to their high dielectric constant, they have the possibility of absorbing electromagnetic energy and of dispersing net energy in the material as heat [104]. The results of this research are confirmed by Zhang et al. [94]. In this regard, the positive effect of $SiO_2$ nanoparticles and microwave treatment on flexural strength is shown in Fig. 10a. As seen in Figure, by adding 3 wt% $SiO_2$ nanoparticles, the flexural strength increased from 19.13 MPa in the control samples to 26.86 MPa in the composites indicating a 40.4% improvement in flexural strength. By using microwave treatment, composites containing 1, 2, and 3% of $SiO_2$ nanoparticles show, respectively, 9.38, 19, and 29.1% improved flexural strength. The flexural strength of composites depends on the bond between their components. The increase in the flexural strength by applying microwave treatment is due to the reduction of the polymer's viscosity with better wetting for the fibers. As a result, the adhesion and resistance in the interphase region between the fibers and the polymer improve [103]. These results are in line with the research by Chavooshi et al. [103]. Due to very small dimensions, mechanical strength, and high length-to-diameter ratio, nano-silica can create strong adhesion in the interphase region with the composite components. The results are consistent with Nourbakhsh et al. [45] and Farsi [48]. Qualitatively, the composites incorporated with 3 wt% $SiO_2$ nanoparticles have the highest elasticity modulus and strength with the respective values of 1116.86 and 25.38 MPa in the presence of a microwave beam.

Figure 11 shows the un-notch impact strength in the composites. From Fig. 11, it is obvious that, in general, there is a negligible difference between the impact resistance in the samples with and without microwave treatment, suggesting a lower impact resistance in the composites treated with a microwave beam. The effect of the amount of $SiO_2$ nanoparticles on the impact resistance of microwave-treated composites shows that the samples without $SiO_2$ nanoparticles have more resistance compared to the samples containing $SiO_2$ nanoparticles, which is probably due to the presence of nano-silica particles and the formation of energy absorption points [45, 48]. Works by Farsi [48] and Hosseini et al. [108] showed different impact resistance effects in the composites since nano-silica is added. The samples treated with microwave irradiation containing 1, 2, and 3 wt% $SiO_2$ nanoparticles reflected 3.9%, 6.03%, and 9.2% impact resistances, respectively, which are less than that of the control samples. By applying microwave treatment and increasing the amount of

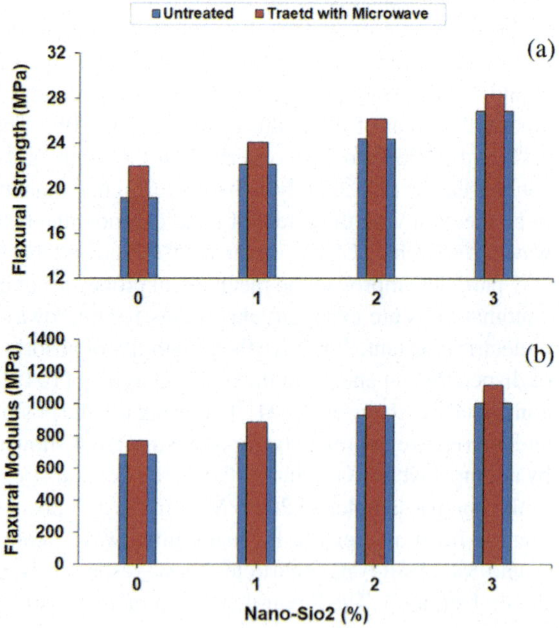


**Fig. 10** Flexural strength and modulus of the WPCs as a function of $SiO_2$ nanoparticles and microwave irradiation

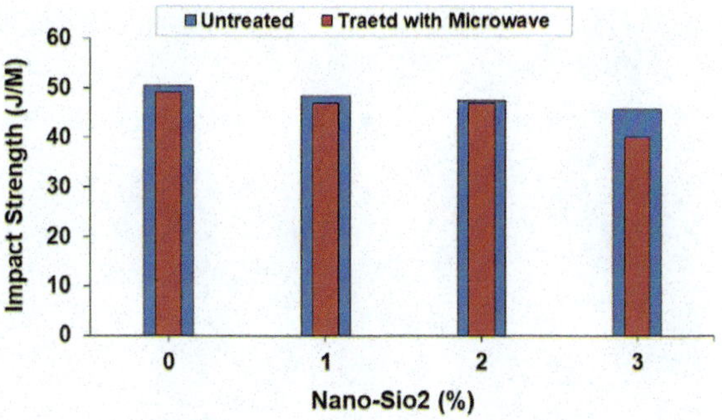


**Fig. 11** Impact strength of the WPCs as a function of $SiO_2$ nanoparticles and microwave irradiation

$SiO_2$ nanoparticles from 1 to 3%, the impact resistance of the composites was reduced by 4.3%, 4.3%, and 18.5%, respectively, in comparison to the control samples and those treated with microwave radiation, which is not consistent with Chavooshi et al. [103]. The un-notch impact resistance shows the resistance of the material against breaking and cracking, suggesting regular cracks occurring at the weakest parts of the composite in the interface between the lignocellulosic material and the matrix. The excessive increase of the $SiO_2$ nanoparticles in the composites can also be considered as the parts with energy absorption and crack development [48]. Therefore, with the increase of $SiO_2$ nanoparticles, the composite impact resistance decreases. These results have been confirmed by Hosseini et al. [108]. Also, the results of the impact test show that the composite without $SiO_2$ nanoparticles and treated with a microwave beam has the highest impact resistance.

Scanning Electron Microscope (SEM) images are suitable tools for justifying the physical and mechanical properties of composites. There is always a relationship between the properties of composite materials and their internal structure. Figure 12a is related to the samples containing 3% $SiO_2$ nanoparticles without microwave treatment, by 10 KX magnification. As it is noticeable, the matrix is dispersed in failure points showing frequent $SiO_2$ nanoparticle particles. In some parts, the wood fibers are separated from the matrix, and instead, empty holes (shown as a red arrow) are seen depicting the poor adhesion between the fiber and the matrix at these points. As seen in Fig. 10a, b, which are related to flexural strength and modulus, compared to untreated samples, the microwave-treated sample has lower resistance with higher water absorption. Figure 12b shows the composite containing 2% $SiO_2$ nanoparticles and treated with a microwave beam with 10 KX magnification. As it can be seen, by performing microwave treatment, the surface of the sample became more uniform and the matrix completely encapsulated the wood fibers, indicating the positive effect of microwave treatment on the reduction of the viscosity of the polymer matrix and the better embrace of the wood fibers and consequently the improvement of the resistance [94]. As shown in Fig. 12b (SEM photomicrograph b) the number of pores on the surface of composites was reduced. At the same time, the water absorption of the sample decreased. Such an issue is more evident in Fig. 12c with a magnification of 10 KX, as can be seen, the surface of the fibers gets covered by the polymeric matrix which is treated and melted by microwave treatment. This led to covering

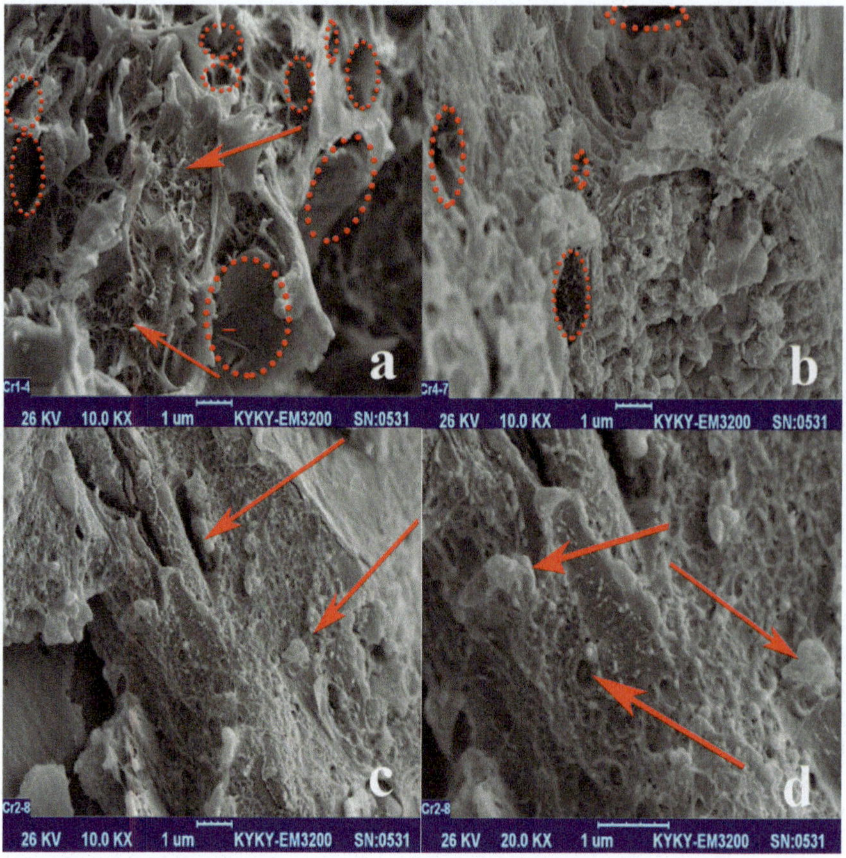


**Fig. 12** SEM photomicrographs of fractured samples of nanocomposites: **a** 3% $SiO_2$ nanoparticles and no microwave, 10000X, **b** 2% $SiO_2$ nanoparticles and with microwave, 10000X, **c** 3% $SiO_2$ nanoparticles and with microwave, 10000X, **d** 3% $SiO_2$ nanoparticles and with microwave, 20000X

and disappearing the holes and pores in the matrix surface. Also, the increase in the amount of $SiO_2$ nanoparticles from 2 to 3 wt% in Fig. 12d with 20 KX magnification indicates the excessive accumulation of nano-silica on the surface, which forms energy absorption and crack development points (see red arrows shown in Fig. 12d), thus reduced impact resistance.

## 7.4 *Conclusion*

In this research, the effect of simultaneous use of post-treatment with microwave irradiation and different amounts of nano-silica on the physical and mechanical properties of WPCs was investigated. Based on the obtained results, the following conclusions were reached:

1. By increasing $SiO_2$ nanoparticles up to 3 wt%, the flexural strength, and modulus of WPCs improved significantly.
2. The un-notched impact resistance of WPCs slightly decreased with the increase of $SiO_2$ nanoparticles up to 3 wt%. Adding different levels of $SiO_2$ nanoparticles to WPCS has led to different positive and negative results in the composites.
3. The use of $SiO_2$ nanoparticles reduces water absorption and the thickness swelling of WPCs, and with the increase of $SiO_2$ nanoparticles, this reduction increases.
4. The post-treatment of WPCs with microwave irradiation increases the flexural strength and modulus and decreases the impact resistance, water absorption, and thickness swelling of composites. SEM images indicate the microstructure surface improvement of the composites with microwave treatment. By applying microwave treatment, the surface of the composite is more uniform, and the fillers are better encapsulated by the polymer.
5. The general results indicate the proper effect of $SiO_2$ nanoparticles and microwave treatment on WPCs. Based on this, the highest amount of flexural strength and modulus, and the lowest amount of water absorption and thickness swelling are related to WPCs containing 3% of $SiO_2$ nanoparticles and treated with microwave radiation. The highest impact resistance belongs to composites without $SiO_2$ nanoparticles and treated with a microwave beam. The SEM images of the samples also confirm these results.

## 8 Summary

This chapter describes the basic, important, and practical concepts of applying microwave irradiation to wood and wood waste products. By reviewing extensive works of literature relevant to the field, a detailed discussion of wood drying and WPCs properties is provided. It was found that microwave treatment greatly influences the properties of wood waste products. It is suggested that new or innovative microwave treatments and methods are demanded to meet the practical application of wood and wood waste products in industries. There is still a lack of experimental investigations evaluating the impact of microwave treatment on wood waste products' resistance to biological agents, fire, environmental conditions, and so on. Finally, the authors discussed their findings for microwave irradiation post-treatment of WPCs by assessing their mechanical, physical, and morphological characteristics. They found that the microwave plays a significant role in the fabrication and wood drying of wood-based composites.

## References


1. Hon DNS, Shiraishi N (2001) Wood and cellulosic chemistry. Marcel Dekker, New York
2. Wagenführ R (2000) Holzatlas (Atlas of Wood), 5th edn. Fachbuchverlag Leipzig, Germany
3. Skodras G, Grammelies P, Kakaras E et al (2004) Evaluation of the environmental impact of waste wood co-utilisation for energy production. Energy 29:2181–2193
4. Mohanty AK, Wibowo A, Misra M et al (2004) Effect of process engineering on the performance of natural fiber reinforced cellulose acetate biocomposites. Compos Part A Appl Sci Manuf 35(3):363–370
5. Bledzki AK, Gassan J (1999) Composites reinforced with cellulose based fibres. Prog Polym Sci 24(2):221–274
6. Global production and trade in forest products in 2020 (2020) FAO. www.fao.org/forestry/statistics/80938/en. Accessed 20 July 2020
7. Forestry Production and Trade (2021) FAO. www.agriculture.gov.au/abares/research-topics/forests/forest-economics/forest-economic-research/global-outlook-for-wood-markets. Accessed 23 Nov 2021
8. Municipal solid waste and construction and demolition wood waste generation and recovery in the united states (2014) Dovetail Partners Incorporated. www.dovetailinc.org/report_pdfs/2014/dovetailwoodrecovery0914.pdf. Accessed 22 Sep 2014
9. Bergeron FC (2016) Energy and climate impact assessment of waste wood recovery in Switzerland. Biomass Bioenergy 94:245–257
10. Clemons C (2002) Wood-plastic composites in the United States: the interfacing of two industries. For Prod J 52(6):10–18
11. Wood Plastic Composite (WPC) Market Size, 2027 (2020). www.millioninsights.com/industry-reports/wood-plastic-composite-wpc-market. Accessed 3 July 2020
12. Haider A, Eder A (2010) Markets, applications, and processes for wood polymer composites (WPC) in Europe. In: Barbu MC, Dunky M, Harper D et al (eds) PTF BPI 2010, Salzburg University of applied sciences October 2010. Processing technologies for the forest and biobased product industries. LIT Verlag, Munster, p 146
13. Medupin RO, Abubakre OK, Abdulkareem AS et al (2017) Thermal and physico-mechanical stability of recycled high density polyethylene reinforced with oil palm fibres. Int J Eng Sci Technol 20:1623–1631
14. Tewari R, Singh MK, Zafar S (2021) Utilization of forest and plastic wastes for composite manufacturing using microwave-assisted compression molding for low load applications. J Polym Res 28(11):1–4
15. Farsi M (2012) Thermoplastic matrix reinforced with natural fibers: a study on interfacial behavior. In: Wang J (eds) Some critical issues for injection molding, vol 1. IntechOpen, Zagreb, pp 225–250
16. Wood Plastic Composite (WPC) Market–Global trends & segment forecast to 2024 (2016). https://renewable-carbon.eu/news/wood-plastic-composite-wpc-market-global-trends-segment-forecast-to-2024/. Accesed 2 June 2016
17. Global Wood Plastic Composites Market-Industry Analysis and Forecast (2019–2027)—By Type, Application and Region (2020). https://www.maximizemarketresearch.com/market-report/global-wood-plastic-composites-market/47079/. Accesed 4 Jan 2020
18. Zaaba NF, Ismail H (2019) Thermoplastic/natural filler composites: a short review. J Phys Sci 2(1):81–99
19. Haque MMU, Goda K, Ogoe S et al (2019) Fatigue analysis and fatigue reliability of polypropylene/wood flour composites. Adv Ind Eng Polym Res 2(3):136–142
20. Kim JK, Pal K (2010) Recent advances in the processing of wood–plastic composites. Springer, Berlin
21. Siddikur Rahman K, Islam MN, Ratul SB et al (2018) Properties of flat-pressed wood plastic composites as a function of particle size and mixing ratio. J Wood Sci 64(3):279–286

22. Adhikary KB, Pang S, Staiger MP (2008) Dimensional stability and mechanical behaviour of wood–plastic composites based on recycled and virgin high-density polyethylene (HDPE). Compos B Eng 39(5):807–815
23. Mankowski M, Morrell JJ (2007) Patterns of fungal attack in wood-plastic composites following exposure in a soil block test. Wood Fiber Sci 32(3):340–345
24. Basalp D, Tihminlioglu F, Sofuoglu SC et al (2020) Utilization of municipal plastic and wood waste in industrial manufacturing of wood plastic composites. Waste Biomass Valori 11:5419–5430
25. Li Q, Matuana LM (2003) Foam extrusion of high density polyethylene/wood-flour composites using chemical foaming agents. J Appl Polym Sci 88(14):3139–3150
26. Bengtsson M, Gatenholm P, Oksman K (2005) The effect of crosslinking on the properties of polyethylene/wood flour composites. Compos Sci Technol 65(10):1468–1479
27. Oksman K, Lindberg H, Holmgren A (1998) The nature and location of SEBS-MA compatibilizer in polyethylene-wood flour composites. J Appl Polym Sci 69(1):201–209
28. Polec I, Hine PJ, Bonner MJ et al (2010) Die drawn wood polymer composites I. mechanical properties. Compos Sci Technol 70(1):45–52
29. Farsi M (2010) Wood–plastic composites: influence of wood flour chemical modification on the mechanical performance. J Reinf Plast Compos 29(24):3587–3592
30. Tasdemir M, Biltekin H, Caneba G et al (2009) Preparation and characterization of LDPE and ppwood fiber composites. J Appl Polym Sci 112(5):3095–3102
31. Gassan J, Gutowski VS (2000) Effects of corona discharge and UV treatment on the properties of jute-fibre epoxy composites. Compos Sci Technol 60(15):2857–2863
32. Yuan X, Jayaraman K, Bhattacharyya D (2004) Effects of plasma treatment in enhancing the performance of wood fibre-polypropylene composites. Compos Part A Appl Sci Manuf 35(12):1363–1374
33. Zlobina IV, Bekrenev NV, Teterin DP et al (2018) Studying the reinforcing microwave modification of aeronautical-purpose structural components made of composite materials. In: Abstract of the 1st international conference on mechanical and materials science engineering, Maharashtra
34. Nozari O, Madanipour M, Farsi M et al (2013) Mechanical properties and water uptake of nano clay/wood flour/LDPE composites after fiber surface mercerization. Cell Chem Technol 47(3):295–301
35. Liany Y, Tabei A, Farsi M et al (2012) Effect of nano clay and magnesium hydroxide on some properties of HDPE/wheat straw composites. Fiber Polym 14(2):304–310
36. Nourbakhsh A, Ashori A (2009) Influence of nanoclay and coupling agent on the physical and mechanical properties of polypropylene/bagasse nanocomposite. J Appl Polym Sci 112(3):1386–1390
37. Lei Y, Wu Q, Clemons CM et al (2007) Influence of nanoclay on properties of HDPE/wood composites. J Appl Polym Sci 106(6):3958–3966
38. Galgalia G, Agarwal S, Lelea A (2005) Effect of clay orientation on the tensile modulus of polypropylene–nanoclay. Compos Polym 45(17):6059–6069
39. Farsi M, Sani FM (2014) Effects of multi-walled carbon nanotubes on the physical and mechanical properties of high-density polyethylene/wood flour nanocomposites. J Thermoplast Compos Mater 27(8):1139–1154
40. Lashgari A, Eshghi A, Farsi M (2013) A study on some properties of polypropylene based nanocomposites made using almond shell flour and organoclay. Asian J Chem 25(2):1043–1049
41. Devi RR, Maji TK (2012) Effect of nano-$SiO_2$ on properties of wood/polymer/clay nanocomposites. Wood science and technology.Wood Sci Technol 46(6):1151–1168
42. Arafa IM, Fares MM, Barham AS (2004) Sol–gel preparation and properties of interpenetrating, encapsulating and blend silica-based urea-formaldehyde hybrid composite materials. Eur Polym J 40(7):1477–1487
43. Lin Q, Yang G, Liu J et al (2006) Property of nano-$SiO_2$/urea formaldehyde resin. Front For China 1(2):230–237

44. Shi J, Li J, Zhou W et al (2007) Improvement of wood properties by urea-formaldehyde resin and nano-$SiO_2$. Front For China 2(1):104–109
45. Nourbakhsh A, Baghlani FF, Ashori A (2011) Nano-$SiO_2$ filled rice husk/polypropylene composites: physico-mechanical properties. Ind Crop Prod 33(1):183–187
46. Roumeli E, Papadopoulou E, Pavlidou E et al (2012) Synthesis, characterization and thermal analysis of urea–formaldehyde/nano$SiO_2$ resins. Thermochim Acta 527(10):33–39
47. Zeng QH, Yu AB, Lu GQ et al (2005) Clay-based polymer nanocomposites: research and commercial development. J Nanosci Nanotechnol 5(10):1574–1592
48. Farsi M (2017) Effect of nano-$SiO_2$ and bark flour content on the physical and mechanical properties of wood–plastic composites. J Polym Environ 25(2):308–314
49. Naik TP, Singh I, Sharma AK (2022) Processing of polymer matrix composites using microwave energy: a review. Compos Part A Appl Sci Manuf 156:106855–106870
50. Jacob J, Chia LHL, Boey FYC (1995) Review—Thermal and non-thermal interaction of microwave radiation with materials. J Mater Sci 30(21):5321–5327
51. Church RH, Webb WE, Salsman JB (1988) Dielectric properties of low-loss minerals. US Department of the Interior. 9194:b1–28
52. Resch H (2006) High-frequency electric current for drying of wood–historical perspectives. Maderas Cienc Tecnol 8(2):67–82
53. Saitou K (2006) Microwave sintering of iron, cobalt, nickel, copper and stainless steel powders. Scr Mater 54(5):875–879
54. Vikberg T, Sandberg D, Elustondo D (2014) High frequency heating in solid wood modification. In: Nunes L, Jones D, Hill C et al (eds) Proceedings of the seventh European conference on wood modification, Lissabon
55. Torgovnikov GI (1993) Dielectric properties of wood and wood-based materials. Springer, Berlin
56. Dömény IJ (2016) Microwave modification of wood, Dissertation. Mendel University in Brno
57. Imenokhoyev I, Windsheimer H, Waitz R et al (2013) Microwave heating technology: potentials and limits. CFI Ceram Forum Int 90(4):41–49
58. Metaxas A (1991) Microwave heating. Power Eng 5(5):237–247
59. Koubaa A, Perré P, Hutcheon R et al (2008) Complex dielectric properties of the sapwood of aspen, white birch, yellow birch, and sugar maple. Dry Technol 26(5):568–578
60. Nuhiji B, Bower MP, Swait T et al (2021) Simulation of carbon fibre composites in an industrial microwave. Mater Today Proc 34:82–92
61. Sgriccia N, Hawley MC (2007) Thermal, morphological, and electrical characterization of microwave processed natural fiber composites. Compos Sci Technol 67(9):1986–1991
62. Farajallahpour M, Layeghi M, Dosthosseini K et al (2017) The effect of layer's moisture content and pre-heating by microwave radiation on physical and mechanical properties of laminated veneer lumber. Iranian J Wood Pap Ind 8(1):39–51
63. Papargyris DA, Day RJ, Nesbitt A et al (2008) Comparison of the mechanical and physical properties of a carbon fibre epoxy composite manufactured by resin transfer moulding using conventional and microwave heating. Compos Sci Technol 68(7–8):1854–1861
64. Torgovnikov G, Vinden P (2009) High-intensity microwave wood modification for increasing permeability. For Prod J 59:84–92
65. Norimoto M, Gril J (1989) Wood bending using microwave heating. J Microw Power EE 24(4):203–212
66. Leiker M, Adamska MA, Giittel R (2004)Vacuum microwave drying of beech: property profiles and energy efficiency. In: Abstract of COST E15 international conference of wood drying, Athens
67. Studhalter B (2005) Temperature and moisture content behavior in microwave heated wood prior to bending Mountain ash (Eucalyptus regnans), Dissertation. The University of Melbourne and University of Applied Sciences, Biel
68. Brodie G (2007) Simultaneous heat and moisture diffusion during microwave heating of moist wood. Appl Eng Agric 23(2):179–187

69. Studhalter B, Ozarska B, Siemon G (2009) Temperature and moisture content behavior in microwave heated wood prior to bending: Mountain ash (Eucalyptus regnans). Eur J Wood Prod 67(2):237–239
70. Larsson BP, Simonson R, Risman PO (1999) Acetylation of wood using microwave heating, Part 1. Studies of dielectric properties. Holz als Roh-und Werkstoff 57(4):259–263
71. Hansson L, Antti AL (2003) The effect of microwave drying on Norway spruce woods strength: a comparison with conventional drying. J Mater Process Technol 141:41–50
72. Makovíny I (2000) Dielectric and electromagnetic characteristics of beech wood. Wood Res 45(3):23–34
73. Binhua S, Xiaoke L, Zhicao S et al (2012) Application of microwave technology on wood processing and product testing in China. J Zhejiang Univ Sc Thecnol 32(5):75–78
74. Larsson BP, Simonson R (1999) Acetylation of solid wood using microwave heating, Part 2: experiments in laboratory scale. Holz als Roh-und Werkstoff 57(5):383–389
75. Treu A, Larnoy E, Militz H (2007) Microwave curing of furfuryl alcohol modified wood. In: The international research group on wood protection, Jackson Lake Lodge, Wyoming, USA, IRG/WP 07-40371
76. Yu L, Gao W, Cao J et al (2009) Effects of microwave post—Modification on the leasing resistance of ACQ-D treated Chinese fir. J Beijing For Univ 31(6):90–96
77. Mori M, Nakazawa S, Norimo M et al (1984) Measurement of internal temperature of wood piece during microwave irradiation by fiber-optical temperature sensor. Wood Ind 39(1):600–603
78. Brodie G (2010) Wood: microwave modification of properties. Encycl Agric Food Biol Eng 1878–1881
79. Balboni BM, Ozarska B, Garcia JN et al (2018) Microwave treatment of Eucalyptus macrorhyncha timber for reducing drying defects and its impact on physical and mechanical wood properties. Eur J Wood Wood Prod 76:861–870
80. He X, Xiong X, Xie J et al (2017) Effect of microwave pretreatment on permeability and drying properties of wood. BioResources 12(2):3850–3863
81. Vongpradubchai S, Rattanadecho P (2009) The microwave processing of wood using a continuous microwave belt drier. Chem Eng Process 48(5):997–1003
82. Zielonka P, Dolowy K (1998) Microwave drying of spruce: moisture content, temperature and heat energy distribution. For Prod J 48(4):77–80
83. Antti AL, Perré P (1999) A microwave applicator for on line wood drying: temperature and moisture distribution in wood. Wood Sci Technol 33(2):123–138
84. Zielonka P, Gierlik E (1999) Temperature distribution during conventional and microwave wood heating. Hols als Roh-und Werkstoff 57:247–249
85. Erchiqui F (2013) Analysis of power formulations for the thawing of frozen wood using microwave energy. Chem Eng Sci 98:317–330
86. Erchiqui F, Kaddami H, Dituba-Ngoma G et al (2020) Comparative study of the use of infrared and microwave heating modes for the thermoforming of wood-plastic composite sheets. Int J Heat Mass Transf 158:119996–212007
87. Singh MK, Zafar S (2021) Wettability, absorption and degradation behavior of microwave-assisted compression molded kenaf/HDPE composite tank under various environments. Polym Degrad Stab 185:109500–109510
88. Singh MK, Zafar S (2020) Influence of microwave power on mechanical properties of microwave-cured polyethylene/coir composites. J Nat Fibers 17(6):845–860
89. Singh MK, Zafar S (2019) Development and mechanical characterization of microwave-cured thermoplastic based natural fibre reinforced composites. J Thermoplast Compos Mater 32(10):1427–1442
90. Li Y, Cheng L, Zhou J (2018) Curing multidirectional carbon fiber reinforced polymer composites with indirect microwave heating. Int J Adv Manuf Syst 97(1–4):1137–1147
91. Thostenson ET, Chou TW (1999) Microwave processing: fundamentals and applications. Compos Part A Appl Sci Manuf 30(9):1055–1071

92. Galindo B, Benedito A, Ramos F et al (2016) Microwave heating of polymers: Influence of carbon nanotubes dispersion on the microwave susceptor effectiveness. Polym Eng Sci 56(12):1321–1329
93. Zhou S, Hawley MC (2003) A study of microwave reaction rate enhancement effect in adhesive bonding of polymers and composites. Compos Struct 61(4):303–309
94. Zhang Y, Cui Y, Wang S (2020) Effect of microwave treatment on bending properties of carbon nanotube/wood plastic composites by selective laser sintering. Mater Lett 267:127547–127551
95. Sutton WH (1989) Microwave processing of ceramic materials. Am Ceram Soc Bull 68(2):376–386
96. Florencio R, Boix RR, Carrasco E et al (2014) Broadband reflectarrays made of cells with three coplanar parallel dipoles. Microw Opt Technol Lett 56:748–753
97. Yuan JM, Fan ZF, Yang QC et al (2018) Surface modification of carbon fibers by microwave etching for epoxy resin composite. Compos Sci Technol 164:222–228
98. Nightingale C (2000) Microwave assisted curing of epoxy resins and composites, Dissertation. University of Manchester
99. Boey F, Lee T. Electromagnetic radiation curing of an epoxy/fibre glass reinforced composite. Radiat Phys Chem 38(4):419–442
100. Yue C, Boey F (1993) The effect of microwave and thermal curing on the interfacial properties of an epoxy/glass composite. In: Abstract of 2nd international conference on the deformation and fracture of composites, Manchester
101. Bai SL, Djafari V, Andreani M et al (1995) comparative study of the mechanical behaviour of an epoxy resin cured by microwaves with one cured thermally. Eur Polym J 31(9):875–884
102. Adeodu AO, Anyaeche CO, Oluwole OO et al (2015) Effect of microwave and conventional heating on the cure cycles of particulate reinforced polymer matrix composites. Int J Mater Sci 4(4):229–240
103. Chavooshi A, Madhoushi M, Saei AM et al (2012) Effect of nanoclay and microwave thermal treatment on mechanical properties of MDF Dust-PP nanocomposite. Iranian J Polym Sci Technol 25(4):323–331
104. Kravchenko OG, Solouki Bonab V, Manas-Zloczower I (2019) Spray-assisted microwave welding of thermoplastics using carbon nanostructures with enabled health monitoring. Polym Eng Sci 59(11):2247–2254
105. Ismaeilimoghadm S, Najafi A. Shahraki A et al (2016) Long-term water absorption and thickness swelling and determination of their characteristics in wood flour/polypropylene//Nano-$SiO_2$ nanocomposite. Iranian J Wood Pap Ind 7(2):241–254
106. Espert A, Vilaplana F, Karlsson S (2004) Comparison of water absorption in natural cellulosic fibres from wood and one-year crops in polypropylene composites and its influence on their mechanical properties. Compos Part A Appl Sci Manuf 35(11):1267–1276
107. Ismaeilimoghadam S, Shamsian M, Kashkoli AB et al (2015) Effect of chemical modification of wood flour on properties of polypropylene-nano $SiO_2$ hybrid nanocomposite. Iranian J Wood Paper Sci Res 30(4):674–689
108. Hosseini SB, Hedjazi S, Jamalirad L et al (2014) Effect of nano-$SiO_2$ on physical and mechanical properties of fiber reinforced composites (FRCs). J Indian Acad Wood Sci 11(2):116–121
109. Sahin H, Ay N (2004) Dielectric properties of hardwood species at microwave frequencies. J Wood Sci 50(4):375–380
110. Agrawal R, Drzal LT (1989) Effects of microwave processing on fiber-matrix adhesion in composites. J Adhesion 29:63–79
111. Hook KJ, Agrawal RK, Drzal LT (1990) Effects of microwave processing on fiber-matrix adhesion. II. Enhanced chemical bonding of epoxy to carbon fibers. J Adhesion 32:157–170

The manufacturer's authorised representative in the EU is Springer
Nature Customer Service Centre GmbH| Europaplatz 3, 69115 Heidelberg,
Germany. If you have any concerns regarding our products, please
contact ProductSafety@springernature.com

Printed and bound by CPI Group (UK) Ltd, Croydon, CR0 4YY
01/10/2026
02225867-0002